Juan Carlos Zutme

So Human a Brain

So Human a Brain

Knowledge and Values in the Neurosciences

Edited by
Anne Harrington

With 12 Illustrations

A Dibner Institute Publication

Birkhäuser
Boston · Basel · Berlin

Anne Harrington
Department of the History of Science
Harvard University
Cambridge, MA 02138
USA

A Dibner Institute Publication

Library of Congress Cataloging-in-Publication Data
So human a brain : knowledge and values in the neurosciences / edited
  by Anne Harrington.
       p. cm.
    "A Dibner Institute publication" — Contents p.
    Includes bibliographical references and index.
    ISBN 0-8176-3540-8 (H : alk. paper). — ISBN 3-7643-3540-8 (alk.
paper)
    1. Neurology — Moral and ethical aspects. 2. Neurology — Social
aspects. 3. Neurology — Philosophy. I. Harrington, Anne, 1960-
    [DNLM: 1. Behavior — physiology — congresses. 2. Brain — physiology —
congresses. 3. Neuropsychology — congresses. 4. Neurosciences —
congresses. WL 300 S675]
QP356.S64 1991
612.8′2′01 — dc20
DNLM/DLC
for Library of Congress   91-33333
                                                                    CIP

Printed on acid-free paper.

Typeset by Lind Graphics, Inc., Upper Saddle River, New Jersey.
Printed and bound by Quinn-Woodbine, Inc., Woodbine, New Jersey.
Printed in the United States of America.

9 8 7 6 5 4 3 2 1

ISBN 0-8176-3540-8
ISBN 3-7643-3540-8

Contents

Foreword

Walter A. Rosenblith

Footnotes to the Recent History of Neuroscience: Personal Reflections and Microstories

The workshop upon which this volume is based offered me an opportunity to renew contact fairly painlessly with workers in the brain sciences, not just as a participant/observer but maybe as what might be called a teller of microstories. I had originally become curious about the brain by way of my wife's senior thesis, in which she attempted to relate electroencephalography to certain aspects of human behavior. As a then-budding physicist and communications engineer, I had barely heard about brain waves, nor had I studied physiology in a systematic way. My work on noise dealt with the effects of certain acoustical stimuli on biological structures and entire organisms.

This was the period immediately after World War II when many scientists and engineers who had done applied work in the war effort were trying to find their way among the challenging new fields that were opening up. Francis Crick, among others, has described such a search taking place in the cafés of the "other" Cambridge, the one on the Cam. At that time the brain sciences, in his opinion, offered much less promise than molecular biology. However, he was sufficiently attracted by what they might eventually have to offer to keep an eye on them, and several decades later his work turned toward the brain.

This was also the period in which Norbert Wiener's cybernetics and Claude Shannon's information theory generated a good deal of intellectual ferment, and not just in Cambridge on the Charles. Wiener, one of the world's leading mathematicians, pulled together ideas, concepts, and mathematical techniques relevant to a broad range of scientific disciplines. Hence, his book's title: *Cybernetics or Control and Communication in the Animal and the Machine*. *Cybernetics* is a term Wiener coined; it derives from a Greek root meaning "steersman." This book and the Macy Foundation series on feedback and related topics stirred a great deal of interest among students of the brain and behavior, and among the developing professions in the areas of statistics, computers, automata, etc. Here – or so it seemed to me at the time – were elements of an exciting program for quite a few years.

In 1946, again through an accident related to my wife's work, I traveled from South Dakota to Harvard University, where Professor S. S. Stevens, then director of Harvard's Psycho-Acoustic Laboratory, took me around the lab and hired me as a research fellow a few weeks later. Under Smitty Stevens's intellectually enterprising leadership the Psycho-Acoustic Laboratory's major wartime contributions had been in the area of speech communication. Now he brought in senior scientists such as Georg v. Békésy (who became the 1961 Nobel Laureate in Physiology or Medicine), the neurophysiologist Robert Galambos, and others to join J. C. R. Licklider and George Miller who were now trying to gain a basic understanding of how organisms process information from sensory stimuli to language.

Four years later I left Harvard and journeyed down the river to M.I.T. where the Electrical Engineering Department and the Research Laboratory of Electronics offered me a home appropriate to my background in physics and communications engineering and to my ambition to use computers for Wienerian *Fragestellungen* to analyze the activity of the brain. When I was asked what kind of a professional title I would prefer, I opted for "communications biophysics." Communications biophysics? What's that? Well, I thought I knew. Anyway, the head of M.I.T.'s Electrical Engineering Department affirmed that biophysics was in the Biology Department's domain. I'm not a biologist, having never taken a course in "real" biology. (That's not quite true, but almost true.) Anyway, the department head advised me to go see Frank Schmitt who was at that time the head of the Biology Department.

After I had shared with Professor Schmitt an outline of my program for the study of the electrical activity of the nervous system, he replied, "Aren't you 200 years too early?" I often gently twit Frank – who a decade later played such a key role in the formation and unfolding of the Neurosciences Research Program – on how fast these 200 years went by.

This was midcentury at M.I.T.: the Mid-Century Convocation on the Scientific Implications of Scientific Progress, which brought Winston Churchill to M.I.T., brought a lot of recognition to an institution whose Radiation Laboratory (radar and all that), Servomechanisms Laboratory, and Instrumentation Laboratory loomed large on the world's technological scene. The mood of the time, but perhaps even more particularly the spirit of M.I.T., is best summarized by the title of Vannevar Bush's report to the president entitled, *Science – The Endless Frontier*. In the next decades M.I.T. would undertake a good many pathbreaking efforts, making use of the equipment and of the innovative techniques that came out of World War II. There also flocked to M.I.T. some of the most fertile young minds from Europe. A fair share of them had been attracted by the Wiener agenda that encompassed both computers and the brain.

Around 1960 the International Brain Research Organization (IBRO) was born. What made the situation favorable for the development of IBRO? Above all, a changed international climate. We were in the period of the Khrushchev "thaw." At a 1958 conference in Moscow, there had been unanimous support for a resolution proposing the creation of an international organization representing the whole of brain research. The plan to create an independent, nongovernmental organization in the area of brain research was welcomed by UNESCO, which has been supportive of IBRO ever since. But IBRO also established links with the World Health Organization (WHO) and the International Council of Scientific Unions (ICSU). When IBRO was founded, it was relatively small and it elected its members. IBRO was initially composed of seven panels, each of which had the label of a more-or-less classical discipline modified by the prefix "neuro." Thus the panels were labeled neuroanatomy, neurochemistry, neuroembryology, neuroendocrinology, neuropharmacology, neurophysiology and behavior, and neurocommunication. This novel multidisciplinary paradigm reflected the common conviction that the brain was not the domain of a single discipline. Since then, IBRO has grown considerably and has transformed itself into a membership organization whose growth parallels the growth of the neurosciences in different countries. In 1988 there were approximately 20,000 IBRO members, with half of the membership coming from the United States. IBRO has, in collaboration with other organizations, set up a number of active programs to stimulate international contacts and to help the development of the study of the brain in the third world.

The growth and progress in the sciences related to the brain—both their basic and clinical aspects—did not pass unnoticed at the highest levels of governments. During the mid-1980s the seven economic summit countries held a series of conferences attended by delegates from the relevant fields. They reviewed the state of research and the therapeutical innovations in the neurosciences but did not overlook ethical issues and concepts.

In July 1989 President Bush signed into law a joint resolution of Congress declaring the 1990s to be the "Decade of the Brain." That resolution estimates that 50 million Americans are affected each year by disorders and disabilities involving the brain. It also estimates the annual economic burden to exceed $300 billion. The last of 20 "whereas's" states that the declaration of the "Decade of the Brain" is designed to focus needed government attention on research, treatment, and rehabilitation, and that the president of the United States is authorized and requested to issue a proclamation calling upon all public officials and the people of the United States to observe the decade with appropriate programs and activities.

Such official documents reflect, at least to some extent, society's current view of the brain, of its role in life and death, of its relation to human intelligence, cognition, behavior, values, thought, addiction, and mental health. At the Woods Hole workshop "So Human a Brain"—and let us hope this is just a beginning—present and future colleagues shared with those in attendance their current views of their explorations into this variegated territory. They tried to inquire into what it is about the brain that makes us human and how knowledge about the brain relates to the rest of human knowledge. Fields that are as bubbling with research as this one are only too often data-rich and theory-poor. Then they fall easy prey to clever and temporarily fashionable metaphors. The thoughtful way in which this volume has been organized should protect us from the temptation of focusing too hard on increasing the market share of our pet metaphors.

Preface

ANNE HARRINGTON

Humanizing Knowledge and Values

In today's world of brain science, having (or being) an enlightened human brain may require a willingness to tolerate a growing professional "confusion about heroism" (to use a phrase by Ernst Becker [1973]). As little as 30 years ago, the image of the devoted scientist–hero portrayed in Sinclair Lewis's *Arrowsmith* was still a powerful one; scientists and laypersons alike could still dream of science as an "endless frontier" (in the words of the famous postwar report by Vannevar Bush [1945]), and still find the vision of rational control over human destiny a noble one. Today, however, even as politicians and scientists join together in proclaiming the 1990s the "Decade of the Brain," a cacophony of disciplinary truths in the human sciences begins to hint at a more complex story; one whose demythologizing collective message is disturbing enough to warrant a serious hearing by the neuroscientific community.

What are the issues? To begin, it is probably fair to say that, for most researchers of the human mind and brain, the imperative to *know,* to increase objective understanding of human thought and behavioral processes, continues to set the agenda — or at least define the frame — for any larger debates about the meaning and imperatives of humanness in the late 20th century. There is a greater or lesser reticence within this group as to how far new knowledge of the brain should or could destabilize traditional humanistic categories of purpose and value, but the logical priority of "natural truth" over "social truth," of "objective" fact over "subjective" value is still rarely questioned. Certainly, few thoughtful neuroscientists deny that values can and do interact with problem selection, design, and theory modeling in the brain sciences, but this awareness seems in practice to have little impact on the actual production and promotion of new knowledge. In general, successful members of the community of brain scientists today continue, more or less consciously, to adhere to what Jacques Monod referred to as an "ethic of knowledge" in science: a commitment to the scientific exploration of nature—including human nature—as a first goal independent of other goals (Monod, 1971).

But is Monod's vision the last word on the matter? In the past 20 years, a different community of scholars, with roots in the history, sociology, philosophy, and anthropology of science, has increasingly begun to argue a very different ethic of knowledge. Here, in varying ways, the claim is made that the old ideal of brain science as a form of objective knowledge about humanness has itself been profoundly destabilized by the fact that brain research is no less an expression of our inescapable humanness than any other activity. For these science analysts, the knowledge-claims of neuroscience are believed to be bound up in complex and unavoidable ways with professional agendas, strategies of knowledge-justification, ethics, and cultural imperatives. While opinions differ within this group as to the implications of this challenge to the Enlightenment ideal of transhistorical, transcultural objectivity, the scholars in question at least agree in their tendency to reverse the emphases of the scientists themselves, and to focus on *values* as at least partly setting the agenda of knowledge.

It is too early to decide whether the perspectives on knowledge and values that set the accent for these two broad communities of scholars are as mutually cannibalizing as they sometimes sound, or whether genuine dialogue and cross-disciplinary enrichment is in fact possible. The jury is still out on this issue in part because there has been a tendency for the different voices to proclaim their different truths in self-reinforcing isolation from one another.

This book derived from a collective effort across disciplines to break free of that mold. In the summer of 1990, the newly established Dibner Institute sponsored some twenty leading scholars from the brain sciences and the social sciences (history, sociology, ethics, etc.) for three days of presentation and intensive debate at the Woods Hole Marine Biological Laboratory.

There had never been a workshop with the broad goals and interdisciplinary focus of this one. Not surprisingly, therefore, our group achieved no final closure and signed no formal peace treaties. Indeed, our exchanges over the weekend were punctuated by often revealing moments of uncertainty, conflict, and incomprehension (cf. the analysis of these exchanges by Fortun and Sigurdsson, this volume). As we persisted, however, we did find that our dialogues had begun to converge around the challenge of what it might mean to "humanize" the sciences of the human brain. In other words, we increasingly discovered in ourselves a willingness to explore alternatives to the pessimism of a Tolstoy, who had envisioned the two imperatives — *what do we know?* and *how should we live?* — as mutually alien categories.

This volume represents both a record of our 1990 interdisciplinary encounter at Woods Hole and a retrospective reflection upon it. The volume is divided into five main parts. The first of these is entitled "Knowledge of and by the Human Brain: Limits and Possibilities." It begins with two attempts from within the neuroscientific community to provide a methodology and knowledge-base capable of tackling the challenge implicit in the provocation: *so human a brain*. We start with an essay on the peculiarly human functions of intentional signing and speech, examined from the perspective of a leading neuroethologist, Detlev Ploog, who admits that ". . . most neuroscientists, regardless of whether they work at the molecular, the systemic, or at any other level of central nervous functioning, ultimately want to contribute to the understanding of the human brain." This first section of the volume grows increasingly self-reflective as it progresses. It is proposed that there are certain "semantic" truths of the mind, such as those revealed by dreams, that are inherently irreducible to the terms of neuroscience (Massimo Piattelli-Palmarini).

This paper is followed by an ambitious attempt by Stephen Kosslyn to use the framework of cognitive neuroscience to conceptualize human selfhood and self-awareness—or, more specifically, "what we mean when we speak of the self, and what would be necessary for the self to change." We are then invited to ponder Paul MacLean's even more radical suggestion, made in the course of an argument that blends brain science with ethical imperative, that the human brain is destined to an "inability ever to achieve certitude of knowledge," including those truths it discovers about itself. MacLean challenges us to consider, in a paraphrase borrowed from Ramon y Cajal, that "the universe is but a reflection of the structure of the human brain." This section of the volume concludes with Jason Brown's proposal that these biologically given limits on knowledge should be of vital interest to humanists, since they interact in crucial ways with the human world of morality and values.

Part 2, "Values and the Nature of the Neuroscientific Knowledge Game," advances the case for linking knowledge and values, but from perspectives and knowledge-bases outside of the brain sciences proper. The main concern here is with (in Rodney Holmes's words) "the kind of knowledge that is sought and found by neuroscience." Given that it will never be possible to research and know everything that one theoretically *could* decide to research and know, what if it turned out that there was a relationship between the kinds of questions scientists actually did decide to ask—the things they wanted to know—and the professional, ethical, and social contexts in which they worked? These and related problems are explored from very different conceptual and methodological perspectives by Rodney Holmes and Londa Schiebinger.

Schiebinger, writing from a feminist historical perspective, notes that "objectivity in science cannot be proclaimed, it must be achieved"; and that, paradoxically enough, the means to this goal must be "to disabuse ourselves of the notion that science is value neutral." In other words, objectivity emerges as a political rather than a strictly epistemological goal. Holmes challenges us to become more conscious of the ethic of research that motivates and indeed defines the neuroscientific community as a profession. He writes:

What are the directions of Neuroscience? Are they to write every grant proposal and to know everything there is to know? Even if it is the latter, then what is knowledge—is it all the possible electrochemical information about neurons, or is there some way of defining our goal that will help us decide what information is worth gathering, or what applications are legitimate or not?

Edward Manier and James Schwartz agree with Holmes and Schiebinger about the centrality of questions of human value to brain science research, but they wonder more about what brain science can and cannot say to those questions of "ultimate concern" that give human beings a sense of existential orientation and grounding in the cosmos. Both authors take their starting point from the critical writings of the novelist Walker Percy, but then proceed to diverge radically in their conclusions, point/counterpoint style. Manier sees Percy as "a neuropsychologically well-informed biblical prophet" whose message boils down to an insistence that "the reductionist project [of the neurosciences] should not be trivialized by pre-shrinking the human world." The "deep sense in which human beings seem 'lost in the cosmos' are data which must be addressed by the human sciences." Schwartz, in contrast, allies Percy with a tendency in the brain sciences to want neatly—diagrammatically—to account for all aspects of human existence, including (or especially) the problem of evil. This sort of sanitizing, he proposes, is sentimentality of the worst kind: in fact, "the Lord's answer [to Job] out of the whirlwind appears to be essentially correct. . . . We cannot hope to make coherent a cosmos that is ethically incoherent. . . ."

The third part of the volume, "Neuroscientific Knowledge and Social Accountability" adds still another layer of tension to an increasingly complex debate. This section sees the problem of knowledge and values in the brain sciences as intersecting most acutely at the point of *praxis,* and challenges us to articulate an ethic of professional responsibility for both the means and the consequences of brain research. Alan Fine examines the justifications and dilemmas associated with research on fetal neural transplantation, a medically promising but highly contentious field, particularly in the United States where the ethics of abortion are so much under fire. Elliot Valenstein warns of inadequate regulation and control on the development and application of new therapeutics in biomedicine and drives his point home with the cautionary tale of psychosurgery. John Durant tackles one of the most controversial and emotive issues in brain science research today: the ethics of experimenting on sentient animals in the pursuit of knowledge that may advance human values. His argument is that our society is increasingly suffering from a dislocation in sensibility on the issue of the moral status of animals, in part because of "conflicting voices" arising out of the sciences themselves. The presentation of this paper at the Woods Hole conference led to one of the most explosive and riveting interdisciplinary exchanges of the weekend. Some neuroscientists vigorously defended their right to pursue knowledge, while slowly coming to concede a need to articulate a moral calculus for deciding what sort of research is acceptable; others found sentimentality and anthropomorphism in many of the claims for "animal mind" that contribute to public sentiment against animal experimentation; and still others found antihumanistic implications in the suggestion that "species-ism" is a form of immorality equivalent to racism and sexism. The reader is referred to the reconstruction of these discussions in the thematically ordered dialogues at the end of this volume.

Part 4 of this volume, "Sociohistorical Perspectives on Values and Knowledge in the Brain Sciences," offers three case-studies in the history and sociology of the brain sciences that raise a range of questions about methodology, language, theory-choice, and epistemology. Here, historical material is employed to focus and elaborate on some of the general theoretical issues and questions that are developed in the earlier sections of the book.

Warwick Anderson explores the social process of "resistances and trade-offs" by which the disorder known as *kuru* underwent a process of epistemological transformation from an anthropologically coherent cultural phenomenon to a neurophysiologically watertight brain disease that would garner its discoverer a Nobel prize. He tells us that "it is on the boundary between a conventional context and another, alien context, that one sees most clearly how each knowledge community struggles to attract phenomena into its own orbit and reinflect them with its own values." Susan Leigh Star uses the 19th-century localizationist model of the brain to ask questions about what a "sociology of the brain" might look like. She begins by proposing that "we can begin to think sociologically about the brain if we recognize the properties of a large zone of negotiation" within which "people attempt to link their experience of a concrete brain with the abstract representations of mind." Her sociological concerns begin to become clear as she continues: "What are the properties of this zone? When did it develop? How has it changed. . . ? Who controls it? Whose experiences are included and whose are excluded? Are there gatekeepers, strategies, technologies, routines, and silences?"

Complementing Star's study of localization theory is my own analysis of the early 20th-century "holistic" reaction *against* mechanistic, localizationist approaches to brain function, particularly in interwar Germany. My concern in this paper is with the way in which disputes over knowledge-claims in the brain sciences of this time period need to be understood as one face of a network of larger disputes over collective cultural and political values. I suggest that

German holistic neurology and psychology, especially after 1918, was not only about discovering or constructing organicist or nonatomistic models of mind and brain; it was also about challenging epistemological rigidity in the natural sciences — and challenging it under the conviction that epistemological reform alone could offer a way out of . . . cultural crisis.

The final part of this volume can be considered an experiment in self-reflexivity. Specifically, it is an attempt to cast some critical light on the clashes of knowledge and values that emerged at Woods Hole as people from different disciplinary worlds attempted — some for the first time — to speak and listen to one another. Entitled "Knowledge and Values across Disciplines: Reconstruction and Analysis of an Interdisciplinary Dialogue," this section of the book consists of two elements.

It opens with a reconstructed rendering of some of the most interesting "live" exchanges and debates pursued within our group over the course of the Woods Hole weekend. Thematically ordered, with a summarizing argument at each new subject juncture for the purposes of orientation, these dialogues take on board the whole range of issues raised by the individual contributions: neurobiological constraints on human knowledge; consciousness and the human brain; the place of ideology in brain models; the feminist critique of science; the "moral rights" of laboratory animals; the ethics of human fetal neural transplant research; and whether the brain sciences lead to a sense of human meaninglessness.

Following immediately on the heels of these dialogues — and critical to a just reading of them — is a meta-level analysis of the dynamics that generally characterized the exchanges at Woods Hole. Aware that attempts to carry out interdisciplinary work are frequently beset with problems, and wanting better to understand the nature of those problems, I invited historians of science Michael Fortun and Skuli Sigurdsson to attend the Woods Hole workshop as "ethnographic" observers of the "tribes" of action. In their paper, which is based on a critique presented to our group on the final Sunday morning meeting, these two scholars have much to say not only about what was discussed but also about what was skirted or left unspoken — the "silences." For example, they were disappointed that

[t]here was . . . surprisingly little attention paid to the technological dimension. . . . Is there an independent technological momentum, and how does that relate to the autonomy of research? Put differently, how much of the tacit knowledge in the neurosciences is encoded in the machine environment . . . ? What assumptions are built into the practices and equipment that neuroscientists use without further thought?

It is worth stressing, though, that the imperative to recognize disciplinary "silences" and to work on overcoming them is incumbent not only on the neuroscientific community but also on its would-be critics and analysts. Thus, one clause in the contract for future interdisciplinary work of the sort begun in this volume might be a commitment by those of us interested in the ethics, culture, and sociology of the brain sciences to invest more energy into opening up the "black boxes" that contain the tacit knowledge and values driving our own analytic technologies, central metaphors, epistemologies, and cultural concerns. We do this, not as an exercise in self-contemplation for its own sake, but so that the critique we direct to our neuroscientific colleagues may be tempered with self-awareness, may ring as clear and true as possible. In an ideal interdisciplinary encounter, there can be no room for any privileged (unexamined) foundation of knowledge and values.

Are the benefits of mutual understanding worth the effort involved in cross-disciplinary communication? We cannot risk anything other than an affirmative answer. It is increasingly clear that Arrowsmith and Jacques Monod represent defunct role models in the complicated world of today's brain science. We need viable alternatives that avoid the extremes of radical subjectivism and of naive scientism alike. Against this background, our interdisciplinary goal of "humanizing" the brain sciences may be understood as a commitment to work together in illuminating the roots joining the tree of knowledge and the tree of life. Ultimately, we may hope that an increasingly sophisticated awareness of interdependence will come to influence both the way we understand the fruits of our neuroscientific efforts, and the way we plan the future horticultural efforts of our society.

Let us have the courage, then, to react to our "confusion about heroism" in creative ways. Let us not close ranks and hunker down in our individual holes of disciplinary certainty, but let us dare instead to be vulnerable, to venture onto the unmapped spaces between the holes, for the sake of a goal and an insight beyond that of our smaller truths. This sort of effort would be a real triumph for the cause of our collective humanness.

Acknowledgments

I am indebted to Everett Mendelsohn, Evelyn Simha, George Adelman, and Stephen Kosslyn for comments on earlier versions of this text.

References

Becker E (1973): *The Denial of Death.* New York: The Free Press.
Bush V (1945): *Science, the Endless Frontier.* Washington, DC: U.S. Government Printing Office.
Monod J (1971): *Chance and Necessity: An Essay on the Natural Philosophy of Modern Biology,* Wainhouse A (trans), 1st American ed. New York: Knopf.

Chapter Authors

Warwick Anderson   Department of History and Sociology of Science, University of Pennsylvania, 215 South 34th St., Philadelphia, PA 19104

Jason W. Brown   Department of Neurology, New York University Medical Center, New York, NY 10021

John Durant   Science Museum Library, The National Museum of Science & Industry, South Kensington, London SW7 5NH, England

Alan Fine   Department of Physiology and Biophysics, Dalhousie University Faculty of Medicine, Halifax, Nova Scotia, Canada B3H 4H7

Michael Fortun   Department of the History of Science, Harvard University, Science Center 235, Cambridge, MA 02138

Anne Harrington   Department of the History of Science, Harvard University, Science Center 235, Cambridge, MA 02138

H. Rodney Holmes   Biological Sciences Collegiate Division, The University of Chicago, 1116 East 59th St., HM Box 25, Chicago, IL 60637

Stephen M. Kosslyn   Department of Psychology, William James Hall, 33 Kirkland St., Harvard University, Cambridge, MA 02138

Paul D. MacLean   Department of Health & Human Services, National Institute of Mental Health, Bethesda, MD 20892

Edward Manier   Department of Philosophy, Reilly Center for Science, Technology and Values, University of Notre Dame, Notre Dame, IL 46545

Massimo Piattelli-Palmarini   Center for Cognitive Science, Massachusetts Institute for Technology, Cambridge, MA 02139

Detlev W. Ploog   Max Planck Institute for Psychiatry, Kraepelinstrasse 2, D-8000 Munich 40, Germany

Walter Rosenblith   Massachusetts Institute of Technology, Room E51-211, Cambridge, MA 02139

Londa L. Schiebinger   Department of History, Pennsylvania State University, 601 Oswald Tower, University Park, PA 16802

James H. Schwartz   Howard Hughes Medical Institute, Columbia University College of Physicians and Surgeons, 722 West 168th Street, New York, NY 10032

Skuli Sigurdsson   Department of the History of Science, Harvard University, Science Center 235, Cambridge, MA 02138

Susan Leigh Star   Department of Sociology and Social Anthropology, University of Keele, Keele, Staffordshire ST5 5BG, United Kingdom

Elliot S. Valenstein   Neuroscience Laboratory, University of Michigan, 1103 East Huron, Ann Arbor, MI 48104

Discussants

Judith Anderson   Woods Hole Marine Biological Laboratories, Woods Hole, MA

Warwick Anderson   History and Sociology of Science Department, University of Pennsylvania, Philadelphia, PA

Jason Brown   Department of Neurology, NYU Medical Center, New York, NY

Jean-Pierre Changeux   Laboratory of Molecular Neurobiology, Pasteur Institute, Paris, France

Terrence Deacon   Department of Biological Anthropology, Harvard University, Cambridge, MA

John Dowling   Department of Biology, Harvard University, Cambridge, MA

John Durant   The Science Museum Library and Imperial College, London, England

Alan Fine   Department of Physiology and Biophysics, Dalhousie University, Faculty of Medicine, Halifax, Nova Scotia

Anne Harrington   Department of the History of Science, Harvard University, Cambridge, MA

Richard Held   Department of Brain and Cognitive Sciences, Massachusetts Institute of Technology, Cambridge, MA

H. Rodney Holmes   University of Chicago, Chicago, IL

Ruth Hubbard   Department of Biology (emeritus), Harvard University, Cambridge, MA

Stephen M. Kosslyn   Department of Psychology, Harvard University, Cambridge, MA

Rodolfo Llinas   Department of Physiology and Biophysics, NYU Medical Center, New York, NY

Paul D. MacLean   National Institutes of Mental Health, Poolesville, MD

Edward Manier   Department of Philosophy/Reilly Center for Science, Technology and Values, University of Notre Dame, Notre Dame, IN

Godehard Oepen   McLean Hospital and Harvard Medical School, Cambridge, MA

Diane Paul   Department of Political Science, University of Massachusetts, Boston, MA

Massimo Piattelli-Palmarini   Center for Cognitive Science, Massachusetts Institute of Technology, Cambridge, MA

Detlev W. Ploog   Max Planck Institute for Psychiatry (emeritus), Munich, Germany

Robert Richards   Department of History, Philosophy, and Social Studies of Science and Medicine, University of Chicago, Chicago, IL

Judy Rosenblith   Wheaton College (emeritus), Norton, MA

Walter Rosenblith   Massachusetts Institute of Technology (emeritus), Cambridge, MA

Londa L. Schiebinger   Department of History, Pennsylvania State University, University Park, PA

James H. Schwartz   Columbia University Medical School, New York, NY

Susan Leigh Star   Department of Sociology and Social Anthropology, University of Keele, Keele, England

Frank Sulloway   Science, Technology Society Program, Massachusetts Institute of Technology, Cambridge, MA

Elliot S. Valenstein   Neuroscience Laboratory, University of Michigan, Ann Arbor, MI

PART 1

Knowledge of and by the Human Brain: Limits and Possibilities

1

Neuroethological Perspectives on the Human Brain: From the Expression of Emotions to Intentional Signing and Speech

DETLEV W. PLOOG

While pondering the theme of this workshop, it occurred to me that most neuroscientists, regardless of whether they work at the molecular, the systemic, or any other level of central nervous system functioning, ultimately want to contribute to our understanding of the human brain. This attempt has produced certain successes at all levels when particular physical or medical questions have been addressed. But if one regards the human brain as the substrate of human behavior, the matter becomes extremely complex. In fact, rapidly accumulating knowledge on the molecular and cellular levels presents an almost hopeless perspective in regard to the possibility of explaining an organism's behavior even when one deals with so-called simple systems, such as that characterizing worms. Without disregarding the relevance of the bottom-up approach in molecular biology or its many success stories in recent decades, nevertheless, if one's goal is to explain behavior via brain mechanisms, one's approach must change to that of a top-down analysis. With this approach, one starts with the behavior of an organism and then proceeds from the analysis of this behavior to an examination of the neural substrates that bring about a specific part of behavior, for example, eating behavior. This approach is taken in neuro-ethology (which does not exclude molecular methods, where applicable). Behaviors that are specific for a certain species or typical for a genus or a family or an order (such as the primates) are first analyzed, and then the causal relationships between such behaviors and their neural substrates are investigated. In attempting to understand the functioning of the brain, one should keep in mind the truth that the pressure of natural selection (which has formed animals, including man) is on the *behavior* of the organism. The selection of adaptive behavior has caused both the further development of evolutionarily older brain structures and the establishment of new structures that together form the modern human brain with all its human faculties, including language.

Regarding human emotions, the neural substrates and mechanisms that underlie emotional expressions such as facial expressions or vocal utterances or emotional speech are of salient importance since, on the one hand, this motor behavior is species-specific and preprogrammed, and, on the other hand, it shows great variability and adaptiveness in regard to the social context in which it is exhibited. In animals and man, this class of motor behavior exercises a dual function: it expresses an emotion and, at the same time, it transmits a message to the recipient. This so-called social signaling has a long evolutionary history in the vertebrates.

To make these points clear, let us consider the head nodding of the green iguana (see Figure 1.1). Among the social signals of this gregarious reptile, head nodding is the most conspicuous one; it is seen in the whole family of Iguanidae (Ploog, 1970). It consists of a sequence of up-and-down movements of the head, highly stereotyped in amplitude, frequency, and time course—a typical "fixed action pattern" (to use Konrad Lorenz's term). The signal is used in a number of competitive social encounters but may also be observed in isolated males in situations lacking a detectable releasing stimulus. Lorenz (1950) termed this the "vacuum activity" of fixed action patterns. The signal undoubtedly has a dual function: it expresses the motivational state of the animal and it sends a message to the recipient of the signal, which may or may not change the recipient's behavior. According to ethological theory, social signals such as head nodding are ritualized behavior patterns, produced by natural selection and based on motivated intention movements.

At this point I should like to introduce the concept of *intentionality* in the communicative act by raising the question of whether the animal has the intention of changing the behavior of the recipient. Does the sender of the signal have a knowledge of the effect that his action can have on the recipient? Of course, we cannot answer this question ourselves, and the animals will never be able to tell us. For the time being let us assume that animals such as iguanas do not have the kind of intention that would involve a purposive transfer of information.

To take this problem a step further, let us now consider another rather more complex example of social signaling. While studying a colony of squirrel monkeys in Dr. MacLean's laboratory back in the late 1950s we found, among other social signals, one that is best described as genital display (see Figure 1.2). It consists of several components: penile erection, lateral positioning of the leg with hip and knee bent, marked supination of the foot, and abduction of the big toe. The signal is used in various types of agonistic, dominance, and courtship behaviors, and is always directed to a conspecific (Ploog et al., 1963). It may be exhibited as early as the first or second day of life, and it is also seen in monkeys displaying to their mirror image (Ploog and MacLean, 1963; MacLean, 1964). (This last observation, by the way, strongly suggests that monkeys, in contrast to apes, cannot recognize their mirror image.) Genital display, then, is a signal that contributes decisively to the formation of group structure and to regulatory processes that I have termed "social homeostasis" (Ploog and Melnechuk, 1970).

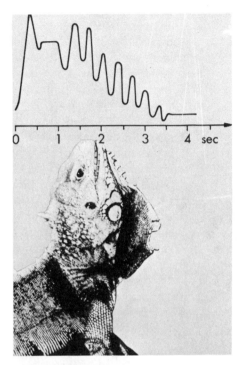

FIGURE 1.1. The head-nodding display of *Iguana iguana,* a South American lizard. Onset (top) and termination (bottom) of the conspicuous social signal. Amplitude, frequency, and time course of the nodding shown in the line above the head. From Ploog (1970).

In regard to the question of intentionality in the communicating act, the following observation is relevant. A juvenile squirrel monkey in late puberty repeatedly addressed "genital display" to the top male of the group, for which he was punished several times. Thereupon he changed his behavior by displaying in the opposite direction so that the alpha male could not see his penile erection, but while displaying he turned his head around so that he could look straight at his addressee. The three human observers

FIGURE 1.2. Genital display, a social signal of the squirrel monkey: a, display at distance; b, counterdisplay at proximity; c, displaying from back of mother at a conspecific on the second day of life; d, 49-day-old male displaying at his mirror image and vocalizing. From Ploog (1974).

who witnessed this behavior the first time could not help laughing. In this case, my previous question about whether the sender of the signal has a knowledge of the effect that his action can have on the recipient is even more difficult to answer. However, the evidence seems to suggest that the juvenile monkey had some sort of knowledge of the effect (and the consequences) of his display.

Since the publication of Charles Darwin's *The Expression of the Emotions in Man and Animals* (1872) we have known something about the evolution of facial expressions and their functions in social communication. Like other social signals, a facial expression also has a dual function: it expresses an emotion and it transmits a message to the recipient of the signal. In man, in contrast to animals, the innate behavior patterns of facial expressions are subject to volitional control learned in the course of ontogeny. The child and the juvenile learn when, where, and to whom such signals are sent. The face expresses not only the affect but also volitionally controlled messages codetermined by social rules. It is often difficult for the observer to distinguish volitionally controlled patterns from uncontrolled emotional ones. And since the understanding of sociocommunicative signals is also based on innate perception mechanisms, the recipient of a message can be misled by volitionally controlled and purposefully applied signals. In this case, we have no difficulty answering the question of the sender's knowledge about the effect that his action can have on the recipient. Intentionality is definitely involved.

Although the behavioral, morphological, and neurological homologies between the human facial expressions and the facial expressions of apes are astounding, I doubt that chimpanzees, for instance, are able to mislead or to deceive their conspecifics by controlled facial expressions, despite the fact that they are able to engage in other deceitful behaviors.

Besides facial expressions, the voice is an outstanding means of social signaling in nonhuman primates and in man. Interest in the evolution and biological foundation of human language and speech inspired our experimental studies on the audio-vocal communication of squirrel monkeys. As with the other forms of communication, vocal behavior serves two purposes, namely, as an expression of emotions and, at the same time, as a communicative signal. Primates have attained the most highly refined means of expressing emotions and of communicating.

How is vocal behavior represented and organized in the brain? Answers to this question contribute in two ways to basic issues in communication research. First, we gain direct access to the brain system that is responsible for the vocal expression of emotions and thereby indirect information about the system that generates and controls emotions. Second, we gain access to a nonverbal communication system that we largely share with the nonhuman primates. This system operates not only in our nonverbal communication system with conspecifics but also in prosodic features in our language — in conversation, in songs without words, in complaining and cursing (Ploog, 1970).

For my purpose here, the results of electrical brain stimulation in monkeys provide enough information for an outline of the system involved in vocal behavior, and hence in the vocal expression of emotions (Jürgens and Ploog, 1970). Surprisingly, only species-specific, that is, natural calls can be elicited by electrical stimulation of the brain sites above a certain level in the pons. The responsive structures reach from subcortical orbito-frontal and temporal structures, through thalamic and hypothalamic structures, into the midbrain and pons and their intimate connections with the limbic system. Not a single vocally responsive site was detected in the neocortex.

The use of several neurobiological methods in combination has enabled identification of a hierarchically organized system that can be described in four levels (see Figure 1.3). The lowest level (I) is represented by the nucleus ambiguus, whose neurons supply the larynx with the innervation required for the execution of vocal movements; by respiratory motoneurons; and by the lateral pontine and medullary reticular formation, which have coordinating functions. The central gray in the lower midbrain and upper pons (II) is the phylogenetically oldest structure for the generation of species-specific calls. This crucial part of the system receives input from motivation-controlling brain structures (III). Electrical stimulation in level II yields species-specific vocalizations in amphibians, reptiles, mammals, and man. Its destruction causes mutism. In the primate, this part in midbrain and pons receives direct input from and is controlled by the anterior cingulate gyrus (IV). This structure is involved in the initiation and voluntary control of the voice, although to a different degree in monkeys and man. Monkeys control their vocal behavior to a rather limited extent in regard to the frequency and amplitude of a species-specific vocalization, but not in regard to the acoustic structure of the utterance (Jürgens and Ploog, 1981). After bilateral ablation of the cingulate area, they lose this control, but are still able to vocalize adequately in the communal situation. In humans, however, clinical work has shown that patients with bilateral lesions of the cingulate area may lose spontaneous speech completely. After a state of akinetic mutism is overcome and the initiation of speech is possible again, the voice sounds monotonous, the speech is aprosodic, and the patient continues to have difficulties in initiating verbal communication (Jürgens and Cramon, 1982).

FIGURE 1.3. Scheme of hierarchical control of vocalization. All brain areas (depicted in coronal sections) indicated by a dot yield vocalization when electrically stimulated. All lines interconnecting the dots represent anatomically verified direct projections (leading in rostrocaudal direction). The dots indicate in (IV) the anterior cingulate gyrus, in (III) the basal amygdaloid nucleus, dorsomedial and lateral hypothalamus, and midline thalamus, in (II) the periaqueductal gray and laterally bordering tegmentum, in (I) the nucleus ambiguus and surrounding reticular formation (the nucleus ambiguus itself only yields isolated movements of the vocal folds; phonation can be obtained, however, from its immediate vicinity). For further explanation, see the text (Jürgens and Ploog, 1981).

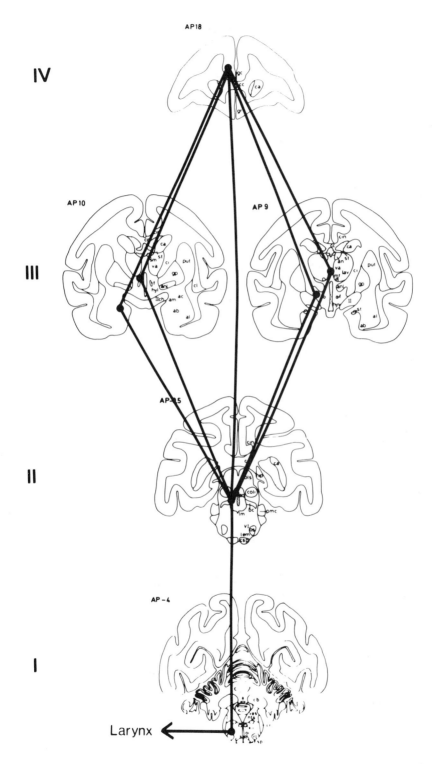

The obvious question about the role of the primary motor cortex in phonation should be briefly addressed. It is well known that this cortex is crucial for the voluntary control of movements. Destruction of the face area in the speech hemisphere in man will result in severe anarthria and aphonia. But a bilateral lesion of this area in the monkey will leave its vocalization completely unimpaired. This profound functional difference is based on different anatomical projections. While in man there is a direct connection between the primary motor cortex and the laryngeal motoneurons — obviously a specific outgrowth of the pyramidal tract in both directions — such a connection is lacking in the monkey. Thus man can exert a direct voluntary control over his vocal cords, but apes and monkeys cannot. This new pathway is clearly a prerequisite for the emergence of social communication in the speaking mode where the acquisition of learned vocal gestures is required. This difference explains why nonhuman primates can learn neither to speak nor to sing (Ploog, 1988a).

It is unfortunate that, when human evolution and the unprecedented increase in neocortex are discussed, concomitant changes in subcortical structures are usually neglected. The limbic system is one of these structures. Since this system is heavily involved in vocal behavior, as just outlined, it is important to know that morphometric analyses of limbic structures revealed more neurons in the human anterior thalamic limbic nuclei than expected in an anthropoid with a brain the size of ours. The increased number of neurons produces an increased limbic input into the human neocortex, especially the supplementary motor area and the premotor and orbitofrontal cortex. The enlargement of the thalamic limbic pool of neurons may represent an increased differentiation of the limbic message being sent to the cortex. Conversely, the enlarged thalamic limbic pool through reciprocal connections may represent an increased ability to activate and control limbic structures via neocortical input. Hence, the specific human corticolimbic connections serve as a basis for the species-typical human motivational system, which is under neocortical control and also cognitively represented (Ploog, 1988b). This means that, for instance, anger, fear, or aggressive impulses can come to mind and be subject to consideration. It thus comprises a very important step in human evolution, as has been stressed by MacLean (1990).

This anatomical excursion makes it quite clear that the subcortical vocal communication system, basic for the emergence of speech, is intimately and reciprocally connected with those neocortical areas that, on the one hand, have this system under control and, on the other, are informed about the processes going on in it. The fact that great emotional excitement makes our voice tremble or can even make us speechless can be explained by these connections. This system may also represent the window that permits self-awareness and verbal reports on our emotions and feelings (Ploog, 1989). Only man can speak about his emotional experiences, and he has words to describe them in every known language. In spite of this, he has no way of knowing whether his individual experience of joy, sadness, or despair is the same as that of his fellow man. Nevertheless, emotions as we know them are inherent in humankind and serve as the basis for empathy and altruism as well as for aggressive and violent behavior.

Just as there are essential differences in the substrates for voluntary control of the voice in human and nonhuman primates, there are also essential differences in the voluntary control of emotions. In man, this control works bidirectionally: emotions can be suppressed and they can also be evoked by creating a mental representation of the motor expression of the respective emotion. Darwin was the first to make use of this typical human reciprocity. For his studies on facial expressions he asked actors to feel a certain emotion and express it facially. There may be a motor matching process of affect perception that works to two ends: namely, to "understand" the emotions of others (empathy) and to evoke emotions in the self.

In regard to human emotionality and the vocal-speech system, I would like to point out one more dimension of the singularity of the human condition, one that becomes apparent in sleep.

If we did not have speech and language, little if anything would be known about the experience of dreaming. We would have to rely on behavioral observations of sleeping organisms such as, for instance, the sleeping hunting dog that vocalizes as it typically would if chasing game, or a sleeping infant crying out and showing the facial expression of fear. From verbal descriptions by people who report about their dreaming we know what dreams, and, for that matter, what unprotected emotions, are like. There is, however, a more direct way to discover emotions and experiences in dreams. Although motor pathways are usually blocked during sleep, there are instances in which the vocal-speech pathway is not blocked and the dreamer reports verbally while he is dreaming. This is the most direct and undiluted access to dreaming experience and thereby to emotions available to a person other than the dreamer himself. There are verbalized dreams without emotional content, which could be classified as "cognitive dreams." But most dreams are colored by emotions, and often they are dominated by distinct emotions strongly expressing a specific motivational state, which sometimes finds its immediate expression in vocalizations such as screaming, moaning, or shouting, or even giggling and laughing. In the sleep-talker's verbal utterances, the whole canon of emotions can be expressed: desire, fear, anger, sadness, and affection, gratifying, triumphant, and successful feelings, as well as feelings of pleasure and happiness. These are all emotions that MacLean (1990) lists as feelings associated with psycho-motor epilepsy, and all of them may be experienced in the dreaming state and verbally reported. From this, we conclude that just as dreams belong to the universals of human experience, so do emotions. They cannot be learned from any model. What is learned is the situations, events, and stimulus conditions that have elicited specific emotional states during an individual's life history. Although emotions (and affects) may be attached to certain external events or personal experiences, the canon of emotions as such is universal for humankind.

The evolutionary changes in the limbic system are, it seems, responsible for the changes in the human motivational system and are, at the same time, involved in the evolution of language. All this means that humans are able to act out their emotions with a species-specific motor system—the speech system. This explains why most aggressive acts are expressed as verbal attacks and why verbal expressions of emotions often lead to relief from the respective emotional state. It may also explain why emotions are often the driving force for cognitive plans that are then acted out independently of the presence of the respective emotional state. In this way, premeditated and planned acts of aggression and violence become possible, a fact that, ironically, distinguishes *Homo sapiens* from his closest relatives and the rest of the animal kingdom.

One way of paraphrasing the meaning of "So Human a Brain" is to ask the question "What is *peculiar* to the human brain?" The approach suggested here for pursuing this question is to start with the behavioral analysis of faculties that are species-specific for humankind—for instance, the control of facial expression of emotions and vocal behavior, language and speech, and other human faculties such as planned intentional behavior, self-awareness, empathy, and altruism—and then to proceed to the specific brain mechanisms that generate or mediate these behaviors. All these human features have their roots or origins in our simian past, but with further development into the human form old brain structures changed in function and new ones evolved. This process has been exemplified here by my discussion of animals' social signaling and differences in the facial and vocal expression of emotions in animals and man, as well as the respective brain mechanisms involved. It turns out that human-specific faculties require newly evolved specific brain mechanisms that rest on older ones. So far these brain mechanisms are identifiable only in bits and pieces, but nevertheless the question as to "what is peculiar to the human brain" is answerable in principle.

References

Armstrong E (1986): Enlarged limbic structures in the human brain: The anterior thalamus and medial mamillary body. *Brain Res* 362:394–397

Darwin C (1872): *The Expression of the Emotions in Man and Animals.* London: Murray

Jürgens U, Cramon D von (1982): On the role of the anterior cingulate cortex in phonation: A case report. *Brain Lang* 15:234–248

Jürgens U, Ploog D (1970): Cerebral representation of vocalization in the squirrel monkey. *Exp Brain Res* 10:532–554

Jürgens U, Ploog D (1981): On the neural control of mammalian vocalization. *Trends Neurosci* 4:135–137

Lorenz K (1950): *Evolution and Modification of Behavior*. Chicago: University of Chicago Press

MacLean PD (1964): Mirror display in the squirrel monkey, *Saimiri sciureus. Science* 146:950–952

MacLean PD (1990): *The Triune Brain in Evolution*. New York: Plenum Press

Ploog D (1970): Social communication among animals. In: *The Neurosciences,* Schmitt FO, ed. New York: Rockefeller University Press, pp. 349–361

Ploog D (1972): Kommunikation in Affengesellschaften und deren Bedeutung für die Verständigungsweisen des Menschen. In: *Neue Anthropologie,* Gadamer H-G, Vogler P, eds. Stuttgart, Germany: Thieme-Verlag, pp. 98–178

Ploog D (1988a): Neurobiology and pathology of subhuman vocal communication and human speech. In: *Primate Vocal Communication,* Todt D, Goedeking P, Symmes D, eds. Berlin: Springer-Verlag pp. 195–212

Ploog D (1988b): An outline of human neuroethology. *Hum Neurobiol* 6:227–238

Ploog D (1989): Psychopathology of emotions in view of neuroethology. In: *Contemporary Themes in Psychiatry,* Davison K, Kerr A, eds. London: Royal College of Psychiatrists, pp. 441–458

Ploog D, Blitz J, Ploog F (1963): Studies on social and sexual behavior of the squirrel monkey *(Saimiri sciureus). Folia Primatol* 1:29–66

Ploog D, MacLean PD (1963): Display of penile erection in squirrel monkey *(Saimiri sciureus). Anim Behav* 11:32–39

Ploog D, Melnechuk T (1970): Primate communication. In: *Neurosciences Research Symposium Summaries* 4:103–190

2

Truth in Dreaming

MASSIMO PIATTELLI-PALMARINI

Introduction

The goal of the workshop, upon which the present volume is based, was to promote an earnest exchange of ideas between neuroscientists and researchers in other disciplines. The "Description of Goals" circulated by the organizers urged us "to explore how modern neuroscience has changed conceptions of what it means to be a human being." I will prudently leave this formidable task to the neuroscientists themselves, but there is something I do want to say about ways in which modern *cognitive science* has changed our conceptions of what it means to *explain* human behavior. Certain general considerations may be seen as relevant by neuroscientists and I will attempt to stress such relevance. The particular departure point I have chosen here is the *logical* problem represented by a scientific explanation of meanings in natural languages. Summarily stated, this is the problem of mapping linguistic sounds onto mental representations that have a definite content, this content (a proposition) having in turn the property of being true or false, or, more generally, of satisfying certain relations of adequacy with respect to real or possible "states of affairs." More on all this in a moment. Adopting a terminology that has become common in linguistics and in cognitive science, the expression "logical problem" suggests an analysis of the basic assumptions that any theory of meaning, reference, and truth is bound to share, and a reflection on the fundamental nature of the task, irrespective of the specific hypotheses one chooses to adopt. Let's look at these points in turn.

The Place of Meaning in a World of Causes

It is a powerful temptation to consider meanings, and our knowledge of them, as causal factors in the explanation of much of what we do, think, believe, and say. Very often, an explanation of our actions and our mental states that is based on what we were *told* is not only legitimate (I mean epistemically legitimate), but constitutes the *only* explanation possessing adequate specificity, plausibility, relevance, and systematicity (Piattelli-Palmarini, 1990). For instance, we surely did not gather in Woods Hole for our symposium because we were driven there by physical forces, or because we were satisfying our primary biological needs. We went there because we were told to meet there. A certain message had reached us, kindly inviting us to be at a certain place, at a certain time, for a certain length of time, in order to responsibly engage in a series of scholarly conversations. The meaning of those messages, and our access to that meaning, given certain shared assumptions, norms, and utilities, is *the* explanation of why we gathered in Woods Hole. Yet, it would certainly be inaccurate to say that those messages were the *cause* of our being there. Nothing, properly speaking, was "the cause" of our being there. We went because we decided to go. Those messages persuaded us to go, they did not "cause" our going. The fundamental problem in the explanation of human action and thought is that a lot of it is not caused *at all*. This is a fundamental reflection that goes back, of course, to Descartes and which has been recently reemphasized over and over by, among others, Noam Chomsky (Chomsky, 1986, 1988b, 1988c).

The deep puzzle is that human behavior is *neither* caused by sensory inputs, *nor* unrelated to them. It is "appropriate to" those inputs and it is selectively "occasioned by" them. An important part of the logical problem of the explanation of behavior, notably including the lawful and systematic dependency of behavior on our mental states (our beliefs, desires, utilities, values, etc.), is one of constructing a scientifically rigorous and truly explanatory link from classes of specific stimuli to classes of specific mental representations and their contents, and via them, to classes of specific "appropriate" behaviors. Everything we know suggests that this link must be *neither* causal *nor* stochastic. This is the chief challenge we have to meet both in cognitive science and in the neurosciences. I must confess that, in spite of awesome recent progress in the understanding of subtle neuronal mechanisms, the volitional component (i.e., the class of properly *un*caused events in the brain) seems to me to still constitute a total mystery for the neuroscientist. Any light, however dim and indirect, one can throw upon these highly elusive components of brain activity represents, almost by definition, a contribution to our "conceptions of what it means to be a human being."

Intimations of a Science of Meaning

Let's take a brief look at certain important conclusions (I do not dare call them "discoveries") reached at the "software" level, that is, in linguistics and in cognitive science. This is the level of analysis of abstract *structures* (not to be equated with "behavior" as commonly defined) and of their *systematic* effects on meaning.

Some of these conclusions are, I believe, worthy of being singled out:

a. Our routine segmentation of the strings of linguistic sounds into salient linguistic units does not match, in general, the purely physical segmentation of the signal. It is not just the case that linguistically salient units (words, phrases, syntactic constituents, clauses, etc.) do not "exactly" or "exhaustively" match the acoustic intervals of speech, the point here is that these two orders of "parsing" often do not match *at all* (Halle, 1990). Let's take the following simple sentence:

(1)   Father sat in a chair

and let's look at its sound-spectrogram, millisecond by millisecond. (This recording was made at the Max Planck Institute for Psycholinguistics in Nijemegen, but many similar recordings are routinely available in many laboratories for a variety of analogous sentences in a variety of languages [Zwitserlood, 1990].) We see that the acoustic intervals do not correspond *at all* to the separation between words. In fact, it is a well-known phenomenon that we are unable to *hear* where a word ends and another word begins in a language we do not understand (Halle, 1985). The purely acoustic parsing would generate

TINA CH AI R

as words of English. It is obvious to any of us, as it is to any child acquiring the language, that they are not. *Why* this is so "obvious," though, is far from obvious. Research in the domain of early language acquisition (as early as one week after birth) has shown that the newborn actively (though, of course, unconsciously) *projects* onto the stream of sounds a very restricted innate set of *highly abstract* hypotheses, rapidly converging on the right ones by a process of selection that bears intriguing resemblances to

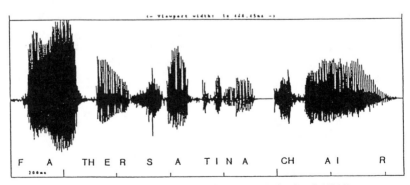

FIGURE 2.1. Father sat in a chair. From Zwitselood (1990).

how some birds acquire the song patterns typical of their species, subspecies, or local population (Mehler, 1990; Mehler et al., 1987).

Much the same *kind* of lesson is to be derived from the process of lexical acquisition and from the intimately related process of the acquisition of syntactic structures. Once again, the purely physical (acoustic) segmentation of the incoming stimuli is *as such* unreliable. The mismatches between "brute" physical segmentation and proper linguistic parsing are so pervasive that we have to conclude in favor of the hypothesis of highly specific, internally (presumably innately) determined sources of active projection onto the stimulus. In the light of overwhelming evidence, we have to reject the traditional instructivist hypotheses of "learning," that is, a molding of the brain/mind by the environment, a transfer of structure *from* the environment *to* the mind (Piattelli-Palmarini, 1986, 1989).

Nobody is claiming that there is a magic process of direct "mind reading." We do basically understand "what we hear." The point to be stressed is that "what we hear," the physically characterized linguistic stimulus, contains enough *cues* for us to construct a family of specific abstract mental representations. It constitutes sufficient *evidence* for us to select the right hypothesis, but it does not "determine" the hypothesis (in this case, a family of abstract structural descriptions of the incoming acoustical string). I want to stress that the notion of cue is quite different from the notion of cause. A cue constitutes relevant evidence only to the sufficiently structured ear. A cue, at variance with a proper causal determinant, does not itself do the work: it constrains and canalizes the work carried out by an active and "prepared" mind.

b. A lot can be said about the systematic effects of abstract linguistic structures on meanings that are in*dependent of context* (independent of storytelling, of appreciation of pragmatic uses, of the beliefs, desires, and utilities of the speaker and the hearer). Evidence for this conclusion is overwhelming and is to be found in a huge technical literature (for reviews, see Chierchia and McConnell-Ginet, 1990; Higginbotham, 1985, 1988, 1989; Jackendoff, 1990). A few selected examples will suffice here.

Lexical Effects

Let's examine the relations between these sentences:

(2)   I persuaded John to go to college
(3)   John (now) intends to go to college
(4)   I (now) intend to go to college

If (2) is true, then (3) must be also true. (4) might also be true, contingent upon (2) *and* some story we can concoct. But the *necessary* link between the truth values of (2) and (3) does not depend on *any* story. It depends only on the respective meanings of the verbs *persuade* and *intend*. It is an analytic dependence and we do not need to know anything at all about John, about the speaker, or the college in question in order to fully appreciate the connection between (2) and (3). Plainly, this is not the case for the *contingent* link (if any) between (2) and (4). This link can *only* be constructed by mounting up a plausible story. Examples of this kind are abundant and quite straightforward for a variety of verb-meanings (Grimshaw, 1990; Tenny, 1988a, 1988c).

Effects of "Larger" Units

Possibly, the most striking examples are offered by cases of mandatory coreference, mandatory disjoint reference, and optional coreference between elements of a same sentence. "His" is preferentially (though not obligatorily) coreferent with "every man" in the sentence

(5)   Every man kisses his mother

but necessarily noncoreferent in the sentence

(6)   His mother kisses every man

"Himself" is obligatorily coreferential with "John" in the sentence

(7)   John expects to support himself

but obligatorily *not* coreferential with "John" in the "larger" sentence

(8)   I wonder who John expects to support himself

Notice that in these, as in myriads of other cases (Chomsky, 1981; Lasnik and Uriagereka, 1988; van Riemsdijk and Williams, 1986), the systematic effects of linguistic structures on meanings are, so to speak, intrinsic. We only have to look at (or listen to) the sentence *in isolation* in order to have access to these effects. We do not need to know anything at all about the speaker, the characters appearing in the sentence, nor about any story that might make these sentences more relevant to our interests and concerns. These facts about meaning are context *in*dependent. Data of great scientific relevance can be collected by examining identical, or analogous, context-independent meaning effects of linguistic constructions in other languages, or local dialects. Which leads me to the next point:

   c. *Significant insights into the systematic mapping of sounds onto meanings* in our language *often need comparative data from* other *languages. Data on "what we mean by what we say" that are simply derived from individual introspection are not* sufficient *for a scientific semantics.* (A corollary, rightly stressed by J. T. Higginbotham, is that we do not have full spontaneous access *even* to the meanings of many terms in our own *private* idiolect [Higginbotham, 1988].)

   This may appear so counter intuitive that at least a few telling examples are required.

In many cases the genitive *of* and the agentival *by* allow for a corresponding adjectival expression that is perfectly synonymous:

(9)   The speech of the president
(9a)   The presidential speech

(10)   The invasion by the Germans
(10a)   The German invasion

but this is not always the case:

(11)   The death of the president
(11a)   *The presidential death

(12)   The arrival of the Russians
(12a)   *The Russian arrival

These are, plainly, not synonyms any more (Giorgi and Longobardi, 1991). The adjectival construction (if at all acceptable) now conveys the idea of a *way of doing* something, it does not specify the agent as in the two previous examples. We have very clear introspective intuitions of which expressions allow this synonymy and which do not, but we would be utterly unable to tell *why*. Nothing in the introspectively accessible *meaning* of these terms allows us to generate the good and the bad examples. Properties of the *verb* from which some of these nominals derive offer a cue. As a first rough approximation, nominals derived from transitive verbs, in the main, tend to allow for this synonymy *(invade, attack, occupy)*, while intransitive verbs, again in the main, tend to disallow it *(die, maturate, arrive)*. Plainly, this cannot be the whole story. Exceptions are too numerous *(insistence,* from the intransitive *insist,* and *allegation,* from the intransitive *allege,* allow it, while *knowledge,* from the transitive *know,* disallows it). Moreover, there are plenty of nominal expressions that derive from no verb at all (at least from no *contemporary* English verb); some allow it *(affluence, predicament, strategy, disease, superiority)*, while others disallow it *(importance, defamation, inception)*. Little wisdom do we derive from the transparently accessible meanings of these expressions. (*The predicament of the Germans* is the same as *the German predicament,* but we cannot say *the presidential importance* to mean *the importance of the president.* Why?) A better, but still insufficient, cue is to pay attention to what the agent has to do, in order to satisfy the adjectival expression. When the action, or property, so to speak, strongly "emanates" from the agent, then the construction is valid *(invade, reach affluence, be superior)*. When the "agentival role" is less prominent, or the agent is the passive receptacle of the property *(death, maturation, arrival, being considered important by others,* etc.), then the construction is invalid (Giorgi and Longobardi, 1989; Grimshaw, 1990; Higginbotham, 1985). To cut a long story short, it appears that *none* of these *intuitive* criteria, which are accessible to us with a moment's reflection, can *fully* explain the phenomenon.

The beginning of a satisfactory explanation has recently come from a comparison between the syntax of noun phrases in the Romance and in the German languages. Interesting asymmetries appear between different families of languages. This intriguing phenomenon has been found to be deeply related to other syntactic, heretofore distinct and prima facie unrelated, phenomena. Just to give a flavor of the explanatory strategy, one has to *also* take into account the reasons why *The my book is ungrammatical in English, while *Il mio libro* is perfectly grammatical in Italian; why in English we do not say *the Callas to refer to the famous soprano (or to anyone picked out by a family name), while in Italian we currently and quite correctly say *la Callas, la Navratilova,* etc. (although their masculine counterparts *il Pavarotti, *il Gianni are ungrammatical in standard Italian and are to be heard—but never written—only in certain dialects). A tentative explanation that commandeers all these superficially distinct phenomena under a single powerful principle relies on technical notions such as those of differently set syntactic parameters, of "specifiers" and "maximal projections," of assigning reference at the level of logical form versus assigning reference at the level of D-structure and surface structure (Belletti and Rizzi, 1981; Burzio, 1986; Giorgi and Longobardi, 1989; Rizzi, 1982). Clearly, nothing of all this is introspectively accessible to us, and there was little hope of arriving at these general, far-reaching, and deep explanatory hypotheses by examining one language only.

The full scientific elucidation of "what we mean by what we say" often consists of unraveling our *tacit* knowledge of *our* language, and this *may* require the full and systematic study of *other* languages as well. We do not get enough semantic evidence just from the introspective analysis of what we mean by what we say, and from how we use certain expressions in our own language. This is a crucial fact about semantics that is often difficult to convey to the philosophers of "ordinary language" in the orthodox analytic tradition (for a thorough analysis, see Chomsky, 1986).

d. It is crucial to analyze what every *speaker of a language* would *say (and not say) in certain* imaginary *and counterfactual situations.* (There is an intimate and systematic connection not only between meaning and the state of affairs that make an expression true—or false—in the real world, but also between meaning and *hypothetical* states of affairs that *would* make the expression true—or false—in "possible worlds " [reviewed in Chierchia and McConnell-Ginet, 1990].)

All attempts to "reduce" meaning to objective statistical co-occurrences of linguistic sounds and physical (or at any rate "public") arrangements of objects have failed (Fodor, 1987, 1990; Higginbotham, 1983). The demise of behavioristic semantics is probably the most exemplary case, but other attempts, for instance, those based on "informational content," could be cited (Dretske, 1981, 1988) One major stumbling block for all these physicalistic reductions of meaning is constituted by the capacity of meaningful expressions in all natural languages to refer to events, including events in the past and in the future, to possible states of affairs, and even to explicitly *im*possible states of affairs (Bennett, 1988; Schein, 1986). Plainly, many such referents cannot *in principle* be characterized physically.

Yet, we have perfectly straightforward understanding of counterfactual conditionals such as: *If you had left home on time, you would not have missed the train.* We also have no problem whatsoever in understanding pronominal and anaphoric reference to *explicitly* nonexistent persons and things: *If you had a sister I would have married* her, *though* she *would have been most unhappy to be my wife. Too bad you never wrote that book,* it *would have been a best-seller* (Fauconnier, 1985).

What makes these expressions *obviously* meaningful is our easy access to mental representations of states of affairs that are *both* relevantly like the real world *and* relevantly unlike it with regard to certain selectively chosen aspects. All serious accounts of how we manage to understand these expressions, and of how we often manage to derive from them important and *true* consequences for our life, must countenance the mental construction of, and reference to, "possible worlds" (also severally called, by different theorists: situations, schemes, scripts, frames, mental spaces, mental models, etc.) (Chierchia and McConnell-Ginet, 1990; Fauconnier, 1985; Jackendoff, 1990; Johnson-Laird, 1983). It appears that we also have to postulate our routine access to a *metric* measuring their greater or lesser "proximity" to the real world (Stalnaker, 1968, 1984). I will come back to this problem when dealing with truth in fiction (Lewis, 1983) and with the semantics of dreams.

How crucial our access to possible and counterfactual states of affairs is for semantics in general can already be seen from our understanding of the meaning of single items in the lexicon. Would it be accurate to say that I "persuaded" someone to do something if I had injected into his veins a psychotropic drug that determines passive subservience to the will of others? Would it be accurate to say that we have "intercepted" today a message sent during World War I and which had then been lost and forgotten? Our robust and quite systematic intuitions about what we would say, or not say, in these imaginary states of affairs is a *crucial* component of semantics. Someone (say, a child or a foreigner) who answers such questions incorrectly thereby shows that he has not understood the meaning of that word.

e. There are specific *tools for a scientific semantic analysis, and there are standard problems that* every *semantic theory must be able to explain.* (It is not satisfactory to construct ad hoc semantic theories that explain some phenomena—possibly generated by weird and infrequent expressions—but that do not even begin to account for other current, garden-variety problems of meaning and reference. It is not satisfactory either—for the reasons explained above—to construct a *mono-lingual* semantic theory, a theory that only holds for, say, English, but that would not be equally applicable to Chinese or Italian) (Chierchia and McConnell-Ginet, 1990; Dowty, 1979; Higginbotham, 1985, 1989; Jackendoff, 1983, 1990.)

Some of the precious tools have been set up through an extensive study of synonymy, intertranslatability across languages, quantification, and the calculus of presuppositions and implications. The typical syntactic and semantic phenomena covered by these in-depth analyses are (the list being, of course, always open to new entries): passivization, relativization, extraction, questioning, nominalization, adverbialization, anaphora, verb tenses, progressives, aspectualization, counterfactuals, antonyms, and opposites. The essential information here is that there is a corpus of highly specific data and well-defined problems, some of which are extracted from a variety of languages the world over, and that every semantic theory these days *has to* be able to tackle. We have already met several examples. I will only offer a few more to show that many terms and expressions in our languages routinely make reference to *events* (Bennett, 1988; Davidson, 1980; Enç, 1990; Grimshaw, 1990; Higginbotham and Schein, 1989; Schein, 1986). This is another physicalistically *non*reducible semantic element that will become important when examining the semantic content of dreams.

We can start from the following two couples of *superficially* analogous sentences:

(13)   It rains, but it does not snow
(13a)   It rains, but it does not matter

(14)   He bought a ticket, but it was too expensive
(14a)   He bought a ticket, but it was too late

Plainly, they are not *really* analogous. In fact, the pronoun *it* in the second member of each couple shows weird referential properties. It refers not to a "thing," but to a *fact* suitably described. As the sentences make it all clear, the pronoun refers to the fact *that* it rains, to the fact *that* he had bought the ticket. Plenty of expressions betray this referential attachment to events, not to persons or things. This includes also certain "silent" elements of speech (technically called "empty categories"), that have no acoustic expression, and therefore no physical "reality" (Chomsky, 1981; Giorgi and Longobardi, 1989; Lasnik and Uriagereka, 1988; MacDonald and Clark, 1987; van Riemsdijk and Williams, 1986). For instance, the Italian word-for-word translation of these sentences *has to* leave the various occurrences of *it* acoustically unexpressed:

(13′)   *Piove, ma non nevica*
(13a′)   *Piove, ma non importa*

(14′)   *Comprò il biglietto, ma era troppo caro*
(14a′)   *Comprò il biglietto, ma era troppo tardi*

These unexpressed (and, in the above Italian examples, strictly inexpressible) syntactic elements share with pronouns a number of crucial properties (and are thus technically called "pro"—lowercase). Both "pro" and an analogous, though syntactically distinct, empty category called "PRO" (uppercase) can refer to events. Striking as it may sound, reference to nonphysical entities (i.e., events) by physically inexpressed linguistic elements is what we *have to* appeal to, in order to fully account for certain routine syntactic and semantic phenomena (Dowty, 1990; Enç, 1990; Hornstein, 1990; Pustejovsky, 1988). For instance, let's examine the differences in meaning between superficially similar sentences such as:

(15)   The boat was built to brave any storm
(15a)   The boat was sunk to collect the insurance

What happens in these cases can be rendered at an intuitive (i.e., nontechnical) level: in (15) the acoustically empty element PRO, which every speaker mentally—and of course unconsciously—inserts between "built" and "to," is interpreted as referring to "the boat" (i.e., a thing), while in (15a) it is interpreted as referring to the sinking of the boat (an event, not a thing).

On the basis of many subtle syntactic and semantic data of this kind, it appears unwise to conclude that certain *public* and "concrete" strings of sounds or characters connect with other, equally public, strings of sounds and characters. This interpretation may appear, at first blush, more prudent and more realistic, because it would "only" acknowledge that overt expressions may connect to other overt expressions (for instance, descriptions). The fact is that this surface-to-surface interpretation would fail to do justice even to the simple data we have just encountered. The correct interpretation is vastly more radical: there are in our languages both overt (acoustically and graphically expressible) and "covert" (acoustically and graphically unexpressed—or even inexpressible) devices, suited to make reference to nonphysical individuals, namely to *events.* These are best conceptualized as *abstract* individuals possessing their own specific identity (raining-there-and-then, my buying a ticket at 8:05, his sinking the boat, etc.). They typically consist of real or *possible* actions, changes of state, happenings, and mental processes *under certain salient descriptions.* Cogent arguments based on considerations of economy, compactness, generality, and on *specific* explanatory hypotheses, militate in favor of the second choice: our "ontology" has to accommodate events and reference to events as perfectly routine devices exploited by natural languages.

Let's briefly consider further kinds of evidence (Tenny, 1988b):

(16)   The paratroopers landed in an hour
(16a)   Paratroopers landed for an hour

In the first case, there is one single event, the landing of the paratroopers, and this event is encompassed within the time interval of an hour. In the second case, there are multiple events, each constituted by the landing of a single paratrooper, and the series of all these separate events is contained within the time interval of an hour. Note that the *physical* characterization of the state of affairs is exactly the same. A quality camera and a reliable watch will give exactly the same data. Yet, the truth conditions of (16) and those of (16a) are different. The syntactic differences generate different descriptions, which successfully pick out different events. There is no other elegant and compact way of explaining how we understand those sentences. Reference to events and different quantifications over events, as routinely found in expressions such as "in an hour" and "for an hour," offer a simple, powerful, and easily generalizable explanation. For instance, verbs of achievement, those that represent actions with a "culmination" (Puste-jowsky, 1988), freely admit the construction "in an hour," "in five minutes," etc. We can say *He built the house in an hour, She solved the problem in three minutes,* but it sounds weird to say **She solved the problem for three minutes.* On the contrary, if she solved many problems (there are many such events) we can say *She solved problems for an hour.* Verbs that do not express actions possessing a culmination behave differently: they freely admit the construction "for," but not the construction "in." *He sang for an hour, She walked for an hour,* but not **He sang in an hour,* **She slept in an hour.* In order to make this "in" construction acceptable we have to *generate* the culmination: *He sang the song in five minutes, She walked to the bridge in an hour.* Reference to events and quantification over events make all these phenomena (and many more) systematically explicable by elegant and general syntactico-semantic theories, applicable to all languages.

There are also cases that vindicate in a particularly compelling way reference to events and *not* to things or sets of things. These are constituted by sentences such as (May, 1985; Schein, 1986):

(17)   Ten boys ate ten pies
(18)   The eating of ten pies by ten boys
(19)   Ten men unloaded five trucks
(20)   The unloading of the five trucks by the five men

(21)   The bushes grow thicker in the middle of the forest
(22)   I resent the approval of all the budgets by some delegates

The gist of the strategy here is to analyze all the possible sets (of pies, trucks, budgets, etc.), and all the corresponding possible sets of boys, men, delegates, etc., exhausting all the combinations of these sets, and of the elements thereof, that make these sentences true or false. The result is that no construction of sets does justice to our syntactic and semantic knowl-edge. Only the construction of events (events of eating pies, unloading trucks, growing thicker, approving budgets) do manage to account for our syntactic and semantic judgments. What is more, all the obvious inferences and presuppositions of all these sentences come out *as* obvious, and the uncertain or hard cases come out *as* uncertain or hard. As they should. This kind of "preservation of inferential ease" does *not* come out from *any* analysis of these cases in terms of reference to sets of objects and/or persons.

These are some of the reasons why the introduction of events, of reference to events, and of quantification over events, has recently contributed to a boom in the explanatory power of theories in syntax and semantics. From this follows the natural conclusion that, therefore, there *are* events in the ontology of linguistic theory, and that we "really" routinely build mental representations of events. This *kind* of conclusion is presently resisted by some philosophers (notably John Searle) but it appears unproblematic to many professional syntacticians and cognitive scientists.

It appears that, if all this is even approximately on the right track, neuroscientists may one day discover the neuronal correlates of these processes. If we really construct events in our mind, and if we really refer to them and quantify over them, then something in our brain allows us to do it. It will be fascinating to discover one day what are the neuronal bases of our mental event-construction and event-manipulation. Something that the neuroscientists would not have looked at, had the "software" scientists not discovered the existence of these phenomena.

A Methodological Caveat

To a hard (or rather "wetware") scientist, all this talk of sentences, intuitions about meaning, reference to nonmaterial entities, and so on, must appear as poor evidence indeed, if not as outright wizardry. It is in part to enlarge the domain of relevant evidence for semantic theories that I recommend also examining dreams and their contents. I said enlarge, not improve or refine. In fact, the kinds of evidence I have just summarily sketched are, and will be for some time, the *central* data for semantics, just as judgments about grammaticality are, and will be, the central data for syntactic theory. A short digression on this topic may prove useful (Chomsky, 1988a, 1988c).

If introspective evidence and judgments about meaning were indeed considered useless as data for science, then nothing I have to say about dreams and their contents would command any attention. Moreover, biologists are (in my experience at least) *particularly* reluctant to accept expressions in ordinary language and intuitions thereon as something even remotely deserving the status of scientific data. They tend to regard pieces of "behavior" as data, but not the abstract structures that underlie these pieces of behavior. Now, it would be quite easy to *superficially* transform the primary data for syntax and semantics into bona fide pieces of behavior. We could just ask the subjects to press a button, or to raise their right hand, if and only if the sentence we present to them satisfies certain criteria (is grammatically well formed, acceptable as such in a standard conversation, synonymous with some other sentence, unambiguous in meaning, etc.). This is routinely done in psycholinguistics, because the *precise* measurement of *time delays* in the responses is often crucial (for a review, see Garrett, 1990). When, on the contrary, the subject is allowed to reflect as much as he or she wants to before issuing an answer, these scientistic paraphernalia become totally unnecessary. We could also place electrodes on the scalp and measure how certain characteristic wave forms correlate with certain syntactic and semantic properties of the input. Once again, this is nowadays done in psycholinguistics (Garnsey et al., 1988; Kutas and Hillyard, 1980) because it is, for instance, important to monitor the timing of different computational *paths* in the parsing of a sentence, when this sentence is flashed word-for-word on a screen, under controlled rates of presentation. Magnetic resonance imaging (MRI) might soon offer still better techniques and it is not totally out of the question that one day these sophisticated recordings might become routine even in the departments of "pure" linguistics. It needs to be stressed, however, that *any* such device will have to be *calibrated,* if it is to be of any use. The initial (and crucial) calibration will *have to* rely on "typical instances" of the linguistic phenomenon we want to monitor, that is, on the standard *intuitions* of a standard speaker–hearer, about certain standard examples. And this, of course, can *only* mean that we have to trust intuitions about sentences and expressions in ordinary language. The initial baseline for any such study will have to be built on the present kind of data, and on the present kind of evidence. The new techniques might well, for all we know, improve the field dramatically and even allow to build, progressively, step by step, different baselines to calibrate ever more refined phenomena. This science-fictional scenario does *not* make the actual data for syntax and semantics irrelevant or obsolete. They will still constitute the proper *evidence for* these explanations and the ultimate *control* for all these findings. No biologist has ever claimed, in virtue of the modern techniques of gene-sequence analysis and polymerase chain-reactions, that it is meaningless to maintain the distinction between insects and mammals, or between neurons and glial cells. We do not expect this "dismissal" to happen in linguistics either.

Since *individual* intuitions about meaning, reference, truth, and well-formedness *are* scientifically relevant data for semantics, I urge that reports on dreams can also be relevant. I will now consecrate the rest of this chapter to show how, and why.

The Content of Dreams as Data for Semantics

In the light of overwhelming introspective evidence, of physiologically monitored sleep-states and well-controlled neuropharmacological effects on sleeping and dreaming, we can safely assert the following (for a truly comprehensive interdisciplinary review, see Hobson, 1988):

1. *Human beings, and many mammals, regularly have dreams*
2. *Many dreams have content* (Michel Jouvet has managed to prove, beyond reasonable doubt, that this is the case even in cats [Jouvet and Michel, 1959].)
3. *Many of these contents are "about" something specific* (persons, places, events, and actions)
4. *The dreaming subject has direct and vivid access to these contents*
5. *Many "characters" and places in the dreams have their full-blown identity.* (Surely identity enough to allow for specific and unambiguous subsequent recall, once the subject is awake.)

If all this is correct (and it *is,* for all we know) then these simple facts *already* have interesting consequences for some semantic theories. I will start by pointing out the semantic consequences that we must derive from the very *fact* that dreams have identifiable contents. In a second phase I will derive other interesting consequences from the *kind of content* that certain dreams can have.

Since dreams have content, then . . .

A widespread misconception of semantics is summarized by the motto "meaning is use." The core idea is that of an *identity* between meaning and the socially admissible ways for us to use linguistic expressions in ordinary life. A variant of this conception, which is notably associated with the later works of the philosopher Ludwig Wittgenstein, equates our understanding of meanings with the ability to successfully engage in communication, and meaning as such with procedures specifying a variety of "language games" (for a critique of this position, see Chomsky, 1986; Fodor, 1987, 1990). Plainly, if dreams have content *while they happen,* and do not *acquire* (pace Norman Malcolm) content *only* post facto, when the subject makes a *public* report of the dream, then this conception is utterly untenable. And, in fact, I think it is, also on many other grounds and for other compelling reasons (Higginbotham, 1983, 1985, 1989).

The contentfulness of dreams strikes me as a further definitive blow to this line of semantic theorizing. I always found very symptomatic that certain philosophers have wished to go out of their way (and out of their depth) in order to *deny* that dreams have content. The most representative piece of such obdurate denial is Norman Malcolm's short book *Dreaming* (Malcolm, 1959), aptly rebuffed by Hilary Putnam (Putnam, 1975a). The philosopher Daniel Dennett has also attempted to demonstrate that dreams may well receive their content in the reawakening phase, in the relatively few minutes during which the sleeper comes back to the real word and reconnects with public life (Dennett, 1969, 1978). There is no physiological plausibility whatsoever to these desperate attempts (Hobson, 1988; Mamelak and Hobson, 1989), but it is indeed symptomatic that philosophers of this meaning-pragmatic tendency are definitely *embarrassed* by the contentfulness of dreams.

I feel entitled to stress that the contentfulness of dreams is per se a further confirmation of the *individualistic, mental-representational,* and *intrinsic* nature of meaning. It discourages us once again from *explaining* meaning in terms of communication and *collective* causal links with the world. Dreams are a sort of semantic "experiment of nature." We observe a subject who lies motionless, isolated from any sensory and factual commerce with the world, *but* who can *still* generate *specific* meanings and have access to them.

It will now repay us to examine what kinds of meanings these can be.

The Bizarreness of Certain Dreams

Table 2.1 is (an abridged and simplified version of) a classification of the bizarreness of dreams painstakingly developed, and routinely used in clinical practice, by A. N. Mamelak and J. Allan Hobson.

The following excerpts are glimpses from a collection of 233 dream reports, diligently written down in a special journal by a subject known as "The Engine Man." I have copied them from Professor Hobson's large corpus of data, to serve here as a quick reminder of the kind of dreaming experiences that each of us might easily summon from personal recollections. I took the liberty of underlining some passages that are particularly relevant to our present semantic revaluation of dream contents. It will be especially rewarding for us to remark how the *identity* of people and places is *maintained,* in the face of striking and flagrant denials of the most essential defining attributes.

Walking South on 14th St., just south of Pennsylvania Avenue. Street was very muddy. . . . A few blocks (about 3) south of the avenue (Pa. Ave.) I turned east, passing behind various buildings none of which seemed large. No one in sight except my companion, *a child* of perhaps 6 to 8 years, *who later turned into Jason but who, at first, seemed like a stranger.* [Hobson, 1988, p. 272.]

TABLE 2.1. Two-stage scoring system for dream bizarreness*

Stage I:	Mark as bizarre physically impossible or improbable items in: A. The plot, characters, objects, or actions B. The thoughts of the dreamer or dream character C. The feeling state of the dreamer or dream character Localize each item of bizarreness in the dream domain (as it is reported)
Stage II:	Characterize the item as exhibiting: A. Discontinuity (change of identity, time, plots, or features) B. Incongruity (mismatching features) C. Cognitive abnormalities (non sequiturs, ad hoc explanations, explicit vagueness)

*Adapted from A. N. Mamelak and J. A. Hobson, *Journal of Cognitive Neuroscience* 1(3):201–221 (1989).

From Hobson's characteristically insightful comments on this fragment of the report, we learn that, in spite of vast incongruences (muddy street, small buildings, etc.), of which the dreamer is *perfectly aware,* both during the dream and after, he *still* pictured himself in Washington, DC, *while the dream lasted.* Let's make a note that in the real Washington, DC, as the subject *knew full well,* such a walk would have brought him straight in front of the White House, but in the dream it does not. Let's also make a note that Jason, the dreamer's nephew, was in fact *an adult.*

The Engine Man then continues in his report:

> It was at the Customs Building, where all animals (except small ones such as cats) *must be registered or declared, weighed, and the proper tax paid. . . .*
> It was a 3-story building, of white stone with "ramps" on outside apparently to enable animals to reach the upper stories . . . We entered the building *somehow (not by means of ramps)* went to the upper stories and looked in several rooms . . . [Hobson, 1988, p. 272.]

For a final assault on the canons of identity, let's read the following:

> Jason went from door to door, pushing each open. Usually we saw two persons in a room. . . . In each of two other rooms was a girl (young lady) in nurse's uniform. Each was talking to small persons, *evidently children* in years *but with aged and deeply lined faces.* [Hobson, 1988, p. 274.]

This is surely enough reporting for our present purposes. What is of interest to us, in view of a reassessment of semantic theories, is the almost unlimited *tolerance* vis-à-vis causal anomalies, contradictions, violations of identity, exceptions to necessary and sufficient conditions, lack of unity of place and time, general vagueness of properties and attributes. Yet, people and places and objects *possess* an identity in the dream. They *are,* in the dream, Jason and Washington, DC, in spite of their manifestly incompatible attributes. In dreams we have another semantic "experiment of nature": the best evidence so far of the ubiquity, and the centrality, in our mental life of what Kripke and Putnam have called "rigid designators" (Kripke, 1980; Putnam, 1975b).

The notion of these rigid designators is, in essence, the following: we have to account for our remarkable capacity to reason, and often to reason *cogently,* in counterfactual situations (Fodor, 1987; Kratzer, 1988). We must also account for our remarkable capacity to change (even drastically) our descriptions of, and beliefs about, a *same* object or person. A certain line of thinking, inaugurated by Aristotle and then refined by Bertrand Russell, had decreed that we pick the referents of our discourse through definite descriptions, by detecting properties that are individually necessary and jointly sufficient to single out the proper referents, and these only. Kripke and Putnam object to this theory of reference, pointing out that reference often *remains* fixed also when the description changes drastically, and when we do not have the faintest idea of what these "individually necessary and jointly sufficient conditions" are. Proper names (Plato, Richard Nixon, etc.) and the terms for "natural kinds" (tiger, lemon, electron, water, etc.) turn out to be typical "rigid designators." We manage to *stably* refer to them in spite of changing descriptions and in spite of changes in their characteristic properties. We understand, in fact, without any problem, expressions such as "Richard Nixon might not have become President," "If Plato lived today, he would like truth-conditional semantics," "If electrons possessed twice the electric charge they possess, matter would be unstable," etc. These different descriptions maintain the same referent. They still are "about" Nixon, Plato, electrons. Yet they project them onto different, counterfactual states of affairs, or different "possible worlds."

In a nutshell, the main objection to a referential mechanism that would be *exhausted* by our competent use of necessary and sufficient criteria lies in the fact that we could not maintain the *same* referents when we drastically *change* these very criteria. Since we *do* often change them, and *still* manage to keep our referents constant, there must be also something else in the referential mechanism.

A rigid designator, unlike a description, *always* picks out the same referent in *every* possible world in which that referent is supposed to *exist*. Proper names, for instance, are the archetypical (though not the only) rigid designators. How they manage to be "rigid," according to Putnam and Kripke, is a complicated story. For us, it suffices to stress that, in their scheme, a crucial component is an *objective* causal link between the term and *actual* true exemplars, bona fide referents, in the real world. This "collective" and "objective" part of the story cannot possibly work for dreams. We will have to rejustify the referential powers of rigid designators *without* the collective-objective component. I think that this needs to be done not only for rigid designators in dreams, but for rigid designators in general, even in plain natural language. This is a most interesting consequence for semantic theory, further corroborated by data on the contents of dreams, on which I cannot delve here.

Another interesting consequence has to do with the *impossibility* to establish a "metric" of proximity between the real world and the worlds of dreams (again, while they last and while they display the meanings and the references they display) (Bennett, 1988; Chierchia and McConnell-Ginet, 1990; Davidson, 1967; Dowty, 1990; Higginbotham and Schein, 1989; Kratzer, 1988; Stalnaker, 1968, 1984). This ought to totally *disrupt* meaning and reference *in the dream,* but it does not. People and places are what they are, do what they do, and say what they say in the dream, *in spite* of the dreamer's utter impossibility of constructing *in the dream* a metric of proximity to the real world. Yet, many dreams *feel* exactly like real worlds themselves.

The worlds of dreams are *at least* possible worlds (I say "at least" because they look mighty *real* to the dreamer while they last) and they appear to be *filled* with rigid designators: persons and places and events that are what they are in spite of drastic redescriptions, and even in spite of downright *contradictory* descriptions.

It is interesting, though, that we seem totally unable to reason cogently in (and about) the contents of our dreams. Unlike fictional contexts, in which we *are* able to reason cogently, by combining truths in the story with truths in the real life (Lewis, 1983), dreams carry all sorts of unredeemable vagueness and bizarreness. We are unable to combine truths in the dream and truths in real life in *any* way that bears consequences in either context. Hobson seems to have a good neurological explanation for this fact (Mamelak and Hobson, 1989). I am inclined to offer a convergent semantic explanation: the problem may well lie in the impossibility of constructing a "proximity metric." Since this metric, and our access to it in the real world, has been shown to be essential to our *reasoning* on counterfactuals (Kratzer, 1988; Stalnaker, 1968), to our understanding of progressives and tenses (Dowty, 1990; Enç, 1990; Hornstein, 1990), to our understanding fiction (Lewis, 1983), and to other subtle semantic inferences (Bennett, 1988), the case of dreams can be particularly illuminating. Rigid designators do preserve their power of reference, but the absence of a metric of proximity baffles the attempts of our dreaming mind to construct minimally consistent fictions. The bizarreness of dreams might be just the result one would expect, given the absence of a metric for degrees of proximity among "possible worlds."

In fact, the divorce between dream and reality does not seem to reside in a lack of meaningfulness, in a lack of referential power, or in a lack of believability. There *are* meanings in dreams, there *is* full-blown reference, and events *are* quite "believable" while they happen. The true divorce is in *any* availability of the content of dreams for *reasoning*. Pace the ancient dream interpreters, there is no truth "for us" to be found in dreams. Unlike even the craziest works of fiction, the content of dreams is *unusable* for lessons bearing on our real life. Pace the Freudians, there is, in dreams, no truth "about us" qua dreamers. The important truth to be found in dreams is *only* of a generic kind. We can derive truths about our mind/brain. Not individually, but as a species.

I will limit myself to offering these hints in order to suggest why and how data from the analysis of dreams *can* be relevant to semantic theory and to cognitive science.

I now conclude with considerations on the content of dreams as data for the modularity of the mind/brain (Fodor, 1983; Garfield, 1987).

The Peculiar "Realism" of Dreams

Dreams, as we have seen, possess a sort of extrinsic bizarreness: they do not match the probability distributions of real life, do not preserve causal inferences, and often do not preserve even logical inferences. But they also have, in my opinion, a sort of *intrinsic* bizarreness. What I find equally stunning is that they, after all, resemble reality *so much*. They should not, according to a theory that makes us the "scriptwriters" of our dreams. How can it be that *in* our dreams, while they *happen,* we:

Encounter places and faces, that we *know* we have never seen, endowed with great vividness and ample details

Suffer from the happening of events we desperately would like *not* to happen

Are anguished from uncertainties and uncertain outcomes, or are greatly *disappointed* by the final turn of events

Are afraid of events and people

Are in admiration of *superior* performances by characters we encounter (perfect native speakers of languages we barely mumble, perfect dancers, pianists, consummate lecturers, etc.)

How can this be? Plainly, nothing of the above *ever* happens to a scriptwriter. These things just *cannot* happen to someone who writes a novel. Yet they happen in our dreams. Therefore, I urge, we are *not* the authors of our dreams. Our brain/mind is, not "us."

It only takes a moment of reflection to realize that these are *some of the best data we can possibly collect on the profoundly modular nature of our brain-mind*. Only a modular brain-mind (Fodor, 1983) possessing a myriad of specialized, bullheaded, domain-driven and cognitively inaccessible (impenetrable) subcomponents can display such features. It displays these modular activities *even* when it is, in fact, *disconnected* from *any* external source of stimulation. Intense, ineliminable modularity, plus the fact that the system cannot tell the difference between externally and internally generated "data," plus the fact it is *bound* to try to make there-and-then the best sense it can of the internally generated data (Hobson, 1988; Mamelak et al., 1989) can explain all this. I do not see any other explanation that even begins to make sense.

Conclusion

There has always been, since time immemorial, a feeling that dreams convey *some* peculiar kind of truth. The long prehistory (and the anthropology) of the analysis of dreams shows that they have been often *interpreted* as being somehow true *of the real world*. True about the future, for instance. With Freud, the locus of truth switched from the external world to the internal world of the dreamer. To Freud, dreams were true of the inner, repressed tendencies *of the dreaming subject*. The modern scientific analysis of dreams (Aserinsky, Dement, Kleitman, Jouvet, Hobson, etc.) has shifted this perspective considerably. Dreams are held to be true *of our brain/ mind,* not of "us" qua persons. What I have been hinting at here is a refinement of this insight. I take dreams to be true of our *modular* brain/mind, and I take them to be able to offer precious data for semantic theory. Since the terrain has been cleared, in linguistics and in cognitive science, for the scientific legitimacy of data about our individual intuitions, the analysis of dream reports becomes also a valid source of evidence for (or against) *specific* semantic hypotheses. In this chapter I have tried to offer some very tentative guidelines for such a novel "modular" and "semantic" approach to the problem.

Acknowledgments

I am especially indebted to Noam Chomsky, James T. Higginbotham, and J. Allan Hobson for providing the primary sources of many ideas that I have freely adapted, at my own risk, to the topic of this chapter. Special thanks to David Pesetsky, who, by asking me to be the commentator of a paper by Hobson and Mamelak at an MIT colloquium, has provided a much-needed incentive to freshen up my long-standing interest in the mystery of dreams. I have carried out researches on which this work is based thanks to the generosity of the MIT Center for Cognitive Science, the Alfred P. Sloan Foundation, the Kapor Family Foundation, the Cognitive Science Society, and Olivetti Ricerca (Ivrea, Italy).

References

Belletti A, Rizzi L (1981): The syntax of "ne": Some theoretical implications. *Linguistics Rev* 1:117–154
Bennett J (1988): *Events and Their Names.* Indianapolis, Ind. Hackett Publishing Company
Burzio L (1986): *Italian Syntax: A Government-Binding Approach.* Dordrecht, The Netherlands: Reidel
Chierchia G, McConnell-Ginet S (1990): *Meaning and Grammar: An Introduction to Semantics.* Cambridge, Mass. MIT Press
Chomsky N (1981): *Lectures on Government and Binding (The Pisa Lectures).* Dordrecht, The Netherlands: Foris
Chomsky N (1986): *Knowledge of Language: Its Nature, Origin, and Use.* New York: Praeger Scientific
Chomsky N (1988a): *Language and Interpretation: Philosophical Reflections and Empirical Inquiry.* Unpublished manuscript. Department of Linguistics and Philosophy, MIT
Chomsky N (1988b): *Language and Problems of Knowledge: The Managua Lectures.* Cambridge, Mass.: Bradford Books/MIT Press
Chomsky N (1988c): *Some Notes on Economy of Derivation and Representation.* Unpublished manuscript. Department of Linguistics and Philosophy, MIT
Davidson D (1967): The logical form of action sentences. In: *The Logic of Decision and Action,* Rescher N, ed. Pittsburgh, Penn.: Pittsburgh University Press
Davidson D (1980): *Essays on Actions and Events.* Oxford: Clarendon Press
Dennett D (1969): *Content and Consciousness.* London: Routledge and Kegan Paul
Dennett D (1978): *Brainstorms.* Cambridge, Mass.: Bradford Books/MIT Press

Dowty DR (1990): *A Semantic Account of Sequence of Tense.* Conference on "Time in Language," MIT

Dowty DR (1979): *Word Meaning and Montague Grammar.* Dordrecht, The Netherlands: D. Reidel

Dretske F (1981): *Knowledge and the Flow of Information.* Cambridge, Mass.: Bradford Books/MIT Press

Dretske F (1988): *Explaining Behavior: Reasons in a World of Causes.* Cambridge, Mass.: Bradford Books/MIT Press

Enç M (1990): *Tenses and Time Arguments.* Conference on "Time in Language," MIT

Fauconnier G (1985): *Mental Spaces: Aspects of Meaning Construction in Natural Language.* Cambridge, Mass.: Bradford Books/MIT Press

Fodor JA (1983): *The Modularity of Mind.* Cambridge, Mass.: Bradford Books/MIT Press

Fodor JA (1987): *Psychosemantics.* Cambridge, Mass.: Bradford Books/MIT Press

Fodor JA (1990): *A Theory of Content and Other Essays.* Cambridge, Mass.: Bradford Books/MIT Press

Garfield JL (1987): *Modularity in Knowledge Representation and Natural-Language Understanding.* Cambridge, Mass.: Bradford Books/MIT Press

Garnsey S, Tannehaus M, Chapman R (1988): *Monitoring Sentence Comprehension with Evoked Brain Potentials.* Paper presented at the 1st CUNY Conference on Human Sentence Processing. New York, N.Y.: CUNY, Graduate Center

Garrett MF (1990): Sentence processing. In: *An Invitation to Cognitive Science: Volume 1. Language,* Osherson DN, Lasnik H, eds. Cambridge, Mass.: Bradford Books/MIT Press, pp. 133–175

Giorgi A, Longobardi G (1991): *The Syntax of Noun Phrases: Configuration, Parameters, and Empty Categories.* Cambridge: Cambridge University Press

Grimshaw J (1990): *Argument Structure.* Cambridge: Mass.: MIT Press

Halle M (1985): Speculations about the representation of words in memory. In: *Phonetic Linguistics, Essays in Honor of Peter Ladefoged,* Fromkin V, ed. New York: Academic Press

Halle M (1990): Phonology. In: *An Invitation to Cognitive Science: Volume 1. Language* Osherson DN, Lasnik H, eds. Cambridge, Mass.: Bradford Books/MIT Press, 43–68

Higginbotham JT (1983): The logic of perceptual reports: An extensional alternative to situation semantics. *J Philosophy* 80:100–127

Higginbotham JT (1985): On semantics. *Linguistic Inquiry* 16(4):547–593

Higginbotham JT (1988): Knowledge of reference. In: *Reflections on Chomsky,* George A, ed. Oxford: Basil Blackwell

Higginbotham JT (1989): Elucidations of meaning. *Linguistics and Philosophy* 12(3):465–517

Higginbotham JT, Schein B (1989): *Plurals.* Unpublished manuscript. Department of Linguistics and Philosophy, MIT

Hobson JA (1988): *The Dreaming Brain.* New York: Basic Books

Hornstein N (1990): *As Time Goes By: The Syntax of Tense.* Cambridge, Mass.: Bradford Books/MIT Press

Jackendoff R (1983): *Semantics and Cognition.* Cambridge: MIT Press

Jackendoff R (1990): *Semantic Structures.* Cambridge: Mass.: The MIT Press

Johnson-Laird PN (1983): *Mental Models. Towards a Cognitive Science of Language, Inference, and Consciousness.* Cambridge: Cambridge University Press

Jouvet M, Michel F (1959): Correlations electromiographiques du sommeil chez le chat décortiqué et mésencéphalique chronique. *Compt Rend Soci Biol* 153:422–425

Kratzer A (1988): *Counterfactual Reasoning.* Unpublished manuscript. Department of Linguistics, University of Massachusetts at Amherst

Kripke S (1980): *Meaning and Necessity* Oxford: Oxford University Press

Kutas M, Hillyard S (1980): Reading senseless sentences: Brain potentials reflect semantic incongruity. *Science* 207:203–205

Lasnik H, Uriagereka J (1988): *A Course in GB Syntax: Lectures on Binding and Empty Categories.* Cambridge, Mass.: MIT Press

Lewis D (1983): Truth in fiction. In: *Philosophical Papers.* Oxford: Oxford University Press, 1:261–280

MacDonald MC, Clark R (1988): Empty categories, implicit arguments, and processing. In: Proceedings of the North East Linguistics Society (NELS), Vol 18, Blevins J, Carter J, eds. Amherst, MA: Graduate Linguistics Student Association, University of Massachusetts at Amherst

Malcolm N (1959): *Dreaming.* London: Allen and Unwin

Mamelak AN, Hobson JA (1989): Dream bizarreness as the cognitive correlate of altered neuronal behavior in REM sleep. *J. Cognitive Neurosci.* 1(3):201–221

May R (1985): *Logical Form: Its Structure and Derivation.* Cambridge, Mass.: MIT Press

Mehler J (1990): Language at the initial state. In: *From Neurons to Reading,* Galaburda AM, ed. Cambridge, Mass.: Bradford Books/MIT Press, 189–214

Mehler J et al. (1987): An investigation of young infants' perceptual representation of speech sounds. *Compt Rend Acad Sci, Paris* 303(2):637–640

Piattelli-Palmarini M (1986): The rise of selective theories: A case study and some lessons from immunology. In: *Language Learning and Concept Acquisition: Foundational Issues,* Demopoulos W, Marras A, eds. Norwood, N.J.: Ablex

Piattelli-Palmarini M (1989): Evolution, selection and cognition: From "learning" to parameter setting in biology and in the study of Language. *Cognition* 31:1–144

Piattelli-Palmarini M (1990): Sélection sémantique et sélection naturelle: Le rôle causal du lexique. *Rev Synthèse,* 4(1–2):57–94

Pustejovsky J (1988): The geometry of Events. In: *Studies in Generative Approaches to Aspect,* Tenny C, ed. Cambridge, Mass.: MIT Center for Cognitive Science, pp. 19–39

Putnam H (1975a): Dreaming and "depth grammar." In: *Philosophical Papers: Vol. 2. Mind, Language, and Reality.* Cambridge: Cambridge University Press, 304–324

Putnam H (1975b): The meaning of "meaning." In: *Philosophical Papers: Vol. 2. Mind, Language, and Reality.* Cambridge: Cambridge University Press, 215–271

Rizzi L (1982): *Issues in Italian Syntax.* Dordrecht, The Netherlands: Foris

Schein B (1986): *Event Logic and the Interpretation of Plurals.* Doctoral dissertation. Department of Linguistics and Philosophy, MIT

Stalnaker R (1968): A theory of conditionals. In: *Studies in Logical Theory,* Rescher N, ed. Oxford: Basil Blackwell

Stalnaker R (1984): *Inquiry.* Cambridge: Mass.: MIT Press

Tenny C, ed. (1988a): *Studies in Generative Approaches to Aspect.* (Lexical Project Working Papers, No. 24). Cambridge, Mass.: Center for Cognitive Science, MIT

Tenny C (1988b): The aspectual interface hypothesis. In: *Studies in Generative Approaches to Aspect* Tenny C, ed. Cambridge, Mass.: Center for Cognitive Science, MIT (Lexicon Project Working Papers, No. 24)

van Riemsdijk H, Williams E (1986): *Introduction to the Theory of Grammar.* Cambridge, Mass.: MIT Press

Zwitserlood P (1990): Language comprehension. In: *Cognitive Science in Europe: Issues and Trends,* Piattelli-Palmarini M, ed. Rome: Golem Monograph Series, no. 1, 55–77

3

Cognitive Neuroscience and the Human Self

STEPHEN M. KOSSLYN

Cognitive neuroscience is the branch of neuroscience that focuses on questions about how memory, perception, reasoning, and so on, arise from the operation of neural circuits. Cognitive neuroscience would contribute to issues about knowledge and values *if* it could explicate ways in which the structure of the brain constrains what and how we can know. In this chapter I describe how the self can be conceptualized within this approach. I argue that cognitive neuroscience provides a framework for conceptualizing what we mean when we speak of the self, and what would be necessary for the self to change.

The cognitive neuroscience perspective on the nature of the self grows directly out of the way we conceive of the mind. This conception is based on the premise that the mind-brain problem can be divided into two parts. One part addresses how "mental" activity (such as memory, language, and so on) is related to the brain. Theories of such activity are often cast in terms of "neural networks," which transform input to produce specific output. Although these networks are thought to capture key aspects of specific patterns of neural activity, they are formulated at a more abstract, functional level (indeed, Kosslyn and Koenig [1992] could illustrate the operation of such networks by describing sets of octopi, which grapple with each other in such a way that when some are stimulated by events in the world, others come to be stimulated or inhibited from responding). The other part of the mind-brain problem addresses the question of how the phenomenological experiences of such mental activity, the *qualia,* are related to brain activity. Cognitive neuroscience focuses solely on the first aspect of the conception of mind; it rests on the idea that mental activity can be understood as brain function (for a detailed development of this idea, see Kosslyn and Koenig, 1992).

Thus, at the core of the cognitive neuroscience approach is the assumption that mental function can be considered separately from phenomenological experience. The study of mental imagery provides a convenient illustration of this approach, given that it has clearly evident functional and experiential aspects. In order to describe the experiments that investigate the functional properties of imagery—and presumably for the reader to under-stand what is being talked about—researchers often rely on the experiential aspects of imagery. Although my work has focused on the functional aspects of imagery (e.g., Kosslyn, 1980), it is impossible to ignore com-pletely the experiential aspects. Participants at the conference, upon which this volume is based, forced me to think about the nature of the introspec-tive aspects of imagery in a new way. Thus, in the second part of this chapter I offer some initial, very tentative speculations about the nature of such experiences and how they relate to the functional state. I offer these admittedly loose ideas simply in an effort to engage thinking in certain directions, not as a proposal for a particular position.

A Cognitive Neuroscience of the Self

In this section I consider two topics: the nature of the self and the means to self-knowledge. I describe a cognitive neuroscience perspective on each.

The Self

If mental events are taken to correspond to brain function, then the self can be understood to arise from an interaction of information in memory, processing abilities, and temperament. In this section I will use the terms "personality" and "self" interchangeably, but note that personality typically refers to those aspects of the self that vary from individual to individual.

MEMORY

Experience shapes one's personality (we speak of "formative years" with this in mind). It does so in at least three ways, all of which depend on information in *long-term* (relatively permanent) memory.

First, experience alters the *contents* of memory. Your favorite pet, favorite expressions and speech patterns, best chess opening, style of driving a car, and method of brushing your teeth—all of this and more are stored in memory. Some of the information in memory affects the preferred method of thinking and problem solving. It is easy to be oblivious to the degree to which memory affects everyday commerce with the world; the fish is the last one to find out about water. As one example, memory plays a critical role in defining one's sense of humor, which hinges on associations evoked by a situation. A large portion of what we mean by one's personality is determined by information stored in memory; if your memory were totally erased, much—if not most—of your personality would be lost.

The contents of memory are understood in two senses: the actual material, and the way it is organized. The organization of memory has proven critical for later recall (e.g., see Glass and Holyoak, 1986). If material is effectively organized, it is much easier to recall because cues can more effectively evoke stored information. Furthermore, the sequences of associations that come to mind depend critically on how memory is organized.[1]

Second, information in memory not only serves to guide one's behavior once a situation has been apprehended, it also colors the way the situation is viewed in the first place. Indeed, much of the emotional content of events is determined by past experience, and the past has a grip on the present only because memory mediates perception and action. I will discuss the role of memory in perception in more detail shortly.

Third, depending on what material is in memory and how it is organized, one will be able to learn and use a given kind of material more or less easily. Experts have a rich memory structure in their domain of expertise, which affords them many hooks on which to hang new information. One can quickly draw analogies, spot similarities-with-discrepancies, and so on, which allow new information to be apprehended easily. ("The rich get richer" definitely applies to learning new information.) Moreover, one can use information more flexibly and creatively if it is properly organized in memory—which for many academics plays a critical role in defining the self.

PROCESSING ABILITIES

Nobody is surprised that people's hands differ in speed and capacity, and nobody should be surprised that people's brains differ in analogous ways. There are two critical ways in which processing abilities differ.

First, people clearly differ in the speed with which they process information (e.g., see Sternberg, 1985). Depending on how quickly information can be processed, one will be more or less able to perform multiple tasks at the same time. Furthermore, the speed of thinking, speaking, and so on are important facets of one's self.

Second, people differ in the amount of information that they can retain in *short-term* memory. Information in short-term memory is immediately available (many speak of it as "being in consciousness"; e.g., see Glass and Holyoak, 1986). A simple test of short-term memory capacity is the digit span, particularly when one is required to repeat a list in reverse order. Short-term memory capacity is important in part because it limits how much information can be consciously considered at the same time, which constrains one's ability to solve problems. This capacity presumably is

---

[1]Many factors affect what we remember and how memory is organized, and some of these factors may be described by psychoanalytic theory and the like. The effects of these factors, however, are ultimately to be understood in terms of what information is stored in memory and how it is organized. The method of free association, for example, reveals something about the way information is organized in memory.

determined not only by innate predisposition, but also by effects of diet, health, age, and so on. To my knowledge, there is as yet no evidence that people differ in the capacities of their long-term memories.

Depending on these processing abilities, one will be able to adopt certain methods of thinking and will be able to comport oneself in different ways.

TEMPERAMENT

People differ in their excitability, patience, and the like. I suspect that such differences in temperament reflect different tendencies to associate particular emotions with information, as well as the "rise time," "peak level," and "fall time" of each emotion. My colleague Jerome Kagan has found evidence that a good portion of this variability is innate (e.g., see Kagan, Reznick, and Snidman, 1988).

If all of one's memory were suddenly eliminated, processing abilities and temperament would be all that is left to define personality.

SELF-CONCEPT

Finally, we must consider one more factor that is critical for understanding the self, namely the role of the *self-concept* in governing behavior. The self-concept corresponds to a special kind of information stored in memory, which functions as a kind of "internal model" that can be used to guide behavior. The self-concept includes information about one's goals, beliefs, desires, values, customary (and not customary) modes of interacting with the world, and so on, as well as information about one's knowledge, processing abilities, and temperament. The self-concept need not be (and perhaps never is) entirely accurate—in any or all of these respects. Nevertheless, behavior can be filtered through these beliefs. For example, one may think of oneself as very ethical, but in fact have the propensities of a rattlesnake. Thus, if one tries hard to "live up to" one's ideals, one's behavior will be very different than if one follows one's impulses (we speak of such behavior as "acting thoughtlessly"). Such "trying hard" corresponds to a kind of internal response, in which information in the self-concept is accessed in the course of information processing and used to guide behavior. In short, the nature of one's self-concept, and the degree to which it plays a role in governing behavior, also must be considered a key aspect of the "self."

Knowing Oneself

This conception of self implies that to know oneself—to build an accurate self-concept—one must fathom the information in one's memory and the way it is organized, and must understand how processing abilities and temperament affect the ways in which stored information leads to emotional and behavioral responses. The task of knowing oneself requires collecting and interpreting information. There are several ways of collecting such information about the self.

Introspection is a form of perception in which instead of observing outer events, one "observes" inner events. Mental images play a critical role in introspection; they are equivalent to percepts in normal perception. Thus, it is critical to examine how accurate this sort of "inner perception" is. The rather lengthy example to follow is intended to illustrate several general points: First, introspection will not allow one to learn about *all* of one's processing abilities. Second, introspection cannot always tell one about the method used in thinking. Third, introspection may not even be a reliable means of recalling what one did, said, or thought. And fourth, there now exist laboratory methods that can simply and reliably provide such information.

Probably the most obvious property of mental imagery is that images are not present all of the time. Images only occur in specific circumstances. For example, if one is asked to decide which uppercase letters of the alphabet have only straight lines and which letters contain any curved lines, images of the letters are likely to be present. These images come to mind only when one begins to consider the question. If asked, people invariably report that introspectively the image of each letter pops into consciousness all-of-a-piece. Work in our laboratory has shown that this introspection is incorrect, at least for some kinds of images. Indeed, not only does one generate images of letters a segment at a time, but there also are qualitatively different methods by which one does so — and some people apparently prefer one method over the others. None of this is evident to introspection.

Let me briefly explain how we discovered these properties of imagery. We studied how people form visual mental images using a task originally devised by Podgorny and Shepard (1978), and shown by them to evoke imagery. Podgorny and Shepard showed subjects a grid in which a block letter was formed by filling in some of the cells. A dot or dots then appeared in the grid, and the subject was to decide whether the dot(s) fell in a cell that was blacked out (i.e., was part of the figure). The time the subjects required to make the decision was measured. These times varied with a number of factors, such as the number of dots and where they fell on the letter. For example, dots that fell on an intersection were judged more quickly than dots that fell on a limb.

Podgorny and Shepard also tested the subjects in a second, related task. In this case, the subjects projected a mental image of the letter into an empty grid, imagining that certain cells were filled. The task now was to decide whether the dots fell in cells that were "occupied" by the imaged pattern. The interesting result is that the decision times varied in exactly the same way in the imagery and perception tasks: the number of dots, their location on the figure, the size of the figure, and so on, affected decision times in exactly the same way when the figure was actually present and when it was only visualized. The subjects did generally require more time to evaluate imaged letters than viewed ones, but that was the only difference between the two tasks. The striking similarity in the variations in decision times was exactly as expected if imagery utilizes the same classification processes used in perception, and these results are good evidence that imagery was used in performing the task.

Kosslyn et al. (1988) altered the Podgorny and Shepard task to study image generation. In our experiments, subjects first memorized a set of block letters in grids, which — in our font — varied in complexity from two (for the letter L) to five segments (for G). The subjects later were shown an empty grid, which had a lowercase letter beneath it. Shortly thereafter, two X marks appeared in the grid. The subjects were asked to decide whether the corresponding block version of the letter — if it were drawn in the grid as previously seen — would fill the cells occupied by the X marks. On half the trials the letter would have covered both X marks, whereas on the other half it would have covered only one. The subjects were told to make their decisions as quickly and accurately as possible.

The key to our methodology was that the two X marks appeared in the grid only half a second after the lowercase cue letter was presented beneath the grid. A half a second is not enough time to read the cue, move one's eyes up to the grid, and finish forming the image before the probes appear. Thus, the time to make the decision in part reflected the time to form the image (see Kosslyn et al., 1988; Roth and Kosslyn, 1988). Although this method cannot tell us the absolute time to form the image (because the decision process affects the time to produce the response itself, above and beyond the image-generation time), it can tell us the relative time to form two different images. Thus, it was of interest that decision times increased with the visual complexity (number of segments) of the block letters. In order to show that this result occurred because more complex forms require more time to image, we needed to show that it was not due to the time to search for the probe marks. This was easy to do; complexity had greatly reduced effects in a corresponding perception condition, in which the subjects evaluated probe marks with the figure actually present.

Furthermore, in another task the probe marks were eliminated, and the subjects were asked simply to read the lowercase cue and form an image of the corresponding block letter in the grid; as soon as the image was fully formed, the subjects were to press a key. Not only did these times increase for letters with more segments, but they did so to the same degree in this task and in the image-evaluation task. Thus, we had reason to infer that differences in response times in the image-evaluation task do in fact reflect differences in image-formation time.

And now let us consider the most interesting aspects of the results. We varied more than the complexity of the stimuli in these experiments; in addition, we varied the positions of the probe marks along the segments of the block letters. We reasoned that if the image is constructed a segment at a time, then some probes should require more time than others before being covered by a segment. And indeed, more time was required in the image-evaluation task when the "farthest" probe mark fell on a segment that people typically draw later in the sequence. (The typical drawing sequence was determined by asking a separate group of subjects to copy the block letters into empty grids, while we covertly recorded the order in which the segments were drawn. We only examined letters for which the order was highly consistent.)

This effect of probe position did not occur in a perceptual-control task, when the letters were physically present. And the effect of probe position did not occur in an imagery task when subjects were allowed to form the image prior to seeing the probe X marks. Thus, the effect of probe position was not due to scanning an imaged pattern in search of the probes. Nor did the effect of probe position reflect the way in which people moved their eyes over the grid; the effect persisted even when subjects fixated on the center of the screen while performing the task. The effect only occurred when subjects did not have enough time to finish forming an image before the probe X marks appeared. In this case, the probes "caught the image on the fly," and subjects required more time if they had to wait longer for the critical segments of the letter to be included in the image. In short, these experiments provided good evidence that images are generated a segment at a time, with segments being included in roughly the order in which they are drawn.

We interpreted these results in the context of research indicating that "what" and "where" are processed by different cortical systems (see Ungerleider and Mishkin, 1982). We extended this analysis to the processing of parts of a single object, when parts are seen with high enough resolution to be individuated. Bower and Glass (1976), Reed and Johnsen (1975), and others have showed that objects often are perceptually parsed into constituent parts, which are stored separately in memory. When this occurs, we hypothesized, the representations of the constituent parts are stored separately from the representations of the spatial relations among them. Thus, we reasoned, when later generating an image, one must activate the representations of parts and spatial relations separately, building up the image a part at a time. The representations of spatial relations are critical in that they specify how the parts are to be arranged in the image.

Does it matter that one is not aware that images are constructed a part at a time? If the neuropsychologists are right, and spatial relations are stored separately from parts, then there is room for error in generating images. If one slips up and uses the wrong spatial relation, or includes the wrong parts, the image will be incorrect. Such errors can lead to self-deception. If imagery is taken as a relatively direct means to self-knowledge, the idea that images can be incorrect is sobering. Furthermore, recent work in our laboratory suggests that there are several distinct methods of generating visual mental images. Depending on how one forms mental images, different information is available, which will shape the patterns of one's thought. Introspection will not allow one to discover any of this. Although it alerts one to the presence of imagery, it does not reveal anything about the processes that form images and it does not allow one to evaluate the accuracy of the images.

PERCEPTION

One also can observe one's own behavior (listen to what one is saying, see one's reflection in a window, and so on), and notice how one is acting in given circumstances. In this case, one has no privileged access to the bases of the behavior, and interprets it using the same methods used to interpret the behavior of others. But there is no such thing as "immaculate perception." Let me provide another, very simple example. Kosslyn et al. (1984) showed subjects ambiguous geometric patterns. The subjects were told that the patterns were constructed by juxtaposing relatively many small adjacent pieces, or by arranging relatively few overlapping pieces. For example, the Star of David might have been described as two overlapping triangles or as a hexagon with a small triangle attached to each side. The time later to form a mental image of the patterns depended critically on how the subjects had construed them initially; the more units (2 versus 7, in the example), the more time was required—even though the end product was the same pattern!

Furthermore, once an image had been formed, subjects found it more or less easy to "see" portions of the pattern, depending on how it was organized. For example, if the star was seen initially as two large overlapping triangles, it subsequently was easy to "see" a large triangle in the image; if it was seen initially as a hexagon with six small triangles, it was relatively difficult to "see" a large triangle.

The point is that depending on how one is "set" to view a pattern, it is organized differently and stored differently in memory. And depending on how it is stored, it is more or less easy later to use the information in different ways. I again have used a very simple example, but such effects are even more dramatic with complex stimuli—which afford many more opportunities for different organizations.

Thus, early in childhood (when these organizational effects first occur), the self begins an active, albeit largely unintentional, role in the further development of the self. One packages and colors the world and its inhabitants, including oneself. Because one typically is not aware of the fact that there are alternative ways of organizing stimuli perceptually, the effects of these so-called "top-down" variables (e.g., see Gregory, 1970; Neisser, 1967) often may not be noticed.

Other people sometimes make observations about one, and report these observations. Indeed, some people even make their living doing this, sometimes within an established framework for interpreting behavior (such as psychoanalysis). All of the problems that apply to self-perception apply here, and then some: such information describes behavior, and the relation between behavior and the self is often indirect and convoluted. Furthermore, in order to interpret how someone else has perceived one's behavior, one must know about their propensities (and/or theoretical perspective, if we are in the context of psychotherapy). And then one must be aware of how their reports are filtered by one's own memory and perception.

Intuition

Finally, one sometimes can sense a pattern that cuts across remembered or current introspections, perceptions, or feedback. If such a pattern can be explicitly formulated (e.g., by being described or visualized), I would treat it as a form of introspection. I will reserve the term "intuition" to refer to unarticulated inferences that can nevertheless be used to guide behavior. Intuition is difficult to characterize, and hence difficult to discuss. I wish to make only one observation. Because of this very problem, it is difficult to build on intuitions. Formulating thoughts explicitly allows one to reason about them, to use them as a component of other thoughts, and so on. Intuitions, because they are vague and often elusive, are of limited use in understanding something as complex as a self.

Conclusions: Limitations on Self-Knowledge

If you accept my premises and observations, a simple conclusion appears to follow: all of the traditional methods of self-knowledge have limitations, and will allow one to understand one's self little better than one can come to understand another person. This form of self-knowledge, although valuable, need not reveal the underlying reasons for one's behavior.

Hence, it would appear that experimental methods developed in cognitive neuroscience one day may have a role in helping people to better understand themselves. Such work would pick up where Freud left off when he used free associations to understand patients' problems. Indeed, this approach would lead us not only to exploit behavioral measures (such as patterns of associations and response times), but also direct observations about the neural events that underlie specific behaviors. Some day it might even be possible to diagnose key aspects of the structure of one's personality by observing which regions of one's brain are relatively active while one is engaging in specific kinds of mental activity.

The "Feel" of Experience

In another sense, true self-knowledge is to be found at the level of phenomenology, of *qualia* — regardless of how well such experience reflects underlying brain events. Some might argue that the texture of one's experience is more important for defining the self than are the contents of one's experiences; the feelings and experiences that accompany mental events may lie at the heart of what we usually mean by "the self."

This consideration leads us face to face with perhaps the stickiest problem in all of science: how to characterize the nature of consciousness. I cannot claim to have an insight into this problem, but I may have a way of structuring it and a glimmer of a direction for further work. Let us begin by considering requirements for a theory of consciousness.

Requirements for a Theory

A coherent theory of consciousness must respect certain requirements. At a minimum, the following criteria must be met.

NONREDUCIBILITY

The characterization of consciousness cannot be replaced by statements about underlying brain events; if it is, there is no reason to posit special properties of consciousness. A theory of consciousness must posit an entity that belongs to a different category than brain events (cf. Ryle, 1949). If it does not, nothing is gained by positing consciousness. So, for example, saying that consciousness is "activity in layer 4 of the cerebral cortex" or the like is not satisfactory; one can perfectly easily imagine such activity without the accompanying experience (but for an alternative point of view, see Crick and Koch, 1990).

UNIQUE ROLE OF CONSCIOUSNESS

If the theory posits a role for consciousness, as opposed to positing that it is merely epiphenomenal, then consciousness must have a unique function. Given that consciousness is a special kind of stuff, it presumably plays a special role; the function of consciousness presumably is not appropriate for a physical state or process. So, for example, positing that consciousness highlights important information will not do: one can posit attentional mechanisms that "flag" important information without needing any special state. A good heuristic is to imagine ways in which a computer can be programmed to carry out a function; if it seems likely that it can be, then the function is unlikely to be accomplished by consciousness.

SELECTIVITY

Not all mental processes are reflected in conscious experience. For example, we have no idea how we actually come to visualize objects, or even to identify objects we see or hear; we are aware of the product of the process — the image or the identity of the object — but not the intermediate steps that lead to it. Any theory of consciousness must specify the characteristics of mental processes that allow them to be accessible to consciousness.

ASSOCIATED WITH BRAIN STATES

There is no question that consciousness is associated directly with brain states, and only indirectly with states of other organs of the body. Consciousness changes when the brain is affected by drugs, and consciousness is often related to functional states in the brain. For example, when one is perceiving an apple, one is conscious of its color and shape. Moreover, brain events precede consciousness, and neural activity must reach a certain level before one is aware of a stimulus; one perceives a skin sensation after the brain has registered it, as indicated by changes in its evoked potentials (Libet, 1987; Libet et al., 1967). Even though consciousness is a qualitatively different kind of thing from brain function, it must be causally connected to it.

CONSCIOUSNESS AFFECTS BRAIN STATES

If consciousness is not epiphenomenal, the relation with the brain must be bidirectional: consciousness affects mental function, and vice versa. It seems to have been this requirement that impelled Descartes to localize consciousness to the pineal gland. At first glance, this requirement is perhaps the most challenging of them all: how can something that is qualitatively different from brain states affect them? But recall that this happens every waking second, when photons and other physical energies affect brain states via the sensory organs.

It is not easy to formulate even a single candidate that meets all these requirements. Rather than stop here, I have tried to sketch out the direction of a viable theory. This approach hinges on the idea that consciousness serves as a kind of "parity check." In computers, there is usually a "parity bit" associated with every "byte" (8 bits, which is the smallest unit in which symbols are represented). The parity bit is set so that the sum of the bits is always odd (or even; the convention differs in different machines). Thus, if parity is disrupted, the machine registers that there is an error in the byte (which can occur if only because cosmic rays occasionally disrupt electronic storage). The fundamental idea is that consciousness arises from an interaction of physical events in the brain with other physical events; it is like a chord, which is either in tune (if parity is preserved) or out of tune (if it is not). The mere presence of a coherent consciousness is an indication that the machine is working properly. Let us explore this idea with regard to the five requirements.

There are many possible variants of Parity Theory. For expository purposes I will sketch out a very strong and a very weak version here. I will begin by focusing on the experience of seeing an object, and later will consider other aspects of consciousness.

Strong Parity Theory

The strongest version of Parity Theory that I can think of posits that consciousness arises from physical interactions between electrical and magnetic products of the brain and ambient electrical and magnetic products in the environment. For example, if the neural systems sensitive to red light are active, parity is preserved if that wavelength is in fact present in the immediate environment. At first glance, this idea seems absurd, but it has some interesting properties. Before we consider some of its critical faults, let us see how well it satisfies the five criteria.

Nonreducibility

On this view, consciousness is something other than actual brain events. It is a kind of resonance between rhythms evoked in the brain and physical energies outside the brain. It cannot be identified with specific brain states because it also depends on the accompanying state of the world. A key assumption here is that neurons fire at different rates depending on the state of the world. Indeed, Richmond et al. (1987) claim that neurons transmit information using frequency modulation, and Gray and Singer (1989) have found that neurons that encode different portions of a stimulus oscillate together at the same frequency. The idea is that these physical events interact with very low power electrical and magnetic fields in the environment, setting up a second-order event, which is consciousness.

Consciousness, on this view, is a physical state, but a different kind of physical state from that which underlies neural information processing. By analogy (and I'm afraid that's all we have to guide us in these murky waters), it is like the beats between two guitar strings, as opposed to the oscillations of the strings themselves.

Unique Role of Consciousness

On this view, consciousness is a direct bridge to the physical energies in the world. As such, it fills a role that cannot be filled by any brain state. Brain states themselves, by definition, are not in direct contact with the world. They are often initiated by physical events in the world, but have no means of checking the accuracy of the concomitant response. Indeed, this observation has led some to worry about the existence of the world at all (cf. Brown, this volume).

SELECTIVITY

Not all neural events have the proper frequency to interact with ambient physical energies. If consciousness is to serve the role of a parity check, it would be most useful for processing that is not "hard wired" and reflexive, which typically is very fast. And, in fact, we seem to be conscious either of the very slow "voluntary" processes that involve many steps, such as those used in reasoning, or of the *products* (not the ongoing process) of other, faster processing. These products correspond to brain states, which linger for several seconds. Thus, although we are conscious of the results of perceptual or imagery processing, we have no awareness of how such processing proceeds.

ASSOCIATED WITH BRAIN STATES

The Parity Theory of consciousness relies on the idea that mental events are produced by the brain. Indeed, on this view consciousness cannot arise without neural activity. The interactions of physical energies that comprise consciousness must follow the production of both the neural and ambient physical events, and hence consciousness must lag behind both types of events.

CONSCIOUSNESS AFFECTS BRAIN STATES

The primary role of consciousness in this theory is as a means of detecting if the brain goes awry. When this happens dissonance arises, which in turn prevents the neurons from continuing business as usual. Thus, it is the failure to obtain a coherent conscious experience that can affect brain states; when a present brain state does not jibe with the ambient physical state, the brain state itself is affected. It is as if a guitar player's hands shake when a wrong note is played, and dissonance vibrates the neck. In this case, the melody stops and is "reset." Different patterns of dissonance affect the brain in different ways, "nudging" it into different functional states. These events are not determined solely by previous brain events, and hence represent an alternative to a strict "neural determinism" (where each brain state is determined by the previous brain state plus the current input).[2]

---

[2]I have often puzzled over the idea of "free will," as opposed to determinism. Determinism implies that a present brain state was caused by prior ones in conjunction with perceptual information. For years, it seemed to me that the only alternative was random brain states, which "pop into existence out of the blue." The present idea, if nothing else, complicates this picture. Indeed, the "dissonance" may become chaotic, and hence the resulting brain state would be impossible to predict — but still systematically related to prior brain states.

PROBLEMS WITH STRONG PARITY THEORY

At first glance, Strong Parity Theory is riddled through with problems, but not all of them are as damaging as they might appear. Some of them are not so easy to dismiss, however, and I sketch them out if only to give the idea a full run for its money.

Contact with Physical Energies

How do subtle physical forces produced by neural events interact with external physical forces? It has been clear since the discovery of the EEG that neurons produce physical forces, and recently it has been found that very low energy magnetic fields can affect biological events (e.g., see Brodeur, 1989). Indeed, there is now convincing evidence that 60 Hz sinusoidal magnetic fields can affect calcium uptake within cells (Walleczek and Liburdy, 1990). Thus, it is at least possible that physical energies produced by the brain interact with physical energies in the world, and that "dissonance" can in turn alter some aspect of brain activity.

However, what about the fact that the frequency of light is many orders of magnitude greater than the oscillations of the brain? It is difficult to see how such widely differing frequencies could interact. Furthermore, the differences in the frequencies of neural activity in different parts of the brain may be trivial compared to the differences between these frequencies and those of ambient physical events. Thus, the idea that such differences underlie the selectivity of consciousness is not plausible.

Interference

If consciousness is an interaction between physical events in the brain and physical events in the world, then a wide spectrum of electromagnetic energies ought to color consciousness. This might be true; I know of no direct tests of the hypothesis.

Another sort of interference is less easy to dismiss: that due to energies produced by the body itself. The heart, intestines, facial muscles, and other organs constantly produce electrical and magnetic fields. These would interact with those produced by the brain, and so would disturb parity resonance.

Memories and Mental Images

We are conscious of events even in the absence of the appropriate ambient physical stimuli. If consciousness is an interaction with ambient physical events, how can we be conscious of something that is not present—as, for example, occurs in mental imagery? The only way around this problem that I can see rests on three premises: (1) the neural state is itself modified by the interaction with physical energies, via resonance; (2) it is this modified state that is stored; (3) the modified state can interact with "baseline" physical energies to reproduce a semblance of the original interaction (and hence conscious experience).

The idea of "baseline" physical energies may allow us to turn a sow's ear into a silk purse: it is possible that the electrical and magnetic energies produced by the body serve this role. Consider an analogy to holography: A hologram can be made by shining a laser through a half-silvered mirror, so that half of the light shines on an object. A photographic plate is placed so that the light bouncing off the object hits it, as well as the remaining light shining directly through the mirror. These two sources of light create interference patterns on the plate, which are fixed via the photographic process. Later, the plate with the interference patterns serves as a kind of filter: if the same kind of laser is shined through the plate, it modulates the light passing through so that it is very much like the light would have been if the object were present; the interference patterns reconstruct the appearance of the object.

In the metaphor, the energies from the body serve as the laser. This activity interacts with the fields produced by the brain (corresponding to the object). Such interaction affects neural activity, and information stored in the brain allows this activity to be re-created later. What is stored, then, is like the hologram. Later, when the neural activity occurs, the energies produced by the body serve as the reference beam, allowing the proper physical interference pattern to be re-created — which produces a conscious experience.

Thus, this form of Parity Theory implies that conscious experience is an accurate reflection of the world, provided that one is not taking drugs or in another state that disrupts the normal operation of the brain. However, the contents of conscious experience only reflect a subset of brain events, and at least some aspects of the world clearly do not influence them. For example, we are not aware of infrared or ultraviolet frequencies when we view objects.

Private Language

Wittgenstein (1952) formulated what he called "the private language problem." If consciousness is distinct from neural states, how can it be stored so that a conscious experience can be identified by comparing it to a previous one? Parity Theory does not rely on processing conscious experiences in any way; one does not compare them to previous experiences directly. Information processing in the brain performs all functions but checking for internal coherence (parity).

Nonperceptual Feelings

What about feeling in love? Where is the physical energy that resonates with the neural state? An important fact about the experience of emotions was revealed in experiments by Schacter and Singer (1962). They showed that the identical physiological arousal was experienced as euphoria or depression, depending on how a subject interpreted the cause of the arousal. In their experiments the subject was injected with a drug that produces an adrenaline rush, but was not told the consequences of the drug. While waiting for another test, the subject was seated with a confederate of the experimenter's, who either was very manic or very depressed—and the subject experienced the effects of the drug differently, depending on the context. In this case, the brain state is a combination of the drug-induced arousal, the perception of the confederate, and processing that identifies a consistent interpretation of the ensemble; the conscious experience arises following a parity check between this entire brain state and the physical situation. Similarly, feelings such as "being in love" presumably would arise from an interaction between a complex brain state and the accompanying physical state. This account is flawed, however, in that it is not clear what "physical states" would be checked against the brain state.

Intentionality

So far, I have discussed consciousness as being conscious *of* an object, event, or state. However, we are also conscious of our relation to objects; depending on our plans, attitudes, and so on, we apprehend objects and events differently. This intentional aspect of consciousness seems to reflect the consequences of attention. Attention is the selective aspect of processing, and it affects virtually all cognitive functions. For example, we attend to an apple (or even simply its color) and not to the table on which it rests, or we attend to the ears of a German shepherd dog we visualize, and not to its nose. Similarly, we attend to one meaning of a word, or one instrument in an orchestra. Depending on what we attend to, different information is processed deeper in the system. For example, Moran and Desimone (1985) showed that neural responses of cells in area V4 that respond to specific stimuli are modulated by an animals' expectations of reward. If a monkey knew it would not be rewarded when the stimulus fell outside a target region, cells that normally responded to the stimulus were "turned off." Thus, depending on what is attended to, different patterns of activity arise in the brain—which in turn may lead to differences in our conscious experience.

Similarly, one's plans also correspond to patterns of activity, presumably largely in the frontal lobes of the brain. Depending on the pattern of activity, one will have different conscious experiences. Indeed, attention works in part by inhibiting some processing, allowing other processing to continue. Hence, the intensity of a conscious experience may reflect the degree to which only a single pattern of activity is present. In short, the fact of intentionality does not seem in principle incompatible with the fundamental claims of Parity Theory.

The most damaging problems, as I see them, are those associated with the time scales of ambient physical energies and those produced by the brain, on the one hand, and those associated with identifying the physical characteristics that define external events, on the other. All of my attempts to deal with these underlying problems have a hollow ring. A weaker form of Parity Theory is much easier to defend.

Weak Parity Theory

The key idea behind Parity Theory is that consciousness is an interaction of physical energies that provides a sign that neural events are internally consistent. Another version of this theory posits that the physical events in question arise solely from the brain. This position has many of the same virtues as Strong Parity Theory, as noted below.

NONREDUCIBILITY

Consciousness is a kind of resonance between rhythms evoked in the brain. Although these energies are produced by the brain, they are not themselves neural events — and hence cannot be reduced to specific brain states.

UNIQUE ROLE OF CONSCIOUSNESS

Consciousness rides herd on billions of neural events that occur every second. Although parity can be checked using patterns of bits in a computer, the brain is far too complex to have a process check the state of each individual neuron. Furthermore, what then would check that process? And what about that process in turn? This possibility of an infinite regress is avoided if the parity check is analogous to consonance or dissonance of the physical energies produced by different parts of the brain itself.

SELECTIVITY

Strong and Weak Parity Theories have the same account of the selective nature of consciousness. Some neural events may simply take place too quickly for the necessary physical interactions to develop.

ASSOCIATED WITH BRAIN STATES

The physical energies that underlie conscious are produced by specific brain states. Thus, consciousness cannot arise without brain states.

CONSCIOUSNESS AFFECTS BRAIN STATES

As in Strong Parity Theory, the role of consciousness in Weak Parity Theory is as a means of detecting when processing goes awry. When this happens dissonance arises, but now the dissonance is between oscillations produced by different portions of the brain. Such dissonance corresponds to a fragmented consciousness, which in turn prevents the neurons from continuing business as usual. Certain diseases (e.g., possibly schizophrenia) may disrupt brain processing so that one is perennially in this state, preventing one from maintaining a coherent line of thought; alternatively, the disease may alter that brain state so much that it resonates incorrectly with physical energies, resulting in an inappropriate conscious experience.

Predictions of Parity Theory

To be worth its salt, a theory must predict new phenomena. Both versions of the theory make at least three kinds of predictions. First, Strong Parity Theory predicts that if all ambient physical energy is eliminated, consciousness should be eliminated. This will not be possible, however, given that the body itself produces electrical and magnetic fields. Nevertheless, this condition establishes a pole of continuum; the theory predicts that as less ambient physical energy is available, consciousness should become increasingly restricted to simple "replays" of previous experiences or should become fragmented.

Weak Parity Theory leads to a similar prediction if ambient physical energies do in fact interact with those produced by the brain. In this case, such interactions are noise, and disrupt the harmonies that would otherwise arise. Thus, removing one's brain from ambient electromagnetic fields should alter consciousness.

Second, Strong Parity Theory posits that consciousness arises only when neural events produce the right sort of physical energies, namely, those that resonate with ambient physical events. Neural events that occur too quickly will not be reflected by a conscious experience, either because they produce physical events of the wrong frequency or there simply is not enough time for the parity resonance to be established. If so, then if neural processing could be slowed down, perhaps by some as yet undiscovered drug, then one should become conscious of mental processes that previously were opaque.

Weak Parity Theory makes a similar prediction, but the relevant interactions are among neural events themselves. If one knew in more detail what patterns of neural activities are produced by different stimuli, it should be possible to make more precise predictions about what sorts of information should become conscious as the system slows down.

Third, according to Strong Parity Theory, the same neural state should lead to different conscious experiences when it occurs in the context of different ambient physical states. Indeed, it should be possible to map out the "resonance space", delineating the dimensions that determine the similarity of the physical events that underlie different experiences. Given this space, it should be possible to predict ways in which combinations of different neural states and physical states will produce similar experiences.

Weak Parity Theory would make a similar prediction if ambient energies interfere with the resonances of the brain. In this case, specific types of ambient energies might interfere with the interactions among brain-produced fields in different ways. Furthermore, if the heart and muscles are affected, consciousness should change. Indeed, if a stimulus (e.g., a loud sound) causes one to startle, which affects the heart and muscles, this should feed back and affect consciousness more generally.

In short, Parity Theory does seems to have some empirical implications, although I can only vaguely sense their broad outlines. However, in order for these predictions to be worth developing in detail — let alone testing — one would have to be sure that obtaining the predicted results would rule out other theories of consciousness. If many theories make the same prediction, it is less interesting to test. At this juncture, I have no idea whether other theories of consciousness would lead to the predictions outlined above.

Conclusions

A cognitive neuroscience approach leads us to treat the self as having two distinct aspects, the functional and the phenomenological. The functional aspects pertain to the representations and processes that affect how we interpret and respond to stimuli, whereas the phenomenological aspects pertain to the texture of experience. The functional aspects of the self are only partially revealed by the conscious experiences that accompany them, and it is unclear how the two aspects, the self are related. I have offered one initial set of speculations merely to illustrate that it may be possible to structure this issue in a way that is amenable to solution.

Acknowledgments

Preparation of this chapter was supported by AFOSR Grant 88-0014, supplemented by funds from the ONR, NSF Grant 90-09619, and NINDS Grant PO1-17778-09. I wish to thank Anne Harrington for thoughtful and stimulating comments; they helped me drag exceptionally vague ideas into the realm of the merely fuzzy.

References

Bower GH, Glass AL (1976): Structural units and the reintegrative power of picture fragments. *J Exp Psychol: Hum. Learn Mem* 2:456–466
Brodeur P (1989): *Currents of Death*. New York: Simon and Schuster
Crick F, Koch, C (1990): Towards a neurobiological theory of consciousness. *Sem Neurosci* 2:263–275
Glass AR, Holyoak KJ (1986): *Cognition*. New York: Random House
Gray C, Singer W (1989): Stimulus-specific neuronal oscillations in orientation columns of cat visual cortex. *Proc Nat Acad Sci USA* 86:1698–1702
Gregory RL (1970): *The Intelligent Eye*. London: Weidenfeld and Nicholson

Kagan J, Reznick JS, Snidman N (1988): Biological bases of childhood shyness. *Science* 240:167-171

Kosslyn SM (1980): *Image and Mind.* Cambridge, Mass. Harvard University Press

Kosslyn SM (1988): Aspects of a cognitive neuroscience of mental imagery. *Science* 240:1621-1626

Kosslyn SM, Brunn JL, Cave KR, Wallach RW (1984): Individual differences in mental imagery ability: A computational analysis. *Cognition* 18:195-243

Kosslyn SM, Cave CB, Provost D, Von Gierke S (1988): Sequential processes in image generation. *Cog Psychol* 20:319-343

Kosslyn SM, Koenig O (1992): *Wet Mind.* New York: Free Press

Kosslyn SM, Koenig O, Barrett A, Cave CB, Tang J, Gabrieli JDE (1989): Evidence for two types of spatial representations: Hemispheric specialization for categorical and coordinate relations. *J Exp Psychol: Hum Percept* 15:723-735

Libet B (1987): Consciousness: Conscious, subjective experience. In: *Encyclopedia of Neuroscience,* Vol. 2, Adelman G, ed. Boston: Birkhäuser

Libet B, Alberts WW, Wright EW, Feinstein B (1967): Response of human somatosensory cortex to stimuli below threshold for conscious sensation. *Science* 158:1597-1600

Moran J, Desimone R (1985): Selective attention gates visual processing in the extrastriate cortex. *Science* 229:782-784

Neisser U (1967): *Cognitive Psychology.* New York: Appleton-Century-Crofts

Podgorny P, Shepard RN (1978): Functional representations common to visual perception and imagination. *J Exp Psychol: Hum Percept* 4:21-35

Reed SK, Johnsen JA (1975): Detection of parts in patterns and images. *Mem Cog* 3:569-575

Richmond BJ, Optican LM, Podell M, Spitzer H (1987): Temporal encoding of two-dimensional patterns by single units in primate inferior temporal cortex: 1. Response characteristics. *J Neurophysiol* 57:132-146

Roth JR, Kosslyn SM (1988): Construction of the third dimension in mental imagery. *Cog Psychol* 20:344-361

Ryle G (1949): *The Concept of Mind.* London: Hutchinson and Co.

Schacter S, Singer J (1962): Cognitive, social, and physiological determinants of emotional state. *Psychol Rev* 69:379-399

Sternberg RJ (1985): General intellectual ability. In: *Human Abilities: An Information Processing Approach,* Sternberg RJ, ed. New York: W. H. Freeman, pp. 5-30

Ungerleider LG, Mishkin M (1982): Two cortical visual systems. In: *Analysis of Visual Behavior,* Ingle DJ, Goodale MA, Mansfield RJW, eds. Cambridge, Mass.: MIT Press

Walleczek J, Liburdy RP (1990): Nonthermal 60 Hz sinusoidal magnetic field exposure enhances $^{45}Ca^{2+}$ uptake in rat thymocytes: Dependence on mitogen activation. *FEBS Lett,* 271:157-160

Wittgenstein L (1953): *The Philosophical Investigations,* Anscombe E (trans). New York: Macmillan

4

Obtaining Knowledge of the Subjective Brain ("Epistemics")

PAUL D. MACLEAN

"Epistemics" is a term I apply to the body of knowledge and collective disciplines concerned with clarifying the nature of the self and the subjective brain (MacLean, 1975, 1990). Epistemics may be regarded as the complement of epistemology because it represents the scientific, subjective view from the inside out, whereas epistemology encompasses the scientific, public view from the outside in.

The introspective process and other considerations lead to the *cerebrocentric* view that, regardless of how it occurs, the entirety of experience and interpretation is dependent on the structure and function of one's own brain. To paraphrase a statement attributed to Cajal, the universe is but a reflection of the structure of the human brain.

We were asked to consider *knowledge, values,* and *neuroscience* with respect to "so human a brain." If knowledge applies to what is known with certainty, then there exists in neuroscience a countless number of things about which one can be reasonably certain. But later on I will point out two principal reasons why, in an absolute sense, there is little likelihood that we can ever achieve certainty in our scientific or other intellectual beliefs. (For this chapter, it will not be necessary to deal with the so-called dualistic nature of brain and mind. For my treatment of that subject, see Chapters 1, 2, and 29 of *The Triune Brain in Evolution* [1990]. Also see my retrospective footnote [no. 3] to the discussion "At the Intersection of Knowledge and Values," this volume, p. 264.)

Nothing is humanly more frustrating than to be caught in the trap of ignorance, and the prospect of never getting out of the trap makes some people dangerous to themselves. The ultimate inability to establish certitude of knowledge, however, need not apply to what we identify as "values" which are, by nature, so relative and unmeasurable that they can never be known with certainty. Hence if one is willing to substitute "belief in values" as a compensation for the inability to achieve certainty of knowledge, one finds that evolution has brought us one compensation that more than any other accounts for "so human a brain." For gaining an appreciation of what that compensation is and how it came about, we must review briefly the evolution and functions of the mammalian brain.

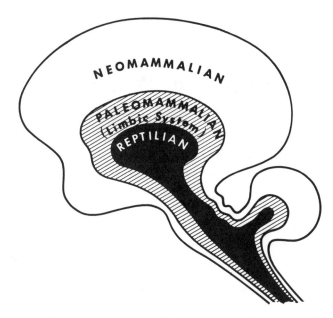

Figure 4.1. Symbolic representation of triune evolution of forebrain and cerebellum leading to human beings. Labeling identifies three intermeshing formations of forebrain that in structure and chemistry reflect a relation to reptiles, early mammals, and late mammals. From Maclean (1967).

Mammalian Origins

In evolution, the forebrain of human beings and other advanced mammals has expanded as a triune structure consisting of three neural assemblies that anatomically and chemically reflect an ancestral relationship to reptiles, early mammals, and late mammals (see Figure 4.1). What this amounts to is an inheritance of three mentalities, two of which lack the capacity for verbal communication.

We derive from an ancient reptilian stock, tracing our ancestry to the mammal-like reptiles that 250 million years ago in Permian times split off from the stem reptiles (see Figure 4.2 [part 1]). Well before the dinosaurs, these mammal-like reptiles had a worldwide distribution when the earth was but one giant continent, now known to us by Wegener's term Pangaea (see

Figure 4.2. Illustration relevant to origin of mammals from mammal-like reptiles (therapsids). 1, Cactus tree, depicting evolutionary radiation of reptiles; 2, Wegener's Pangaea; 3, carnivorous therapsids illustrating mammal-like features; 4, A and A', therapsid jaw bones (quadrate above and articular below) that became incus and malleus of mammalian middle ear (B). Adapted from MacLean (1990).

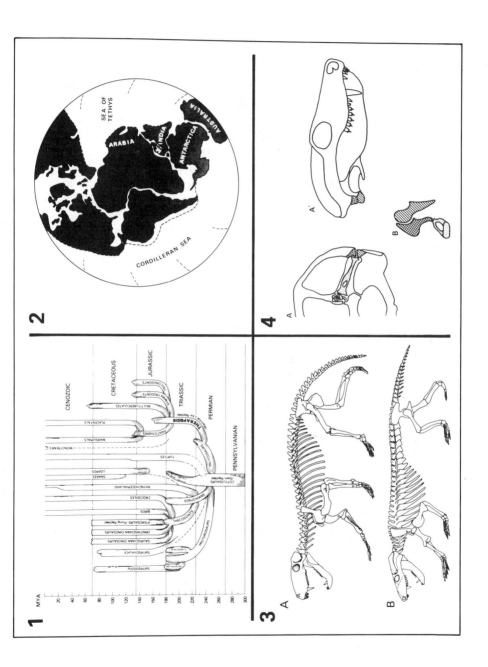

1

MYA

CENOZOIC
CRETACEOUS
JURASSIC
TRIASSIC
PERMIAN
PENNSYLVANIAN

20
40
60
80
100
120
140
160
180
200
220
240
260
280
300

MONOTREMES
PLACENTALS
PANTOTHERES
MARSUPIALS
MULTITUBERCULATES
TRICONODONTS
DOCODONTS
THERAPSIDS (Mammal-like Reptiles)
TURTLES
PELYCOSAURS
COTYLOSAURS (Stem Reptiles)
RHYNCHOCEPHALIANS
SNAKES
LIZARDS
CROCODILES
EOSUCHIANS
BIRDS
PTEROSAURS (Flying Reptiles)
THECODONTS
ORNITHISCHIAN DINOSAURS
SAURISCHIAN DINOSAURS
PROTOROSAURS
ICHTHYOSAURS
PLACODONTS
NOTHOSAURS
PLESIOSAURS

2

SEA OF TETHYS

ARABIA
INDIA
ANTARCTICA
AUSTRALIA

CORDILLERAN SEA

3

A

B

4

A

A'

B

Figure 4.2 [part 2]). The more advanced mammal-like reptiles are referred to as therapsids because of a single, large temporal opening resembling that of mammals.

It is a matter of great interest that there were several lines of therapsids all showing directional evolution toward the mammalian condition. As illustrated in Figure 4.2 (part 3), they had gotten well up off their bellies so that the legs, doglike, supported from underneath. There was a reduction of the phalangeal formula, with the number of digits corresponding to our five fingers and toes. The jaws and teeth were acquiring a mammalian character, while a secondary palate made its appearance. As diagrammed in Figure 4.2 (part 4), two small bones of the reptilian jaw joint — the articular and quadrate — were becoming smaller, as though preparing for their final migration to become the malleus and incus of the mammalian middle ear. And most notable, there were several signs that the therapsids were developing from a cold-blooded to a warm-blooded condition.

One of the forebears of the therapsids was so lizardlike as to be called Varanosaurus after the monitor lizard, of which the giant Komodo dragon would be a striking example. There are no existing reptiles directly in line with the therapsids. Hence, because of the similarities of the lacertilian skeletal features and auditory apparatus to the mammal-like reptiles, we have used lizards in our comparative neurobehavioral studies.

In analyzing the behavior of lizards, one identifies more than 25 special forms of basic behavior also observed in birds and mammals. What brain structures would account for the same kinds of behavior in all three classes of terrestrial vertebrates — reptiles, birds, and mammals? Histochemical methods have made it possible to identify corresponding structures of the forebrain in all three classes. In all three, the structures correspond to the so-called basal ganglia of the forebrain. They are also referred to as the striatal complex because of their striped appearance. In a comparative context, I refer to them as the protoreptilian complex, or, for short, as the R-complex.

As is well known, the basal ganglia of the forebrain have been tradition-ally regarded as slaves of the motor apparatus, having, so to speak, no mind of their own. If purely motor in function, it is curious that large cavities may be present in the R-complex without any apparent loss of motor function.

The main elements of the behavioral profile of terrestrial vertebrates might be compared to the main peaks of two mountain ranges seen from a distance. In one range are behaviors linked together in the *daily master routine* and *subroutines*. In the other range are features of four types of displays used in social communication. In lizards, the four types are referred to as signature displays, aggressive displays, courtship displays, and submissive displays.

Our comparative neurobehavioral studies on animals as diverse as lizards and monkeys indicate that the R-complex orchestrates the performance of *displays* used in *social communication.* Moreover, there is both experimental and clinical evidence that the R-complex is essential for an intelligence linking together the behavioral repertoire of the daily master routine and subroutines.

In observing lizards, one gains the impression that they are about as self-centered animals as one will ever see. Lizards of most species go it alone from the day of hatching, being equipped to do everything they have to do except to procreate. For the first few months of life, hatchlings of some species must hide in the deep underbrush or, like the young of Komodo lizards, take to the trees for the first year of life so as to avoid being cannibalized by their parents or other adult lizards. Most lizards do not vocalize and for the young that is fortunate because, if, like infant mammals, they were to let out a separation cry, they might call attention to themselves and become prey of their parents or other lizards.

The Paleomammalian Brain (Limbic System)

With the evolution of mammals there comes into being the primal commandment, "Thou shalt not eat thy young or other flesh of thine own kind." The evolutionary transition from reptiles to mammals is characterized by the development of three cardinal forms of behavior identified as (1) nursing, conjoined with maternal care; (2) audio-vocal communication for maintaining maternal-offspring contact; and (3) play. Such developments appear to have depended on the progressive elaboration of the primitive cortex enveloping a large convolution which Broca called the great limbic lobe because it surrounds the brainstem (see Figure 4.3A). This cortex, together with its primary connections with the brainstem, comprises the so-called limbic system, a term I used in 1952 (MacLean, 1952) as a substitute for the old name rhinencephalon that implied primarily olfactory function. As illustrated in Figure 4.3A, it was Broca's special contribution to provide evidence that this lobe is found as a common denominator in the brains of all mammals.

On the basis of experimental and clinical evidence, it can be inferred that the limbic system derives information in terms of emotional feelings that guide behavior required for self-preservation and preservation of the species. The limbic cortex and associated nuclei can be subdivided into three main subdivisions. As diagrammed in Figure 4.3B, the two older subdivisions associated with the amygdala and septum are closely related to the olfactory apparatus. The amygdala division is organized around the mouth and is primarily involved in feeding and the search for food, as well as the fighting and defense that may be required in obtaining food. As opposed to these oral-related manifestations, the septal division has been shown to be involved in primal sexual functions and in behavior conducive to sociability and mating. The third division comprising the limbic cingulate cortex and its thalamic connections represents a later development. Experimental findings indicate that it is this thalamocingulate division that is primarily involved in the evolution of family-related behavior — namely, nursing and maternal care; audio-vocal communication for maintaining contact; and play. Significantly there is no apparent counterpart of this division in the reptilian brain.

Case histories of patients with psychomotor epilepsy provide the most compelling evidence (and indeed the only subjective evidence) that the limbic system plays a prepotent role in thymogenic (emotional) functions. Indeed, one might say that this cruel form of epilepsy provides more windows for insights into mechanisms underlying human subjective experience than any other clinical condition. This disorder has been variously known as psychomotor epilepsy, temporal lobe epilepsy, and limbic epilepsy. Since 1970, it has been diagnostically labeled as *complex partial seizures* because of the frequent absence of generalized convulsions. This common form of epilepsy may result from certain kinds of infections or head injuries that are prone to affect limbic cortical areas at the base of the brain. Sometimes with a slowly progressive scar there may be a lapse of several years before an epileptogenic focus develops. Such foci trigger epileptic storms that tend to spread within, and be confined to, the circuitry of the limbic system.

With the aura at the beginning of an epileptic storm, the patient's mind may light up with one or more feelings that in one case or another involve the entire range of affective feelings. The affects may be regarded as agreeable or disagreeable feelings that "impart subjective information that is instrumental in guiding behavior required for self-preservation and preservation of the species" (MacLean, 1990, p. 425). The scheme in Figure 4.4 shows the world of affects with the equator separating the agreeable from the disagreeable affects and the longitudinal lines the division into basic, specific, and general affects. There are no neutral affects because, by definition, affects are either agreeable or disagreeable (pleasant or unpleasant). The *basic* affects apply to gradations of feelings associated with the basic bodily needs such as those arising with the need for food, water, and air, the need to expel, and so on. The *specific* affects are agreeable or disagreeable feelings conveyed by specific sensory systems and include the so-called aesthetic or cultural affects. It is the *general* affects that apply to what we generally regard as emotional feelings. They are called general because they may pertain to individuals, situations, or things. Unlike the basic and specific affects, they are not dependent on particular gateways to the sensorium and may occur and persist as the result of mentation. In both human beings and animals, the general affects can be identified with six main forms of behavior denoted as searching, aggressive, protective, dejected, gratulant (triumphant), and caressive. Corresponding words

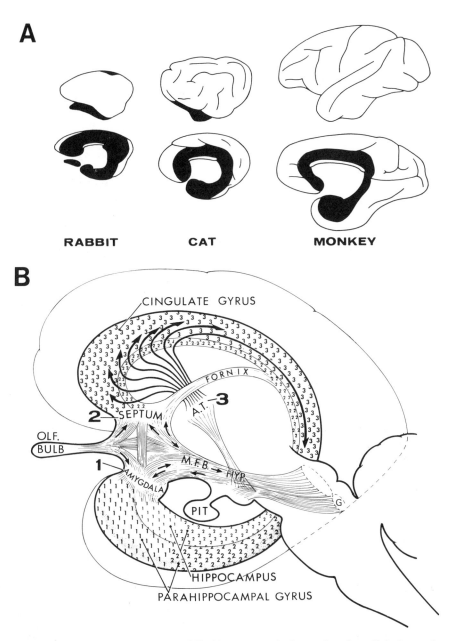

A

RABBIT   CAT   MONKEY

B

CINGULATE GYRUS

FORNIX

A.T. **3**

2 SEPTUM

OLF.
BULB

1 AMYGDALA M.F.B. → HYP.

PIT

G

HIPPOCAMPUS

PARAHIPPOCAMPAL GYRUS

FIGURE 4.3. Anatomical aspects of limbic system. A, Lateral and medial views of brains of three well-known mammals, illustrating that cortex of limbic lobe (black) surrounding brainstem is a common denominator in all mammals. Neocortex shown in white. After MacLean (1954). B, Diagram of three subdivisions of limbic system. See text for explanation. *Abbreviations:* A.T., anterior thalamic nuclei; G, nuclei of Gudden; M.F.B., medial forebrain bundle; PIT, pituitary; OLF, olfactory. After MacLean (1973).

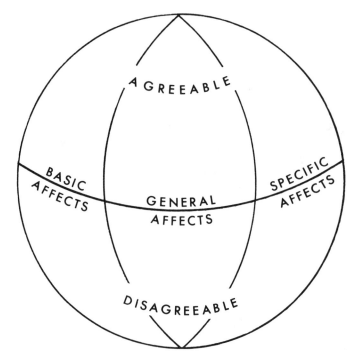

Figure 4.4. Scheme for classifying affective feelings. See text. From MacLean (1970).

descriptive of these affective states are (1) desire, (2) anger, (3) fear, (4) dejection, (5) joy, and (6) affection.

In limbic epilepsy, the general affects range all the way from intense fear to ecstasy. I will give just one illustration, selecting one that applies to feelings that make the condition of being a mammal so painful — namely, feelings of separation, or feelings of loss and sorrow. It is that of a factory worker who described his aura to me as a sensation in the pit of the stomach that gave him a feeling of sadness and wanting to cry. This feeling was followed by a welling up of tears and a sensation of hunger.

In connection with such ictal affects, it is of fundamental epistemic interest that affective feelings experienced during the aura may include eureka-type feelings of discovery; feelings of enhanced reality; feelings of conviction that what one is experiencing at the moment is of utmost importance; or feelings of revelation of what the world is all about, of the truth, the absolute truth. Significantly, these feelings are free-floating, being attached to no particular thing, situation, or idea.

Before further comment on the epistemic implications of this material, it is necessary to complete the description of a patient during a psychomotor seizure. With the spread of the epileptic storm, the lights go out, so to speak, in the limbic system, and the patient may engage in simple or complex automatisms and have absolutely no memory of what happened. Yet, the patient's behavior may involve appropriate cognitive neocortical functions such as that of an engineer who took his train through a long succession of red and green lights into a New York station, or that of a physician who performed a breech delivery with no recall afterward of what had taken place.

Why is there an inability to recall anything that happened during the automatism? The memory of ongoing experiences depends on a person's having a sense of self, a sense of individuality. Otherwise there exists no subjective "matrix" to which memories can become attached. Introspection reveals that a sense of individuality depends on an integration of information from the internal and external environment. The neocortex is primarily oriented toward the external world, receiving inputs largely from the auditory, visual, and somatic systems. The limbic cortex, on the contrary, is closely tuned in with the internal environment through inputs from visceroceptive systems, and, as our microelectrode recordings have shown, is also supplied by inputs from the exteroceptive systems. Elsewhere I have explained the reasons for inferring that a bilateral spread of an epileptic storm within the limbic system interferes with the integration of information from the internal and external environment and thereby obliterates a sense of self and the registration of ongoing experience (1990, pp. 512–516). A "functional ablation" of this kind would temporarily have the same effect on the registration of ongoing experience as an actual ablation or disease of such pivotal limbic structures as the hippocampus.

Descartes said, "I think, therefore I am." But limbic epilepsy indicates that there is a precondition, and that is, "I feel affectively, and therefore I am." Now it is one thing to have this primitive mind to assure us of the reality of the self, and beyond that to assure us of the reality of food that we eat or people that we meet. But where do we stand if we must depend on this same untutored mind to assure us of the truth of our ideas, concepts, and theories? In the intellectual sphere, it was as though we were being continually tried by a jury that cannot read or write.

Neocortical Function in Regard to Knowledge and Values

We must consider another profound uncertainty in knowledge that can be ascribed to the neocortex, which, as just mentioned above, is primarily oriented toward the external world. Reference to Figure 4.3B illustrates that in contrast to the expression "common denominator" used to describe the limbic cortex, the neocortex (shown in white) would be like an expanding numerator. In computer language, the neocortex might be compared to the progressive enlargement of a central processor providing an expanding memory and intelligence for increasing the chances of survival.

To the best of our knowledge, logic depends on neocortical functions. The use of symbolic logic leads to the conclusion (and it was Gödel's special contribution [1931] to prove it in terms of number theory) that any complex logical system cannot avoid contradictions because of self-reference. As Bronowski has commented, self-reference "creates an endless regress, an infinite hall of mirrors of self-reflection" (1966, p. 7). Moreover, since the subjective brain is solely reliant on the derivation of immaterial information, it could never establish an immutable yardstick of its own; it must forever deal with a nondimensional space and a nondimensional time for which it must arbitrarily set standards of its own.

In expounding neurological positivism, Vandervert (1990) has commented upon "science's erroneous placement of the world outside the skull" (p. 9). He would contend that knowledge of ourselves and the world around us depends solely on the "algorithmic organization of the brain" (1988) — algorithms that are to the brain what software is to the computer. Although it is necessary for us to keep probing in order to discover things about our environment, we can never hope to discover more about it than is provided by the brain's built-in neural networks that function algorithmically to achieve solutions.

Moreover, although inherent algorithms may allow us to design this or that instrument to obtain more precise measurements of environmental happenings, the precise measurements must be played back to the brain for analysis and interpretation. This realization illustrates the "pathetic fallacy" of unlimited confidence in instruments of precision.

All things considered, it would appear that because of (1) the matter of limbic delusion and (2) the other matter of self-reference, we are obliged to remain in the state of continuing ignorance. But there is, as I said in the introduction, a compensation that could prove to be beneficial to all living things. And this compensation derives from the quite recent evolution of the granulofrontal cortex that functions to provide what we regard as incomparable values.

What is the granulofrontal cortex? It is the cortex distributed over the forward half of the frontal lobes and characterized by a thin, but distinctive fourth layer of small granule cells. In Figure 4.5, I have used two types of skulls to illustrate my suggestion that the expansion of the granulofrontal cortex may have taken place in one of two contemporaneous human subspecies evolving during the past few million years (MacLean, 1990, pp. 553–556).

The letters A, B, and C identify, respectively, Dart's gracile form of *Australopithecus africanus;* Leakey's handy man, *Homo habilis;* and *Homo sapiens sapiens.* In the next row are successively the robust form of *Australopithecus* (X), *Homo erectus* (Y), and the Neanderthal human (Z). It is the popular view that the robust form (X) evolved from A; that *Homo erectus* evolved from the robust form; and that modern humans derived from *erectus.* But in X, Y, and Z, note the heavy buttressing around the

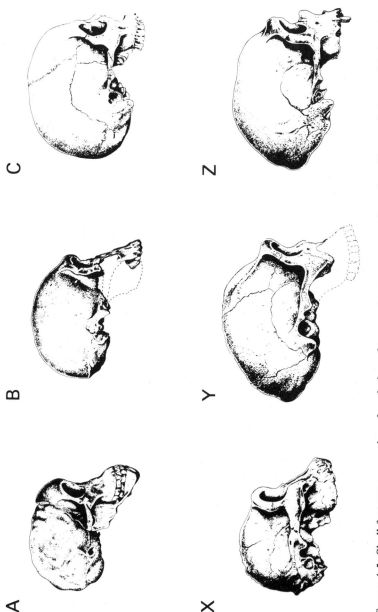

FIGURE 4.5. Skull features suggestive of evolution of two contemporaneous subspecies of human beings, with those in A, B, and C being especially relevant to site of expansion of granulofrontal cortex. See text for explanation. In answer to the questionable use of immature skull in A, an adult skull could have been shown instead. From MacLean (1990).

67

eyes; the sagittal ridge in X and Y; and the occipital bun in Y and Z. Note also the low, rather flat crania. Hence, I would suggest in airplane terms that the erectus is a stretched form of robustus, and that *Neanderthal* is a stretched form of *erectus*. In contrast, note the rounded occiput of the skulls in the first row and, in particular, the rounded frontal region. Note in particular the high brow of *Cro-Magnon* as compared with the low brow of *Neanderthal*. It is in the region of the high brow that the granulofrontal cortex has expanded.

Unlike other disciplines, the field of medicine is in the unfortunate position that advances in knowledge largely depend on an analysis of conditions resulting in disease and suffering. All the more for that reason, there is the obligation to wrest from human misery information that will help to prevent future suffering. If there was anything positive to come out of the medically bleak days of frontal lobotomy, it was the confirmation that the frontal granular cortex is involved in anticipation and planning. In regard to emotional experience, this is a matter of utmost importance because the capacity to anticipate and plan multiplies interminably the number of things that can induce either concern and anxiety or joyful anticipation. Anxiety may be defined as the unpleasant affect associated with the anticipation of the unknown outcome of future events. It was this symptom that, more than any other, was alleviated by frontal lobotomy.

But there is an undesirable consequence of being relieved of anxiety that is well illustrated by another kind of case. I refer to the case of a young man who, because of an injury suffered early in life, grew up largely devoid of granulofrontal cortex (Ackerley and Benton, 1948). Good looking and mannerly, he was dubbed "little Lord Chesterfield." But, he was always getting into trouble because of the inability to foresee the consequences of his behavior. At age 20, he was arrested for committing what was regarded as a premeditated crime. Although tests showed that he had an excellent memory for both past and ongoing events, his physicians defended him by saying that he was incapable of committing a premeditated crime because he could neither plan *nor* remember the sequential steps involved in carrying out a crime. It was as though premeditation required not only the ability to plan but also the step-by-step memory of what is planned, or as one might otherwise say, a "memory of the future" (Ingvar, 1985).

The expression "memory of the future" inclines one to ask, "How can there be a memory of something that has not yet happened? Isn't the correct word *planning*, rather than memory?" Planning, however, is comparable to a blueprint in the drawing stage, not the blueprint itself. In a football game it would usually be ineffectual to invent plays while the game is in progress. What is important is to remember the coach's blackboard drawing of where each player is to go and about how many steps are required to get there.

In primates, the evolution of both the granular and agranular frontal cortex went hand in hand with a surge in the development of neocerebellar systems. It has been known for many years that the cerebellum is involved in goal-directed behavior such as bringing the finger to the tip of the nose or making it possible for a baseball player to catch a high, fly ball. In extrapolating the path of the ball, it was as though the player had a built-in calculus. In a paper appearing in 1974, Robert Dow challenged the traditional view that the neocerebellum is involved only in motor functions. In the introduction of the present chapter (p. 60), I made reference to *directional evolution*. Dow did not use the expression directional evolution, but he did lay great emphasis on the contrasting difference between the development of neofrontocerebellar systems in the monkey and the great enlargement of these systems in the anthropoid apes and in the human brain where the structures appear almost tumorous in proportion. Indeed, because of the appearances of a new part of the dentate nucleus, one might almost say there had been an actual *turn* in the directional evolution of primates.

In 1986, Dow published a paper with the Leiners entitled "Does the Cerebellum Contribute to Mental Skills?" (Leiner et al.). They report their clinical studies indicating that just as the cerebellum participates in skillful motor performance, it may also contribute to the skillful manipulation of ideas.

In an extension of this suggestion, I have reviewed clinical and anatomical findings compatible with the hypothesis that the great development of neofrontocerebellar systems in the human brain may have provided, inter alia, neural circuitry for mathematical algorithms, including those used in planning and prediction (MacLean, 1990, pp. 545–552).

Now, finally, we come to a consideration of the first-time evolutionary appearance of a value that, with the educational honing of sensitivities, has the potential of providing a compensation for the inability ever to achieve certitude of knowledge. Along with anticipation and planning, the granulofrontal cortex is known to be involved in feelings of empathy and altruism. In this respect, it is significant that the granulofrontal cortex is the only neocortex that receives a large input from the great visceral nerve, that is, the vagus nerve. Presumably this visceral input helps to provide "insight" requisite for identifying with the feelings of another individual. One might conjecture that algorithms for altruistic feelings grew out of interconnections of the granulofrontal cortex with the thalamocingulate division of the limbic system involved in parental and other family-related behavior. Through generalization, a concern for the future welfare of one's own family might extend to the community family, to the regional family, to the worldwide family, and, eventually, to the family of all livings things. For the first time in the known history of biology, the granulofrontal cortex with its capacity for generating concern for the suffering of all living things, represents a 180 turnabout that could affect a turnabout of what has heretofore seemed a vicious life-death struggle, *long recognized as the struggle between good and evil*.

In the words of T. S. Eliot, we might imagine that "The whole world is our hospital," and we might continue with Howard Sackler's comment in regard to Semmelweiss that "somehow to intervene, even briefly, between our fellow creatures and their suffering or death, is our most authentic answer to the question of our humanity" (quotes from Martin Gottfried, 1978).

Acknowledgment

I wish to thank Francie Kitzmiller for typing this manuscript.

References

Ackerly S, Benton A (1948): The frontal lobes. *Res Publ Assoc Res Nerv Ment Dis* 27:479–504 (reprinted in 1966 by Hafner, New York)

Ackerly S, Benton A (1966): Report of case of bilateral frontal lobe defect. In: *The Frontal Lobes,* Reprint. New York: Hafner Publishing Company

Broca P (1878): Anatomie comparée des circonvolutions cérébrales. Le grand lobe limbique et la scissure limbique dans la série des mammifères. *Rev Anthrop* [2]1:385–498

Bronowski J (1966): The logic of the mind. *Am Scientist* 54:1–14

Descartes R (1637): Discourse on the method of rightly conducting the reason and seeking for truth in the sciences. In: *Philosophical Works of Descartes,* Vol. 1, Haldane E, Ross G (trans). New York: Dover

Dow R (1974): Some novel concepts of cerebellar physiology. *Mt Sinai J Med* 41:103–119

Gödel K (1931): Über formal unentscheidbare Sätze der Principia Mathematica und verwandter Systeme. *Monatsh Math Phys* 38:173–198

Gottfried M (1978): Program comments on Sackler's play *Semmelweiss.* Washington, DC: Kennedy Center for the Performing Arts

Ingvar D (1985): "Memory of the future": An essay on the temporal organization of conscious awareness. *Hum Neurobiol* 4:127–136

Leiner H, Leiner A, Dow R (1986): Does the cerebellum contribute to mental skills? *Behav Neurosci* 100:443–454

MacLean P (1952): Some psychiatric implications of physiological studies on frontotemporal portion of limbic system (visceral brain). *Electroencephalogr Clin Neurophysiol* 4:407–418

MacLean P (1975): On the evolution of three mentalities. *Man-Environment-Systems* 5:213–224. Reprinted in: *New Dimensions in Psychiatry: A World View,* Arieti S, Chrzanowski G, eds. New York: John Wiley and Sons, 1977

MacLean P (1990): *The Triune Brain in Evolution: Role in Paleocerebral Functions.* New York: Plenum

Vandervert L (1988): Systems thinking and a proposal for a neurological positivism. *Systems Res* 5:313–321

Vandervert L (1990): Systems thinking and neurological positivism: Further elucidations and implications. *Systems Res* 7:1–17

5

Morality and the Limits of Knowledge: A Neuropsychological Meditation

Jason W. Brown

> Nature has neither core
> Nor outer rind,
> Being all things at once.
> It's yourself you should scrutinize to see
> Whether you're center or periphery.
>
> — Goethe

This book will be exploring some of the ethical implications of neuroscience research, issues such as biogenetic factors in mental health and personality, sexual dimorphism in the brain and the feminist agenda, vivisection, fetal transplantation, and so on. I believe that these issues, important as they are, as much by their occurrence as by their outcome, herald a still greater revolution in our thinking about the nature of the human mind, one that is not contingent on innovative technologies (which are mere distractions) or new findings (which are never as new as they seem to be). Methods and data are not theory-neutral. They take on import by virtue of the concepts from which they arise. It is my thesis in this chapter that those concepts that matter in our struggle to understand morality as a product of the "human brain" are those that relate to the dominant or consensus theory on the nature and limits of human knowledge.

Currently, the most popular approach to knowledge and its natural limits is rooted in some variety of machine intelligence or functionalism, a theory that in turn is linked to mosaic and/or modular concepts of brain organization, what Bunge has dubbed the lego theory of the mind (Bunge and Ardilla, 1987). However, this is not the only possible framework for reflecting on the complex relations between knowledge and morality. This chapter reviews a competing theory of mind/brain, that of microgenesis, and explores the import of this theory for some problems in the philosophy of mind. I will argue that microgenetic theory has the potential to cause a radical transformation of our understanding of the nature of consciousness, introspection, access to private knowledge, free will, value, and what it means to be human.

Microgenesis: A Synopsis

Microgenesis is the hypothesis that evolutionary and, to a lesser extent, maturational process recurs in the unfolding of the mind/brain state, in fact, that a single process in different time frames mediates phyletic, ontogenetic, and microgenetic growth.[1] Put differently, the process of cognition is the growth of organic form. This is the changing configuration of the present, irrespective of the duration over which the growth occurs, the experience of duration, in any event, being mind-dependent. The staged unfolding of microgenesis recapitulates evolutionary structure, as well as configural aspects of ancestral behaviors. The theory does not entail stacking (terminal addition) of behaviors in mature organisms. For example, drives and rhythmic motility, not archaic repertoires, are engaged early in the microgeny.

The transitions from one level to the next and the links to evolutionary process are revealed through a study of symptom formation in patients with brain damage. Impaired structures display, in symptoms, the normal processing at the damaged point. The theory is an interpretation of pathological symptoms with damage at different levels (moments) in the hierarchy (unfolding). The finding that disorders of language can be understood in the same way as disorders of action, perception, affect, and memory led to the hypothesis of a common or unitary framework underlying all aspects of cognition. Even right-left differences, for example, holistic and analytic strategies, are interpreted as the entrainment of successive levels in a single system, for example, whole to part (context-item) or syncretic to featural transformation (Brown, 1990b).

Evolution builds up a stratified system of distributed levels over which mind is elaborated. In this system, early phyletic stages correspond to early stages in the cognitive process, whereas recent stages in phylogeny mediate late or end stages in cognition. The theory assumes an unfolding reiterated in milliseconds over this structure in an obligatory direction from depth to surface. There are reciprocal connections between successive levels and across homologous ones, within and between the hemispheres. In language and perception, there is a progression from word or object meaning to phonology and object form. In action, there is a passage from axial motility in body space to fine movements in object space. The structure of the microgenetic transition—the mind/brain state—is surrounded by tiers of sensory input, which constrain but do not enter the object formation, and tiers of motor output, which are read off the developing action (see Figure 5.1).

The unfolding lays down a series of three layers: subconscious cognition, the self and mental content, and the object world (see Figure 5.2). The self arises as a preliminary or preobject in the surfacing of subconscious

---

[1]This is a brief sketch of the theory. The reader is referred to Brown (1988, 1989, 1990a, 1991) and Hanlon (1991) for further material.

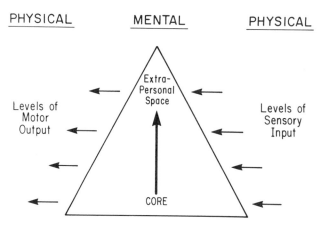

FIGURE 5.1. Sensory input at successive moments in the unfolding of the brain state constrains a configuration to model the external object. Similarly, motor keyboards discharge at sequential moments in the action structure, the discharge folding into the movement. This pattern of an emerging mental state interfacing with, but insulated from, a physical sensorimotor surround is reiterated at successive stages.

cognition, fractionating into concepts, images, and other mental contents which then exteriorize as external objects. The awareness of objects and mental content is a derived product emerging out of the underlying self-concept that stands behind and is distributed into the awareness content. Put differently, the self that looks out at external objects is an early stage looking at a late stage in the same microgeny.

All acts and objects grow out of the self, and the self is the conceptual core of the personality. The affective content of the primitive self-concept is the will to survive. One can say, the will as a core affect, and the self as a core concept, are given together as a single inherited form. The will distributes into the drives and these distribute into partial "affect-ideas" that accompany the percept development outward to build up the affective life of objects. The different forms of emotion are moments in a continuum with parcellation, qualitative change, and diminishing intensity over levels

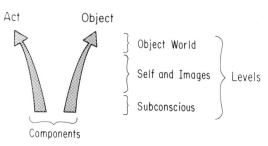

FIGURE 5.2. The microgenesis of the act and object formation lays down levels in the mental state.

in the elaboration of intra- and extrapersonal space. Thus, an archaic emotion such as will develops into drivelike states such as fear, which contain the nucleus of what later become affectlike states such as shyness, timidity, or humiliation. With this outline of microgenesis now complete, we can move on to consider the impact of the theory on the concept of the self, introspection, agency, and value.

Introspection and the Self

Most of us would define introspection as a state in which we can look into and describe the contents of our minds, and we would probably agree with William James (1890) that a belief in the faculty of introspection is "the most fundamental of all the postulates of Psychology." If we define *introspection* as the self examining its own mental contents, such as images and ideas, it can be contrasted with *exteroception,* or perception, which is the self examining objects in the world. We assume that the self in introspection is the same self as that in exteroception, so the difference is the content — an image or an object — that the self happens to be looking at. If the self is the same self in both states, the difference between introspection and perception will be the difference between an image and an object. Of course, the image is a content in the mind and the object is a content in the world, so our understanding of the difference between image and object is dependent on our understanding of the relation between mind and externality.

Consciousness is the reference point mediating knowledge of this relation. The consciousness of objects provides direct knowledge of the relation between the self and the external world. The consciousness of ideas or images provides direct knowledge of the relation between the self and other mental contents. Introspection and perception are states of the self looking at ideas or objects, while consciousness is the relation between the self and those contents. Introspection is often considered a "higher" form of consciousness than the consciousness of objects because the self is aware of the conscious state. For some, this is the crux of the problem, the state of being conscious that one is conscious. But consciousness of self is just an instance of introspection in which the content of the introspecting state is the idea of a "self-that-is-conscious."

For introspection, there has to be a self and there has to be something for a self to think about. As William James said, it is the fundamental belief, the basis of the *Cogito* of Descartes. Reichenbach (1954) remarked that "the discovery of the *ego,* of the personality of the observer, is based on inferences of the same kind as the discovery of the external world." This is explained by the fact that the self is a precursor of private *and* public space, an image from which all the images of introspection and the part-objects of the world develop. The nature of introspection is approached through the composition of the self and the relation of the self to images and objects, as well as the link between image and object. As these relations clarify, we obtain a better understanding of self-knowledge and of the knowledge we have of the external world.

The clinical material (Brown, 1988) relating to these topics shows that the transition from an image to an object is a transition from one mental space or level to another. An image is a type of preparatory object. An object is a completed image. The degree to which we know, or recognize, or are aware of an image depends on the degree to which the image objectifies.

The Self in Awareness

The object is preceded by the image and the image develops out of memory and the experiential store of the personality.[2] The self is a stage in the unfolding prior to the image, arising as a configuration in subconscious memory. The different aspects or expressions of the self are not isolated components but ideas the self pours out. The unity of the self derives from a position at a depth beneath analytic consciousness. Self, image, and object are stages in a process of creative becoming.

Hume argued that an attempt to discern the self in introspection always leads to a particular perception and not the self, that is, there is no self apart from the bundle of perceptions occurring during introspection. This does not mean the self is a composite of perceptions or a logical construction. It is what one would expect of a self that develops *into* ideas and objects with the source of the self subconscious or beneath access to that plane where the perceptual content individuates.

Introspection

Two oppositions are set up in object development: between viewer and external objects and between viewer and internal representations, with the latter anticipating the former. Both phenomena result from the outward development of percepts, a stage of mental (verbal, perceptual) imagery preceding that of external objects. As the object draws outward and becomes independent, its internal or subjective phase persists as a separate mental space in opposition to the external world. This is the basis for exteroception, the belief that objects impinge on mind. Introspection is

---

[2]Microgenesis entails that the mind/brain state is an *intrinsic* or endogenous series constrained by sensory (physical) input to model the external world. Learning is the persistence (i.e., the increased probability of recurrence or the enhanced role in subsequent unfoldings) of phases in the brain state that are reconfigured by sensation. The theory does not deny the existence of an objective world, only direct access to a world other than what is mirrored in mental representation.

attenuated exteroception, mind looking at its own preliminary object representations.

This leads to the paradoxical conclusion that introspection is a precursor to object awareness even though object awareness is a more primitive function.[3] There is a difference, however, between an awareness of objects and an awareness in which a self apprehends an object field. Externality and object independence require a self. Piaget thought that object awareness was an early stage in the ontogeny of self-awareness, but this is not the same awareness as when an observer is an agent distinct from the objects being viewed.

Introspection is a state in need of an explanation, not a method of psychological investigation. There are problems in the traditional depiction of the self, there are problems with the self's access to mental content, and there are problems with the reliability of verbal reports (see below). These problems do not just arise through the inability to verify mental states and their verbal descriptions or through the potential for bias or dissimulation by the subject. There is a deeper problem inherent in the very nature of introspection that defines the limits of self-knowledge. Knowledge of other objects has to do with the way that representations emerge and the nature of intrapsychic content. A fuller understanding of the process through which representations develop is likely to erode the "commonsense" belief in the existence of a self that scans mental content, and vitiate theories of mind and self built up instead on metalinguistic data.

Knowledge

The external world develops out of systems of experiential and conceptual knowledge. Introspection draws on this knowledge, which is implicit in the perception. Reflection on an object experience is a journey to the depths of the object formation, uncovering potential contents in the substructure of the object that were given up for the sake of the one object that developed.

In the withdrawal from the object the image rises to prominence and the feeling of self in relation to image is heightened. There is still an object; otherwise the individual would be dreaming, but the focus is on the preliminary content. This content has an imagelike quality. The relative richness of the content is contingent upon how early or unresolved the stage is. The play of ideas and images in introspection, the rising up of novel and

---

[3]The emergence of introspection at a penultimate phase in the microgeny corresponds with the specification of the language areas (Wernicke, Broca) within generalized or integration cortex, that is, from a stage antecedent to the primary "sensorimotor" cortices. This is consistent with the evolutionary concept that new growth occurs as an outcropping of earlier, less highly adapted (more homogeneous) form-building layers rather than as an addition to the (up-to-that-point) most recent evolutionary phase.

creative forms, the knowledge brought to bear on the object experience, are phenomena that emerge in the recurrence of the perception when stages bypassed in the earlier traversal are reclaimed.

We all have the impression that reflections on an object lead to the revival of prior events through a linkage of memories. I can search out a past that an object calls to mind, a body of stored knowledge that can be selected out of memory. I can introspect on the content of the knowledge and choose the information that is required. But what is the "I," the knowledge, the search, and the revival? The I is the self that comes up. Searching is an active way of describing the passive revival that "coming up" refers to, and revival is the derivation into awareness of what remained below the surface in the structure of a prior object. Knowledge consists of this content. One can say that access to knowledge is the depletion in a concept of those potential objects out of which the final object developed.

Limits of Knowledge

The depth and scope of knowledge available to the self narrow down as the theoretical issues clarify. The only direct, primary, unmediated knowledge is the content of consciousness in the present moment. The real object, the thing-in-itself, the sensory information it conveys, and the brain state that receives this information are all part of the physical world, extrinsic and indirectly known only as an inference about the origins of mental states. Since an object in the world is perceived as a representation in the mind, the perceptible world around a mind is part of the mind that perceives it. The real object cannot be penetrated through the representation, which is the limiting point in the mind's knowledge of the world. The inability to know the physical object is part of the desolation of the privacy of knowledge, but what is sacrificed in the recognition of our loss of the object is compensated for in the expansion of mind to include the object representation.

The healthy mind does not extend into the surrounding world but instead engages that world as a spectator. The intuition that object space, the surface of mind, derives from formative layers giving rise to the self-concept, carries with it the implication that self and object are part of the same sheet of mentation and that object representations are directly known as part (products) of the self-representation.

In sum, the self is elaborated in a capsule of mentation — the mental state — stretching from a subconscious core to a surface representation of the world. There is no access to physical states around this capsule. The world on the other side of our representation of it is an inference about the origins of that representation. The self is laid down as a phase in the mental state. The world passes out of the self, a self that is a product, not an agent, constructed each moment in the generation of objects that are its own derived content. The feeling of agency and the illusion of an external world belie the reality of a self that is deposited in the reiteration of the mental state, and a world that is a continuation of the mind of the viewer. The solipsistic conclusion is not that the world does not exist, nor that mind alone exists, but that one cannot access any event beyond its momentary mental representation.

Awareness is a relation between levels in the mental state. It stands in relation to what-is-given-to-awareness, not as something apart, but as that part of the representation developing in the mental state of that moment. The depth of meaning and the surface image, the knowledge and memory that objects call up, the object itself, are part of the same unfolding. That part of mentality that fails to elaborate awareness is excluded from the scrutiny of the introspecting state. This content is the *major* part of the representation, including the potential content given up in the realization of the final representation, to the extent that it cannot be revived on later reflection.

Choice

The account of the self as a product, and consciousness as a relation between phases in the microgeny, has consequences for our understanding of choice and responsibility. Our system of morality, not to mention our beliefs, intentions, and sense of autonomy and control are based on the assumption of an active decision-making self and an awareness that can scan a menu of competing options. This assumption carries with it the full weight of common sense and is part of the folk psychology of the self. Persuasive and incontrovertible as it may seem, this *assumption* is inconsistent with microgenetic principles.

The nature of choice is at the heart of the contemporary moral crisis. We are besieged by choices from every direction and required to "take a stand" on every one of them. To be indecisive is to be uncommitted, and lack of commitment, in a society where strong opinions come so easily, is perceived as a weakness of character. The issue, however, is the freedom of free choice, the elucidation of which in relation to brain function is essential for a scientific understanding of moral behavior.

Each of us has to make his or her way in society, finding some position in the stream of opposing values. Often it is difficult to decide what that position should be. A knowledge of the facts is not always helpful.[4] In most cases the choice comes down to a bias toward the freedom of the individual or the values of the society. Take the pro-choice and pro-life positions on abortion, euthanasia, and suicide: in each instance, the argument for the rights of the individual depends on free choice, the belief that an individual should decide how she or he wishes to live (or die).

---

[4]As Hume argued, reason is a faculty of determining what is true and what is false but is completely dumb on the question of what to do (Watson, 1982).

For most people, free choice is a value but the thing that is chosen is conceived as a preference,[5] like a preference for clothing or ice cream. To treat a value as a preference is to defuse it of its affective content and shift the debate to a less contentious level. The course that is chosen may reflect a value or a preference, but the freedom to choose is considered a fundamental right that society must respect. The concept of choice is a value independent of what the choice is about, even if one disagrees with what is chosen. For example, I may disapprove of pornography but grant the individual's right to decide. Choice is important, I believe — not so an individual can "freely" choose (taking us to the question whether there is "free choice" as commonly understood) — but because morality depends on some degree of indecision. Morality begins when self-interest hesitates.

When rights are conceived in terms of choices, self-interest tends to predominate and a restriction on one's own freedom, barring injury to others, is perceived as a restriction on the freedom of all. On the other hand, the good of society is vague and culture-relative, and lacks motivation in those societies where divine authority no longer counts as argument. The result is that the idea of a collective good is subjected to a debilitating scrutiny while the idea of individual liberty is taken as a given.

Ultimately, the tension, in modern societies, between individual responsibility and shared values has its roots in our ideas about the nature of value and choice. The way that value is assigned to concepts is a theory on the affective link between the self and its concepts or objects. The role of choice in decision making is a theory of agency and free will. These are topics that should be addressed by neuroscientists as well as by philosophers. What is the nature of the self? How are values formed? Are they conscious and freely acted upon or are they more like unconscious habits that drive behavior from below? Is choice the exercise of free will (i.e., does an agentive consciousness intercede in behavior?) or is consciousness informed after the fact in a behavior that is preset by conceptual processes beneath the surface of awareness?

Free Will

What we call "free will" concerns the interval between an idea and an act, not the external constraints on an action.[6] If I think I will move my finger at a certain moment and then move my finger, the microgenetic interpretation is that a conceptual core gives rise to both the idea — I will move my

---

[5]Unlike a value, a preference does not carry a strong affective charge: it is less generic and in a more superficial relation to the personality (self-concept).

[6]See Brown (1989) for the clinical material on which these conclusions are based, specifically the elucidation of the action structure, the unfolding from axial and rhythmic midline systems and body space through a phase of personal memory and core concepts to an articulation in the fine distal musculature in extrapersonal space,

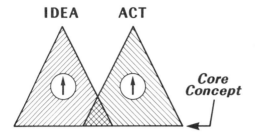

FIGURE 5.3. The decision to act and the action are generated out of the same core.

finger — and the actual finger movement. The core bridges both of these surface representations. Since the core is beneath consciousness, the microgeny of the idea informs the actor of the impending movement while the movement empties the concept of its remaining content. The point is, idea and act are driven by subsurface content. The idea does not cause the action, it is only a forewarning (see Figure 5.3). On this theory, actions are internally driven ("voluntary"), but the feeling of *decision* is a deception created by several factors, including the order of idea and act (which is really a problem of succession and time awareness), the precedence of the self in the unfolding sequence and its opposition to other mental and external contents, and the flow of mind outward toward the world. Our feeling of freedom in action parallels our feeling of the independence of objects "outside" in the world, objects which in fact arise out of concepts in mental space.

Voluntary action (understood in the microgenetic sense) is only one form of volition of which the self is capable. Thoughts and images are also under voluntary control. There is a different relation of the self to the various types of images. Dream images and hallucinations arise spontaneously and overpower the self which then becomes a victim of its own imagery. Memory images are "looked up" voluntarily, and thought images can be "manipulated" at will. There is an effect of effort on eidetic images. Objects appear independent of the viewer but are not independent. Similarly, these states of volition, passivity, and detachment are not states applied to the image (as commonly understood) but are generated as aspects of a larger state in which the image or object in question is also embedded. The feeling of volition, and the sense of a self that is engaged in a voluntary behavior, depend on the degree of objectification in the unfolding of concepts.

Indecision is a sign of conflict and conflict is accompanied by anxiety. The relationship between these phenomena has been documented in perceptgenetic studies by Sander (1928), Smith and Danielsson (1982), and others. These studies have shown that irresolution in a forming (subconscious) percept carries a strong affective charge that dissipates when the

the perceptual basis of action in awareness, and so on. For a discussion of external constraints on action in relation to free will, see Berofsky (1966).

object clarifies. Anxiety and conflict characterize states of incomplete object formation. Put differently, the object is incompletely selected out of its subsurface conceptual base. The multiplicity of potential or preobjects (i.e., indecision) is the other side of a lack of object specification.

This is the case even when the competing objects (propositions, ideas) are all entertained in consciousness. The inability to choose is the failure of any one object to gain the ascendency. On this view, indecision points to unresolved or preliminary cognition. The greater the indecision, the richer, or more diverse, the concepts guiding the action. Indecision or choice enhances the feeling of volition because the conceptual underpinning of the action has to be realized through many intermediate steps before finally discharging into the action. The enrichment of the self-concept preceding the action accentuates the indecision and increases the feeling of agency.

The self receives this lack of resolution and interprets it as a choice among competing goals. The final zeroing in on a target action, the choice that is made or the value that is satisfied, points not to conscious selection but to selection *into* consciousness. The reasons we give for our choices or our value judgments, therefore, are not true motivations but justifications of an unconscious valuation.

Responsibility

If the self is a way station in the progression toward a goal, and if ideas in consciousness are only an advance notice of inchoate acts, actions cannot be attributed to conscious deliberation and choice. Ideas do not initiate acts but are messengers of acts in preparation. What, then, is the role of choice in decision making, and specifically, what is the meaning of responsibility?

Since the self is generated out of the experience of the individual, good and bad, and all the acts and ideas of that individual are generated out of the self, whatever he or she does is that individual. Personality and self actualize in every behavior. The only criterion for whether a behavior is responsible is whether the action expresses the self fully and completely. The action then reveals what sort of self gave rise to an action of that type. The self must assume the burden of being the self that it is. From an intrapsychic standpoint, responsibility is the judgment we make of character, of the way the person, the self, is constituted. How a given character fits a society determines how well it is adapted to that society.

What does this mean for the role of choice in moral decision? As a society develops, there is growth and change in the education and experience of its members, and thus a change in the concepts underlying behavior. This is how society—the local or the collective society—shapes character. When we hold an individual to a moral standard, we indicate the distance between his or her behavior and that deemed acceptable by the society. We provide a scorecard on adaptibility or fit. If we assume that choice plays a role in behavior, we assume the individual can alter his behavior to meet societal criteria. But character changes with a change in the concepts underlying behavior, accomplished through instruction and learning; behavior does not change by fiat.

Value

Moral behavior requires an enlargement of the self-concept with a greater inclusiveness of those objects toward which the self is directed. The will has to be redistributed into other affects as the self-concept differentiates. In a mature self, survival no longer requires an immediate discharge into the drives (hunger, sexual, and so on) but incorporates a variety of secondary outlets, many indirect and muted paths — including even the altruistic denial of the self — as valid means to the goal of preservation. Self-preservation, in other words, becomes — in a mature self — contextualized.

Value is feeling invested in concepts that are important to the self. What happens is that the growth of the self is accompanied by a redistribution of will. In maturation the self is gradually articulated by partial concepts and their attendant affects. Such concepts enrich the mature self and the affect that fills them creates value. The self is defined by these concepts. I am the objects I need and love. The grief that ensues when an object is lost, or the compassion felt for an object that is injured, depends on how large a share of the self the object consumes.

A value, therefore, develops when part of the self is charged with feeling. When there is injury to a value, there is injury to the self. A sociopath has a narrow self-concept. He is guided by only one principle: self-interest. There is no compassion because the self does not contain the concepts in which other individuals have a share.

In contrast, the self that is a mother or father is a self that is defined by a role and a family. We say that a person is his or her work. Without the family or the work, the identity, the self, of such an individual is lost. We are not speaking of an external relation. The self incorporates and is a product of these objects. "Parent" and "profession" are intrinsic features of the self-concept, with the result that the loss of the object is like an excision of mental structure.

Feelings are part of valuation and values are the basis of morality. In the development of a concept the internal assignment of value has to externalize and become a stable part of the culture. The concept leaves the self as an object. The private experience of valuation becomes a shared concept and the self submits to the constraints imposed by that concept. As Tom Nagel (1979) put it, "morality depends on the objective assertion of subjective values."

Clearly, there is a two-way flow in the growth of value. On the one hand, personal concepts with their affective content exteriorize as shared values and gradually solidify in the morality of the culture. On the other hand, the prevailing morality penetrates the maturing self-concept and articulates the self with concepts independent of those primary ones motivated by will. The goal of socialization should be to combat the insularity of the primitive self. This is achieved by the assimilation into the self of values that are enriching, for then the self will fear their loss and struggle for their survival and for the opportunity to share and participate in their life. A society should promote choice, not to provide the individual with competing outcomes but to enrich the self-concept with claims other than those of self-interest.[7]

The Good Life

For most people the goal of life, simply, is happiness. The good life is a happy life. Chance and illness play a role in unhappiness, but unhappiness is most often the result of personal conflict. The unhappy individual is a hotbed of incompatible pursuits. The internal conflict that leads to unhappiness is the tension of a self that is a composite without clear direction. This tension, and the indecision or irresolution that it reflects, is displaced to the environment and interpreted as a struggle between the self and the limits on its freedom.

In the same way, moral decisions are perceived as a struggle between the will of the individual and the constraints of the culture. For example, the contest over abortion is experienced as a conflict between the wishes of the individual and the limits imposed by the society. In such instances, the constraints on action represent personal values that have been externalized and now serve as rules of conduct, antagonistic to the will of the individual. The real conflict is (or should be) one of an *intrapsychic* struggle between competing values.

Our faults, Shakespeare said, are in ourselves, not in our stars. The feelings that accompany unhappiness are created by the constitution of the self and the incomplete microgeny of components in the self-concept. The depressive, for example, is centered in the past, and is therefore characterized by an inability to go on with life. The inward turn, the revival of (proximity to) memories, the passive and helpless attitude, the pervasiveness of the mood—all are part of the withdrawal. A feeling of inadequacy may arise from the difficulty in coping with an unhappy situation. The inadequacy is the judgment of a discrepancy between goals and capacities, both of which are self-perceptions. The loneliness of the unhappy person is usually not due to abandonment by others but to a lack of sociability in the

---

[7]Echoing J. S. Mill (1874) who wrote, "The object of moral education is to educate the will."

self. The more the self is restricted by internal conflict, the more the microgeny is fixed at a preliminary phase, the less active, decisive, and assertive the self, the more unhappy life becomes.

Happiness is proportional to the completeness of self-expression, but primitive (noncontextualized) self-expression is just the appetite of the will and the drives. The primitive expression of the will leads to a vulgar type of happiness that occurs at the expense of the self-expression and happiness of others. When self-interest is motivated by the drives, generosity and compassion are the first to be sacrificed. The *quality* of the happiness that is achieved depends on the richness of the self that is expressed.

Thus, the construction of a good and moral life does not depend on logic and argument in maturity but on a nurturing development that articulates a self not crippled by dissent and sufficiently diverse to contain concepts that overlap with those of others — ideally, a self for which the happiness of others is a goal. The self at birth is a unitary construct of dispositions charged by will. The fate of this construct and the affect that fills it are determined by the conditions in which the self develops; that is, the family and the society. The child is shaped by these conditions to an adult with values. It is no small task to determine how this shaping ought to proceed, though it is clear that the aim of rearing should be to embed the self in a rich and varied social fabric. Only then will the self-expression of the individual, and his happiness, be achieved in the context of the society as a whole.

This socialization of the self is the balance that is attained between the will and the society. The adaptation of the organism to society is the fit between the primitive expression of will and its refinement by environmental pressure, a process of adjustment that is the nucleus of moral behavior. This conclusion suggests a basis for morality not only in intrinsic brain and mental process but in the evolutionary concepts of variation and fitness. Evolution requires a balance between creativity and elimination. New forms are generated and the unfit are pruned by a hostile environment. In human terms, one has to go beyond fitness in the sense of reproductive success, which (sociobiology not withstanding) is its biological product, not its definition. Fitness is contextual, thus relativistic. Fitness is not superiority but a balancing act on the road to self-actualization. In a perfect adaptation one finds a type of harmony or order. The closer one comes to a perception of this order, the nearer one approaches truth, the idea of the good, and an intuition of the mind of nature.

Acknowledgment and Afterword

Thanks to a critique of this paper by my brother, Professor Richard Brown (Brown, 1987), I am reminded that the solipsistic bent of microgenetic theory begs much of social-historical thought. For the social theorist, history and cultural diversity are the primary sources for an understanding of the self and moral behavior. After all, we seem to be conscious *over time*. We have a personhood that is stable, that stands behind and interacts with objects. Humanity depends on social interaction. From an intrapsychic perspective, however, history, like memory, is the shaping of the now in the momentary mind/brain state. Self and world are in a process of continuous re-creation. Mind invents and remembers itself, and others, for the moment in which the self is alive. *Pace, frater!*

References

Berofsky B, ed. (1966): *Free Will and Determinism.* New York: Harper and Row
Brown JW (1988): *Life of the Mind.* Hillsdale, N.J.: Erlbaum
Brown JW (1989): The nature of voluntary action. *Brain Cogn* 10:105–120
Brown JW (1990a): Psychology of time awareness. *Brain Cogn* 11:1–21
Brown JW (1990b): Overview. In: *Neurobiology of Higher Cognitive Function,* Scheibel A, Wechsler A, eds. New York: Guilford Press, pp. 357–365
Brown JW (1991): *Self and Process: Brain States and the Conscious Present.* New York: Springer-Verlag
Brown RH (1987): *Society as Text: Essays on Rhetoric, Reason, and Reality.* Chicago: University of Chicago Press
Bunge M, Ardilla R (1987): *Philosophy of Psychology* New York: Springer-Verlag
Hanlon R, ed. (1991): *Cognitive Microgenesis: A Neuropsychological Perspective.* New York: Springer
James W (1890): *Principles of Psychology.* New York: Holt
Marcel A, Bisiach E, eds. (1988): *Consciousness in Contemporary Science.* Oxford, England: Clarendon Press
Mill JS (1874): The freedom of the will. In: *An Examination of Sir William Hamilton's Philosophy.* New York: Holt, Rinehart and Winston, Vol. II
Nagel T (1979): *Mortal Questions.* Cambridge: Cambridge University Press
Reichenbach H (1954): *The Rise of Scientific Philosophy.* Berkeley and Los Angeles, CA: University of California Press
Sander F (1928): Experimentelle Ergebnisse der Gestaltpsychologie. In: *10 Kongres bericht experimentelle Psychologie,* Becker E, ed. Jena, Germany: Fischer
Smith G, Danielsson A (1982): *Anxiety and Defensive Strategies in Childhood and Adolescence.* New York: International Universities Press
Watson, G. (1982): Free agency. In: *Free Will.* Oxford, England: Oxford University Press

PART 2

Values and the Nature of the
Neuroscientific Knowledge Game

6

What Is the Ethical Context
of the Neurosciences?

H. Rodney Holmes

The title of this chapter is a question: what is the ethical context of the neurosciences? Neuroscience is a profession. Since the aim of this chapter is to describe the *ethical* context of this profession, the approach will not be a pure sociological analysis but an ethical analysis. It aims to bring a certain kind of clarity to our interpretation of the context of neuroscience by analyzing the values, goals, and purposes of the profession. These values, on the one hand, bind neuroscientists together and define the context of the profession; and, on the other hand, they strain the coexistence of individuals with competing values within the profession, and of the profession with other groups with conflicting values. Ethics, as an intellectual enterprise, does not try to provide an absolute hierarchy of values nor to design an ideal society of moral coherence. Unlike the social sciences, its aim is not to define, classify, and generalize about the role of morals in society or about the facts, points of view, and actions taken in specific cases of moral conduct. Its aim is to clarify the understanding of moral existence, that is, to develop a clarity that enables more responsible activity in a social world.[1]

This chapter will develop the position that neuroscience as a profession is founded on fundamental principles, that is, *ethics,* the violation of which undermines the activity of neuroscience itself. To achieve this aim, I turn to classic works not to argue from a basis of authority but to select relevant thinkers who still provide the terms in which we reflect on contemporary problems. The first works were written around religious ideas. The scriptural bases of those ideas is not my concern in this chapter: rather it is what has happened to the ideas themselves. Those ideas have become secularized in the literal sense of the term. The ideas have been converted from ecclesiastical to civil use. They are built into the *saeculum,* into the order of the modern professionalized world. It no longer matters so much to the professional world whether people believe them, and the religious formulations may not be useful as explanatory models for contemporary culture. But although they may have lost their importance as independent religious ideas, they have become important as explanations of how the social world came to take the form that it currently does. Moreover, as threads woven into the secular fabric, the explication of the ideas will aid in understanding the moral terms as they are used secularly, and hence in understanding the moral basis of the secular, professional order itself. My method in this chapter is to explicate these terms and how they were used by relevant thinkers to develop moral and political theories that in some instances define the modern systems in which we function, and in others addressed problems fundamentally the same as those emerging in the practice of modern neuroscience. By this understanding I hope to find some clarity for help in articulating and critically evaluating the ethical principles and moral codes of modern neuroscience.

It is necessary to give brief explications of these systems because the authors addressed their problems systematically, and that is why the terms they used are still valuable today. Specific problems were solved precisely by developing complete systems in ways that analogous problems today cannot even be articulated if they are raised merely on an ad hoc basis. I hope to show in the following pages how these authors contributed to the formation of concepts and modes of thought that are indispensable if we are to see our way clear through certain modern problems and dilemmas.

The first section of this chapter presents a case study and mentions several other current dilemmas that illustrate complexities of conflicting values in contemporary neuroscience. The chapter then turns to classical sources that explain how professions are ethically grounded and what critical principles may be raised to evaluate the ethical conduct of neuroscience. With each author, some of the most obvious analogies with contemporary issues are raised. The first work marks the turn away from the ancient world in which moral laws were understood to be absolutely the same in all times and places, in the way that fire is understood to be hot. In John Calvin we will see how "moral law" is not fixed in all times and places, but depends for its interpretation on particular circumstances. The second work literally provides the terms of political freedom which are in the opening words of foundational political documents learned by every schoolchild in the United States. John Locke raises the issue of the relation of the voluntary society to its civil government, and he finds that the society must make its own laws and have ample material means to remain free and to achieve its purposes as a society. The final works by even more modern social thinkers focus on the moral existence of the individual and the profession in the modern liberal society. Durkheim and Weber agree that there are necessities, but they find these moral necessities in the external and material forces of the society upon the profession, and between the professional organization and its individual member. Hegel critiques empiricism and radical rationalism to show that the ethical principles or laws derived by those methods are abstracted from reality, are incomplete, and therefore miss the unity and inner necessity that are central to ethical life and morality. Taken together, these classical sources raise criteria for clarifying the ethical context from perspectives both internal and external to the profession. But, more importantly, this method of approaching values in neuroscience will open even more significant questions about the nature of knowledge from the profession of neuroscience within its cultural context.

A Case Study

In late 1987, Edward H. Oldfield and Robert J. Plunkett of the National Institute of Neurological and Communicative Disorders and Stroke became the focus of conflicting values in science and society. They were denied federal approval for support to implement a promising new procedure to treat Parkinson's disease using tissue from aborted fetuses. Experiments by others using animals indicated that grafted cells from fetal brains might compensate for some of the functional loss associated with adult neuronal death in Parkinson's disease. They submitted a research protocol for grafting healthy cells from the brain of an aborted human fetus into the brain of a patient suffering from Parkinson's disease. According to the research protocol, a woman who had already decided to have an abortion, in a different hospital, would be offered the opportunity to donate fetal tissue for the procedure. This donated tissue would be grafted into the brain of a Parkinson's patient who had given informed consent for the procedure. This protocol was approved by the relevant ethics committees and by the director of clinical research for the National Institutes of Health. A Parkinson's patient entered the hospital and consented to the procedure. But just before surgery, the director of the NIH halted the procedure until the Department of Health and Human Services gave its approval, an approval that never came.[2]

Shortly afterward, the assistant secretary for health ordered a moratorium on federal funding for all experiments involving fetal tissue from elective abortions. This moratorium, a compromise offered in lieu of an executive order by President Reagan banning all such experiments, was offered to give the scientific community an opportunity to draft its own regulations, and with the offer came a deadline. During this time the neuroscientific community honored a voluntary moratorium on this research regardless of the funding source, with two notable and vocal exceptions.

The NIH convened an interdisciplinary panel to examine the ethical and moral questions. In its first meeting the panel set for itself the initial standard that there must be unanimous consent by panel members on its recommendations, although this was subsequently amended. The NIH interdisciplinary panel met again on September 14, 1988, in preparation for its final report. In its published report of that meeting "the panel did not recommend that research be halted on fetal tissue within the context discussed, although the recommendation of the committee is not binding, and an additional assembly of the panel will probably occur before the final recommendation to an NIH advisory committee is made in November."[3] During the November meeting of the Society for Neuroscience, a member of the Human Fetal Tissue Transplantation Research Panel reported to the society members that the difficulty in reaching unanimous consent in the panel had delayed the process past the deadline. In the meantime a research team in Colorado announced that it had used its own funds to transplant tissue from an aborted fetus into a Parkinson's patient.

The next year the NIH panel members reached their decision. They concluded that transplantation of fetal tissue *was* permissible public policy if certain safeguards were enacted. Following the guidelines developed in Sweden and generally accepted in Europe, the panel recommended certain procedures to ensure that the prospect of helping a sick person would not persuade a woman to have an abortion. However, in November 1989, Secretary of Health and Human Services Louis W. Sullivan announced a permanent ban on federal support for transplanting tissue from elective abortions. Furthermore, President Bush has made it a condition that the vacant post of Director of the NIH be filled by an individual opposed to abortion.[4]

This anecdote illustrates a clear conflict of interests between certain special interest political groups and the civil government against neuroscientists, several disease foundations and the American Association of Medical Colleges. The conflict is one of ethical absolutes. Other ethical concerns are raised by other external groups: scarcely an issue of the *Neuroscience Newsletter* or *The American Physiologist* has been printed in the past two years that did not contain major pieces concerning pressure from various groups they call "animal rights activists." National newspapers and even *Readers' Digest* carry stories about break-in incidents and harassment of scientists at the nation's medical research institutions. The question of honesty in science has been raised most recently about a Nobel laureate in molecular biology who is under investigation by a congressional subcommittee for fraud.[5] And following the release of a report by the General Accounting Office on NIH-funded research just before the workshop that prompted this chapter, Colorado Representative Patricia Schroeder voiced "serious questions about funding," and various groups have leveled accusations of "male science" against the National Institutes of Health.[6] Indeed it is clear that every modern neuroscientist is powerfully motivated by values drawn from more than one way of understanding the world. The task is to articulate a professional ethic of neuroscience that provides a full account of the human brain that is at once informed by the scientific tradition and at the same time deeply responsive to a wide range of human goods and experiences.

Within the Profession

The values of science derive neither from the virtues of its members, nor from the finger-wagging codes of conduct by which every profession reminds itself to be good. They have grown out of the practice of science, because they are the inescapable conditions for its practice. Science is the creation of concepts and their exploration in the facts. It has no other test of the concept than its empirical truth to fact. Truth is the drive at the center of science.[7]

Ours is not the first time in history that the aims and values of special interest groups and society have conflicted. What rolls easily off modern American tongues, "separation of church and state," is a resolution in principle of centuries of bloody conflict between people bound together by absolute commitment to achieving goals in service to a common idea of an uncompromised good with other people similarly bound and committed. Since there was no formula available for resolving conflicts of absolute values other than by death, moral theologians and philosophers had to develop ways of expression and regulation of special groups *within* rather than *against* civil society. John Calvin and John Locke explicated principles of internal connections between the conduct of members of religious groups and their common goals of divine purpose in ways that illuminate how any professional group is defined, directed, and regulated by its internal values and purposes.

John Calvin's political writings do not appear in an extended formal treatise, but appear distributed in his biblical commentaries and *Institutes of the Christian Religion*.[8] To recognize that the immediate political context was the Geneva of the early 16th century is to know that government was by elective assemblies; hence, citizens in general shared civic responsibility. Discerning Calvin's writings to be theological texts founded upon scriptural conceptions of the relations of God to man and consequently of man to man, is to appreciate the magnitude of the moral problems with which he was dealing. Calvin had no mean task in trying to connect great and absolute metaphysical truths of divine and natural law to particular and circumstantial applications. By making moral laws dependent on the context, Calvin moved ethical theorizing into the modern era.

Calvin defined two realms, the spiritual and the temporal. Separation is maintained in that neither legislates for the other. Yet these two realms are not entirely separable in practice since the civil government protects, supports, and provides necessary conditions for the worship of God. The most important conditions include all the things needed for maintaining the peace and promoting justice. Furthermore, the two realms are related foundationally by the way each is connected with justice. Judicial or civil law is a political constitution of rules of equity and justice by which citizens can live together in peace. These particular laws differ according to the needs of various times and place. But these laws are consonant with the laws of God, that is, natural or moral law, since equity or justice is "natural," is engraved on the minds of all men as conscience, and is the end of judicial laws. Similarly for ecclesiastical or ceremonial laws, their end is the "preservation of that love that is enjoined in the eternal law of God." The particular laws, such as the Decalogue, are neither ends in themselves nor given to all people, but are witnesses to and particular enactments of the primal law of God, that is a law of love, a natural and popular sense of common justice that "is the same for all mankind."

For Calvin, both civil and ecclesiastical laws are applications defined by particular circumstances, and never to be in conflict since both are in the service of the natural law, which is to be understood as justice:

What I have said will be more clearly understood if in all laws we properly consider these two things — the constitution of the law and its equity, on the reason of which the constitution itself is founded and rests. Equity, being natural, is the same to all mankind; and consequently all laws, on every subject, ought to have the same equity for their end. *Particular enactments and regulations, being connected with circumstances, and partly dependent upon them, may be different in different cases without any impropriety, provided they are all equally directed to the same object of equity.* (Institutes XVI, emphasis added.)[9]

In words that sound strikingly modern, Calvin has considered the most unconditional concepts in moral philosophy to provide a solution to a major quandary in applying ethical principles. He has shown that "moral law" is not fixed in all times and places, but depends for its interpretation on particular circumstances. Furthermore, he has provided the first critical principle for evaluating the conduct within a particular discipline: justice. This contextual application of ethical principles and this critical principle form his key to the peaceful coexistence of the special group and its greater society.

John Locke expanded these ideas about how voluntary associations are defined and regulated internally, and about the proper relationships between associations and the society. Locke's political context was a period of civil war, and of sometimes near-anarchy following the civil war as the monarchy was reestablished in late-17th-century England. This political context was inseparable from the connections of factions within the Anglican Church with the civil political powers against the non-conforming and dissenting Baptists, Independents, Presbyterians, and Quakers. Amid this turmoil, Locke was politically exiled to Amsterdam for advocating religious and political toleration regardless of one's particular beliefs. Thus the major problem of *A Letter Concerning Toleration*[10] was to resolve how individuals may freely associate, define their group, and regulate themselves according to a common purpose without imposing upon, or without imposition by, the authorities represented in civil law. His theory and very words permeate the major documents that founded the United States, and his thought on how to solve particular problems that arise when individuals freely associate for their own purposes in a liberal society is applicable to analogous situations that arise in this society, which was founded in part on his words.

Locke's central thesis appears in several of his extended political treatises. Civil government is formed and exercised to ensure the basic human rights: life, liberty, and possession of property ("outward things such as Money, Lands, Houses, Furniture, and the like"). The fundamental right is the right to one's own body, that is, to life itself and to "health and indolency of body." The rights to property derive from that fundamental right since in working, whether working the land for planting or the clay for sculpting, one is mixing one's self with the material. Hence the product is an extension of one's body, and one thereby derives the right to determine its fate. Liberty is the basic freedom to dispose of one's life without coercion from outside. Hence the most basic infringement upon this right would be corporeal coercion. These three basic human rights are equated with the "Civil Interests." Hence civil government is "bounded and confined to the only care of promoting these things," and its only legitimate use of power, force, or coercion is in their service.

Liberty, a basic human right and civil interest, is developed in the *Letter* to explicate the formation of the voluntary society and the proper relationships between voluntary societies and the civil government.[11] Locke finds that voluntary societies, for example, professions, are formed by the joining together of individuals of their own accords to achieve a common purpose. Members of the society function together to achieve the commonly desired result. Therefore, the activities that are appropriate to the society are those that are directed toward achieving the society's commonly held purpose or desired result. It follows that the proper regulation of the members and their activities is not by command, coercion, and penalties, but by persuasion, and use of light and evidence to change opinions.

Locke finds that voluntary societies are held together by laws. Since the society is voluntary, freedom and spontaneity are its first necessary conditions, and it necessarily follows that the right of making its laws belongs only to the society itself. Since the society is joined to produce a commonly desired result, the primary criterion of its laws is to make the conditions of the society's activity such that are necessary to the purpose of the society.

The laws of the voluntary society become definitional and foundational. They are definitional because laws, in the service of the common purpose, state what it means to be this particular society. Conditions set for society membership are only those that are necessary to the purpose of the society. Societal laws are foundational since they establish the groundwork for the freedom and spontaneity to perform the activity necessary to achieve the purpose for which the members joined together. Undermining these laws restricts freedom and spontaneity, compromises the conditions, and hampers that activity — in short, violating the laws enfeebles the society itself.

From this it follows that the laws of a voluntary society are restricted to the affairs of the society. One voluntary society cannot bind its laws upon another voluntary society, nor upon the civil society at large. Since its affairs are restricted, the voluntary society can demand no civil consequences of breach of societal laws. Because of the differences in purpose, methods, and laws, the voluntary society that transforms into civil society undermines its own purpose and activities. In short, the voluntary society is founded upon two conditions that are its ethical principles: its free and voluntary nature, and its purpose. Undermining either principle destroys the institution itself.

Understanding the concept of the purpose, the *telos,* of the voluntary society, shows why the civil government needs and supports voluntary associations within its civil society. From classical times the idea of "the good" has been implicit in any notion of the telos or purpose. Locke connects civil government, a good life, and moral actions by the activities of voluntary societies. The concern of civil government is personal life, liberty, and possessions, and by assuring these things, civil government provides the necessary conditions for the good life. In Locke's political system it is necessary to hold and use property to exercise one's liberty in accomplishing one's ends. Since "some depraved individuals do prey upon the fruits of others' labors," individuals associate together to preserve their possessions, liberty, and strength in order to accomplish what they want. Since the particular associations are defined by common purposes, it follows that particular societies must have certain things to achieve their good ends — whether those ends be "knowledge for the sake of knowledge" or new medical applications that can make the blind see and the lame walk. Removal of these possessions removes the tools necessary to achieve the good ends. Therefore, the voluntary society makes laws and persuades its members to support the proper acquisition and use of those tools to achieve its ends, and the civil government enforces the conditions necessary to guarantee the right of free use of the property.

Locke has shown us many new things about the meaning and ethical grounding of voluntary societies such as a profession of neuroscience. Freedom and spontaneity are its first necessary conditions. Purpose to produce good effects is the reason for its members voluntarily to assemble and for the civil government to support it. Possessions are the necessary material means to achieve its ends. The professional society regulates itself by persuading its members by evidence and reason. Its self-made laws are directed to assuring the activities that achieve its purpose. They include only laws that are necessary for that purpose. Tools and other material conditions are necessary for the activities of professional societies. This is certainly true in modern neuroscience, and as we shall see, this need for material possession becomes not only an internal necessary condition of the society, but also an external force that gives it shape.

Calvin and Locke provided the terms for what it means to be a voluntary society within a political context, and for how to distinguish between separate societies with separate aims within one political context. Each laid a foundation and constructed a framework for how a voluntary society functions and regulates itself. For Calvin it was based on a critical principle of justice, and for Locke it was based on a critical principle of freedom. These moral principles were derived and applied internally in these systems in ways that help to address some of the present conditions in modern neuroscience insofar as it is a profession that is formed, joined, and sustained by its members voluntarily. These conceptions help to illuminate the moral ambiguity surrounding the groups who did not honor the scientific society's voluntary moratorium in the case study, both in terms of the telos and the activity of the society. Practically, they offer ways of articulating the purposes and the regulations of the society, and they indicate the practical advantages of addressing these questions systematically from a theoretical basis. But even while deriving the values and regulations internally, these two systems have already alluded to the external conditions in a way that requires connections between the explicit purposes of voluntary professions like neuroscience and the values of the politically free society.

Outside the Profession[12]

These multiple ties which link us to each other and to the group in which we take part are not so obvious that all we need to perceive them is a somewhat developed taste. Having received, therefore, no other education, we are necessarily led to deny their existence; that is to say, to see in the individual an autonomous power which depends only on itself and to see in society a simple framework of relations of all these independent forces. This is why whoever undertakes to react against this superficial simplism and to recall the true place of the individual in society runs up against rather lively feelings and prejudices.[13]

When a federal funding agency, such as the National Institutes of Health or National Institutes of Mental Health, issues a "Request for Application" to study a specific question or disease, it is not difficult to line up the interests of society with the desirable effects neuroscience wishes to have on its political society. When funding is offered for specified problems it is easy to see that the values of society often are appropriated by science to a significant measure. What is much more subtle and difficult to see is how the cultural context informs the core values of science, articulated as David Baltimore's basic scientific research for the sake of knowledge,[14] J. Bronowski's drive for truth,[15] and S. Chandrasekhar's aesthetic appreciation of truth.[16] These subtleties are the topics of writings by Emile Durkheim, Max Weber, and Marx and the Frankfurt school, and they focus upon the very motivation of science itself. In this section we will see in

Durkheim how knowledge or science for its own sake is a value supplied by modern liberalism, and how that value is incorporated in a professional ethic.

To see these cores of science from this externalist point of view, one focuses on Durkheim's definition of the *intellectual,* and from that definition traces the connections to the core value of liberalism, the individual:

Let us note in passing that this very convenient word [intellectual] has in no way the impertinent sense that has so maliciously been attributed to it. The intellectual is not a man who has a monopoly on intelligence; there is no social function for which intelligence is not necessary. But there are those where it is, at one and the same time, both the means and the end, the agent and the goal. In them, intelligence is used to extend intelligence, that is to say, used to enrich it with new knowledge, ideas, or sensations. It thus constitutes the whole of these professions (the arts and sciences), and it is in order to express this peculiarity that the man who consecrates himself to them has quite naturally come to be called an intellectual.[17]

If it is true that the means and ultimate end of the intellectual also characterizes the scientist, then it characterizes the same core value of science that scientists themselves express. This places reason above authority as a core criterion, and presumes the right of the individual to decide questions for oneself. The right of the individual seems then to be inalienable, and there emerges from this characterization an implicit basis for a schism between the individual and society, between the individual scientist and society. But Durkheim finds that such radical separation, or individualism, is quite rare, and that the ethical justifications are so morally impoverished and simplistic that they have few serious adherents.

There is another form of individualism of much more concern to Durkheim and to us. It happens when this core value becomes religious (in Durkheim's use of the term). Duty becomes only that which concerns one personally or professionally in order to seek out only that which one's individual conscience requires, and which one shares with one's fellow professionals. When this definition of duty becomes the sole criterion defining good and evil, it takes on a stamp of religiosity and is considered "sacred" in Durkheim's ritual sense of the word. Whoever attacks this core value "inspires in us a feeling of horror analogous in every way to that which the believer experiences when he sees his idol profaned. Such an ethic is therefore not simply a hygienic discipline or a prudent economy of existence; it is a religion in which man is at once the worshipper and the god."[18] Ultimately Durkheim lays this form at the feet of Kant and Rousseau and calls it a "cult" with "its primary dogma the autonomy of reason and its primary rite the doctrine of free inquiry." Rhetorically he asks, "if all opinions are free, by what miracle will they be in harmony?"[19] Systematically he retorts in *Principles of 1789 and Sociology,* the individual, and the core value of free inquiry, is more a *product* of society than its cause.

Durkheim offers several objective means of testing the source of the core values. If the principles are themselves seen as scientific doctrine, then they must be subjected to scientific critical method. *Is it true,* according to the sources cited above, that science proceeds best when left unrestricted to advance knowledge for its own sake; that scientists provided freedom for dissent and independence for originality are motivated by and will discover Truth; that the tenacious pursuit of Truth and the appreciation of Beauty join to create the moment of scientific discovery? For Durkheim's scientific method, "the only way to answer these questions is to confront the reality of facts with the prescriptions which are supposed to include them."[20] The answer is: not necessarily. The connections between these posited causes and their effects are not fixed, but vary by country and national temperament. Furthermore, abstracting these guiding principles from science will show that they motivate the entire society as well as the profession. Since it is thus demonstrated that the core, guiding principles are not unique to the scientific profession but are ubiquitous to the society, they cannot be said to be self-generated by the profession. No positive correlation can be demonstrated between these core values and their predicted effects. Indeed "good facts" are discovered by those who do not understand these to be core values. Thus we cannot conclude that these values are the *causes* of good science. What must be concluded is that these core values were created neither *by* science nor *for* science; rather they result from the very practice of life in society.

The consequences of the demonstration provided by Durkheim are of the deepest significance for anyone who might want to defend the core idea of "science for science's sake" in a purely internalistic way. It is important to determine the source of these core values because the judgment brought to bear on those principles will change completely. If it can be demonstrated that the core value of science is not an independent consequence of its own nature but a consequence of the political society and its values, then the implications for understanding the sources of its ethics and moral codes, and even for understanding the sources and interpretations of its scientific theories, are fundamentally changed. Neuroscience thus conceived, at the same time that it represents naturalistic theories about the brain, sees only a succession of abstract propositions when it does not consider the social conditions in which they are produced. Society's guiding needs and aspirations also determine the theoretical and technical outline of the work that is undertaken.

Durkheim's professional ethics focus on the inevitable conflicts between the guiding needs and aspirations of individuals, professions, and society.[21] Moral facts consist of rules of conduct that have sanction, that is, consequences of conforming or not conforming to rules of conduct already laid down. There are two kinds of moral rules: those applicable to everyone and those applicable to the individual. Between these two extremes is a kind where diversity is most marked: moral rules of the professions. Professional ethics are a plurality of morals found in every society that operate on parallel lines. They have a distinctive feature marked by "the sort of unconcern with which the public regards it," an unconcern that arises precisely because professional ethics are not common to all members of society. This fact points to a fundamental condition without which no professional ethics can exist: *The group protects its members.* Professional ethics are worked out as dialectic tensions of sanction and protection between the society and the profession, and between the profession and the individual.

Durkheim traces the history of the business professions from Roman guilds, through the Middle Ages, to the late 19th century to show the dialectic and the roots of professional ethics. These stages represented a progression from a time when Roman guilds were forms of virtual enslavement of individuals by the government through the present time when the profession may not hold any claim of sanction upon the individual. This development corresponded to shifts from the explicit to implicit binding of societal needs; shifts from governmentally, to guild, to individually imposed standards of work; and shifts from government, to guild, to individual empowerment of the profession. As the explicit role of the government decreased, there increased a need of the individual member for support and protection by the profession from outside forces.

Also found in this history are the moral roots of the profession. The guild was first a religious collegium, celebrating particular religious rites together. Later the guild became a *sodales,* a brotherhood whose leaders were father and mother. Now the religious and family functions extended to the financial support of some of its members (e.g., aged poor who had fallen upon hard times). Moral codes were written that bound apprentices to masters, but also specified that they could be discharged only for just reasons. Regulations governing the quality of work were enforced upon the members to guarantee to the public a standard set by the guild. With time these sanctions became oppressive, and the guild of the Middle Ages had to transform itself into the modern profession. What was retained was the ethical character of the guild: its structural roots were in family-like loyalties and in religious ritual (i.e., communal articulation and celebration, and individual affirmation of commonly held values). The point for Durkheim is that since these facts clearly demonstrate that in the past professions functioned as moral spheres, it can be concluded that they are certainly capable of functioning as such today.

The nature of the professional sphere in terms of the conflicts about sanction and support depends upon the group itself. The more coherent and interactive the group, the greater is its professional strength, and the more numerous and binding are its moral rules over individual members. Moreover, the larger the group, the greater is the potential that individual and group interests may conflict.

This conflict is rooted in the very core value spoken of before: individualism. The individual may not be sensitive to social interests precisely because they are exterior to the individual: they are interests of some thing that is not one's self. What seems to be needed is a system that functions to bring these interests to the mind of its individuals and oblige respect for those interests so as to maintain the community. This system is thereby a system of morals. It has a transcendent nature (i.e., its scope and power transcend the individual) that has been expressed as fundamental principles of ethics deriving from a divine source. The larger the group, the more exterior and abstract will seem the group interests to the individual, and the perspective of the individual will be comparatively limited. Because of this difference in perspective, the aims of the individual will become antisocial in the absence of rules that prescribe actions that conform to collective aims. This system is the profession. And because of its role, Durkheim finds that no professional activity can be without its own ethics.

In summary: Durkheim found the scientific professional ethic to be a part of the liberal tradition. The profession is a social structure based on ideals of individualism and intellectualism. As such, the core value of science is not the glorification of the individual self, but of knowledge in general. Hence the profession does not address itself to the individual member, nor does society address itself to the individual profession, but rather to the "human person" and to impersonal and anonymous aims, respectively. Thus the focus is diffuse, soaring far above individual consciousness. Within the inadequacy of the individual conscience as an ethical system also lies the two-tiered conflict of professional ethics. The individual is both bound and empowered by the profession according to common aims, just as the profession is bound and empowered by society. On the other hand, the individual finds decided advantage in taking shelter under the roof of a collectivity that ensures peace, and benefits by the rules of conduct that prevent wrangling and endless friction with professional colleagues. These conditions would undermine the conditions of peaceful interchange looked for when individuals first sought association as an adjunct to free inquiry. Thus knowledge for its own sake, the core value of professional science, is concluded to be a living out of the core value of a liberal society, and it is to the good of society that the profession regulate its own morals. The profession, with its professional ethic and codified morals, is needed as an interface between the abstract and diffuse needs and aspirations of society collectively, and the concrete and focused needs and aspirations of the scientist, individually. Consequently, professional neuroscience as a matrix of core values, applied purposes, and two-tiered individual conflicts, must be understood and assessed not in terms absolutely peculiar to it, but ultimately in terms of its interdependent life in liberal society.

A Critical Principle

Our look at internal and external determinants of values in professions has prepared us for a deeper examination of the aims, the mores, and the justification of the conduct of neuroscience and the neuroscientist. What are the underlying moral principles that define the "ethical context of the neurosciences?" What is the necessity, the critical principle by which particular laws of the profession can be justified?

Some modern ethical theorists draw a distinction between *ethos* or *mores* and *ethics* or *moral principles*. Ethos or mores are the way we actually live our lives. They reflect the way we think it is right to behave. Ethics, or moral principles, are the formal philosophical principles that justify the practical life. How are we to come to an understanding of the critical principle, by way of empirical study of mores, by rational analysis of ethics, or by some other way?

Hegel criticized the empirical method generally for focusing on a single aspect of a problem and holding it up for measurement and description to the exclusion of understanding the whole. What is missed is the unity, the inner necessity. Montesquieu looked at the law in England and the law in France, and empirically described what was common to both. The flaw was that this snapshot-in-time was an abstraction, a distortion of the reality of moral law. Reality for Hegel was not an empirical account of how things accidentally are at this moment, but a timeless role in history. Reality is developmental, not empirical. It is not accidental, but absolute with an historical absolute necessity.

On the other hand, Hegel found Kant to be equally abstract. He distorted reality by not looking at the content of ethical life but only at the formal framework itself: the Categorical and Moral Imperatives. For Hegel, purely formal ethics is as mistaken as an empirical approach, therefore the two approaches must be combined to understand the inner moral unity that is real and to articulate it as a critical principle.[22]

It is time to call the question: Why look to these theologians, philosophers, and sociologists for guidance in articulating an ethic, a real moral unity, of neuroscience, particularly when few neuroscientists would accept the metaphysics of their systems? The reason is that each of these authors offers guidance by expressing an idea of some kind of necessity that links together the commands and prohibitions when we establish a code of ethics for a profession.

These are not simply an arbitrary collection of rules that happen to be observed by members of the profession. In the good old times everybody had a good notion of how the good doctor, good lawyer, or good scientist was supposed to behave. Today it is rather clear to us that we need some kind of a justifying framework, a "necessity," to identify what elements belong in a particular code, such that anything in the code can be justified in relation to its connection with that critical principle. Lacking this, one does not have an ethic: all one has is a collection of rules, a laundry list of things that people ought and ought not to do. This insight was shared by Durkheim and Hegel.

But what is the basis for, and where do we locate, that necessity? John Locke and John Rawls locate that necessity in the agreement among people who come together to make a society: You and I agree on a purpose. Hegel located it in the universal metaphysical process, an historical process that is working itself out. But if we do not agree that it is there, suppose that the development of society itself imposes a set of constraints on morality. Max Weber asked, "Why was it wrong for so many years to take interest on a loan, but suddenly it was okay?" The answer—the material development of the society. One does not choose one's own goals; the social system chooses the goals for one. Marx, Durkheim, and Weber find the source of the critical principle, the necessity of a profession in material reality and in the ideas and purposes that attend the material forces of the civil society at large. Indeed, the direction of inquiry and the resources of science have been shaped and enabled by a Kennedy's dedication to space exploration and to battling cardiovascular disease, and by a Nixon's desire to cure cancer. Perhaps now the aims of society and the profession will line up to reap the benefits of Bush's "Decade of the Brain." Somewhere connected to the activity of brain research the necessity should be found.

A second important question: What have we learned about critical principles that will help us regulate and direct our profession? The critical principle that supplies the justifying framework of our professional society is to be found not in separate and abstract facts or ideas, but in the unity of our profession as practiced, enabled, and constrained by purposes and tools that are at the same time internal to the profession and externally provided. Properly identified, the critical principle of neuroscience is not a set of universal and abstract principles of morality to be applied in any possible society. These professional standards properly identified and understood are not arbitrary and reactionary, but can be shown to be necessary to the life of the profession, they form the *ethics* of the neuroscientific community. Again, the insight of Hegel was to link universal abstract principles of moral rationality to particular objective life of a person or professional society. When we finally figure out what these principles really are, they will probably look more like Durkheim's social laws than Kant's categorical imperative or Aquinas's eternal law.

Together Durkheim and Hegel reveal the ultimate significance of the knowledge and values of neuroscience. If the most foundational, sacred thing about the society is the society itself, then the neuroscientific activity is found within the matrix of the most sacred element of society. Society gives to neuroscience a particular sanction that empowers and directs the scientific process toward knowledge about the human brain, the physical locus of what it means to be human. Study of the human brain is both the means and the end, the agent and the goal of that sacred element. The values implicit in that study itself and the knowledge derived from it are connected at once to the inner necessity that defines and unifies neuroscience and to the sanction that society gives to neuroscience.

Understanding neuroscience and its professional values this way transforms the way of evaluating the cases enumerated in the first section of this chapter. The way is first to ask: What is the necessity that drives this neuroscience? What is the critical principle of this discipline? Is it an issue of informed consent, of abortion in the use of fetal tissue, or is it a fundamental respect for human beingness that provides a set of ethical standards?

What are the directions of neuroscience? Are they to write every grant proposal and to know everything there is to know? Even if it is the latter, then what is knowledge—is it all the possible electrochemical information about neurons, or is there some way of defining our goal that will help us decide what information is worth gathering, or what applications are legitimate and what are not? It may be good for scientists to cure Parkinson's disease and for scientists to get all the possible knowledge about the brain, but "What does it mean to be human?" is such a central question to the wider society that our findings transcend our professional interests. A society intrinsically interested in the brain comes with its own goals, preconceptions, and values. The best way that our neuroscience can enlighten the most profound questions of what it means to be human is not to negate those goals, preconceptions, and values, but somehow to reintegrate our knowledge with those of the time in which we live concretely. These are tricky and tough questions to which I do not pretend to have the answers.

Conclusions

Although it is pretentious to claim to answer these kinds of questions, this chapter is suggesting an approach for addressing them. It calls for an understanding of a critical principle that is particular to contemporary neuroscience. This requires a disciplined reflection upon the activities and values that is enlightened by sources such as those used in this chapter, and an articulation of a critical principle that is sharpened and tempered by conversations such as those occurring at the conference that prompted this volume. I understand this critical principle to be a real, inner necessity, that unites the ethos and moral principles of the profession. It grounds and justifies the conditions, activities, and direction of neuroscience. Ultimately, the critical principle determines the value and the kind of knowledge that is sought and found by neuroscience. With Hegel, I find elements of truth in the solutions to the particular problems offered by each of the systems enumerated in the second and third sections of this chapter, but I also find that each of these, or for that matter any other, articulation is itself a distortion of the critical principle, the inner necessity that is real. Although the approach is limited both by the kinds of distortion of any empirical approach and by the limits of looking at problems analogously, it does offer the terms in which similar problems have been debated as well as indispensable insights into the formative sources of our core conceptions and modes of thought. These insights deepen the discussion, but also show the complexity of formulating statements about these core values and ways of understanding.

The need in articulating a critical principle of neuroscience is not merely to identify it as a particular mode of thought, but to understand how the profession and each individual member lives in a context of competing frameworks simultaneously. A neuroscientist has at once many loyalties to very different value systems. Unfortunately, in a context of historical ignorance, it is all too easy to caricature each of the value systems that simultaneously do impinge on and inform our lives. Such readings make the targeted critical principle shallow, one dimensional, and easy to repudiate. Narrow proponents of any single value system can point at particular elements discussed in the second and third parts of this chapter and dismiss them out of hand with little intellectual difficulty. On the other hand, it is very hard to articulate a coherent, multidimensional statement of what gives meaning and unity to one individually, and certainly to a modern profession like neuroscience.

In this chapter I have tried to offer a depth dimension for our understanding of the moral existence of the neuroscien*tist* by tracing several relevant threads of thought from their original sources. These texts served as points of origin for sets of ideas that should not be understood as ideas that somebody else had about a foreign situation, but as the historical origins for ideas that make up various facets of the modern context. I have tried to show through the systems that were explicated that, insofar as neuroscience is a functioning profession, it has moral roots and functions as a moral sphere. Its members have a sense of purpose that is ethical and moral to a great degree. Neuroscience, since it is a part of a greater society and is supported by that greater society, is grounded on basic principles that are fundamentally consonant with what society values. Nevertheless as each of the case studies shows, there are all sorts of moral ambiguities with which the profession and its individual members must come to grips. It is my prescription that the profession will be better equipped to deal with these moral ambiguities in a manner that is consistent with its own core values as well as other values in society by disciplined reflection upon those values.

Disciplined reflection by neuroscientists will be aided by understanding the roots of these values as they were explicated in sources like those treated in this chapter. Perhaps the way is first by internalist systems that draw upon the intuitive knowledge of the purposes and actual activities, that is, the inside sources of the scientist, to offer ways of articulating the ethos and moral principles. On the other hand, neuroscience does not exist in a moral, political, or financial vacuum. Modern problems and dilemmas reflect a kind of complexity and moral ambiguity that are better understood with the help of externalist systems that show the connections between the values of society and the values of neuroscience.

It is also my opinion, based on these same externalist systems, that it is not necessary for those in other professions to be radically skeptical about the basic moral principles of the natural sciences. Readers of Thomas, Bronowski, Snow, and Chandrasekhar may understandably not be totally persuaded by their written descriptions of what motivates scientists. If it is true that any statement is empirical and is contextually conditioned, then even the most eloquent statement about the neuroscientific ethos or laboratory life is necessarily inadequate to the task of communicating what about it is real.[23] On the other hand, if the directed activity itself is an expression of a particular purpose, then the empiricist does have some access to neuroscience's basic moral principles by a social avenue. That avenue provides access through an understanding that professional activities are particular forms of societal values. The movement can be understood as a conditioning from general values and purposes of society to particular values and purposes of professions, finally to those values that are internal to one profession but not to another. Having arrived this way at the particular aims and values of the internalist systems of voluntary associations, the claims by Locke and Calvin that those internal values are ultimately consonant with civil values may be more palatable.

Interdisciplinary approaches such as this conference may play key roles in clarifying the moral issues in specific cases such as those enumerated in the first section of this chapter. It is not hard to see how neuroscience could benefit from outside perspectives. Furthermore, from our own perspective, we know of internal structural problems not commonly known to outsiders such as occurrences in our particular method of peer review of grant proposals and research papers. We might be helped in implementing a better way of achieving our ends by social scientists who have shown the single-blind system to be unjust and most open to abuse, and have devised better methods. Knowing first hand of the problems in large laboratories, we may want to reassess the proper size of science on the one hand, or more vigilantly conform to our mores and defend our harassed members on the other. But ultimately, the questions being asked by modern neuroscience are so important and so central to what human beings are all about, that we cannot presume that we can derive the answers by taking over a set of imperatives from some other discipline or some other set of social concerns.

Most significantly, interdisciplinary conferences like this one are most fruitful for deepening the very questions of value and meaning. We hold these conferences precisely because the information and technology coming out of modern neuroscience make these times interesting and exciting for anyone seriously interested in human beings. These workshops help neuroscientists to better understand the core values and to better justify and direct the activity of neuroscience by the articulations and considerations encouraged by the dialogical process. Conversations such as these almost certainly promote the reintegration of the ethics and information of neuroscience with other cultural conceptions and values.

Walter Rosenblith suggested at the opening and closing sessions of this conference that to reach the goals of this conference we need to establish an *ongoing* dialogue among members who will dedicate serious attention to the questions articulated in the titles of the conference sessions. Neuroscience as an objectifying system of inquiry into the brain is changing the way we think, and what we think about many humanistic conceptions. Whether many penetrating formulations are discarded as mere "folk psychology" or are integrated within the new system so as to optimize the value of its knowledge may well depend upon that serious attention.

Acknowledgments

The author is a beneficiary of a Cross-Disciplinary Study Award from the Ethics and Values Studies directorate of the National Science Foundation. The author is indebted to Robin Lovin, Dean of the Theological School of Drew University, for his patient and invaluable guidance.

Endnotes

1. For more on this distinction see James M. Gustafson's writings, such as his introduction to H. Richard Niebuhr's *The Responsible Self* (New York: Harper, 1963):6–41.
2. T. Beardsley, "Aborted Research. Ideology Seems to Have Put Some Medical Advances on Hold," *Scientific American* 262, no. 2 (1990):16. The first section of this chapter was written using Beardsley's article and my notes of the Public Affairs Forum on Neural Transplants at the Toronto meeting of the Society for Neuroscience in November 1988. Thus the reader might note that most of this account was drawn from sources within the scientific profession.
3. D. W. Barnes and R. E. Stevenson, "Meeting Report: Human Fetal Transplantation Research Panel," *In Vitro Cell Developmental Biology* 25, no. 1 (1989), 6–8.
4. Bernadine Healy, MD, a cardiologist and former research director at the Cleveland Clinic, has been nominated and confirmed Director of the National Institutes of Health. Dr. Healy was a member of this NIH panel. See J. Palca, "Healy Nominated," *Science* 251 (1991):264; see also D. Koshland, "The Choosing of the NIH Director," *Science* 246 (1989), 981; and C. Holden, "OSTP Fills Life Sciences Post," *Science* 251 (1991):158.
5. David Baltimore recently retracted a paper that was found by the NIH to contain fraudulent data, and he offered an apology to the postdoctoral researcher who challenged the paper, Margaret O'Toole. See Philip Weiss, "Conduct Unbecoming?" *New York Times Magazine* (October 29, 1989):41–95; David P. Hamilton, "NIH Finds Fraud in *Cell* Paper," *Science* 251 (1991):1552–1554; and David P. Hamilton, "Baltimore Throws in the Towel," *Science* 252 (1991):768–770.
6. On these and related concerns, see "Women in NIH Research" in *Science* 250 (1990):1601; and in *Science* 251 (1991):159. NIH Director Dr. Bernadine Healy has been praised for her positive steps in correcting these situations in David L. Wheeler, "Scientists Draft List of Priorities for Research at NIH on Issues

Critical to Women's Health," *Chronicle of Higher Education* 38, no. 4 (September 18, 1991):A9–A10.

7. J. Bronowski, *Science and Human Values.* (New York: Harper, 1956):77.

8. A collection of these writings can be found in *John Calvin on God and Political Duty,* ed. John T. McNeill, (New York: Macmillan, 1950).

9. McNeill, *John Calvin,* 64–65; the emphasis is mine.

10. John Locke, *A Letter Concerning Toleration.* 1689. ed. James H. Tully (Indianapolis: Hackett, 1983).

11. As with Calvin, the particular association is religious. However, Locke, who is more interested in the coexistence of several religious denominations, grounds his system in political and moral philosophy rather than in scripture. This makes it easier for the careful reader to extrapolate his principles to other kinds of voluntary associations.

12. This section is my formal response to the only comments that time permitted for this paper, those by Susan Leigh Star. She pointed out that the paper as presented gave the most time to John Locke and the implications of his system. This presented a strong internalist bias. Moreover, since the quotations used to represent an externalist position were drawn from the conclusions of Durkheim, Weber, and Marx without explicating any of their systems, the externalist points were unsupported and hence lost. Leigh Star was quite correct. I am grateful for the opportunity to respond more adequately to her perceptive criticism by sketching the framework of one of those systems. Fortunately the reader will also have the opportunity to see in other chapters in this volume how specific values and questions in neuroscience have been influenced by their social and political contexts.

13. Emile Durkheim, "les principles de 1789 et la sociologie," *Revue internationale de l'enseignement* 19 (1890):450–456; reprinted in *Emile Durkheim On Morality and Sociology,* trans. Mark Traugott, ed. R.N. Bellah (Chicago: University of Chicago Press, 1973):40.

14. David Baltimore, "Limiting Science: A Biologist's Perspective," *Daedalus* 107, no 2 (1978):37–45.

15. J. Bronowski, *Science and Human Values* (New York: Harper, 1956).

16. S. Chandrasekhar, *Truth and Beauty: Aesthetics and Motivations in Science* (Chicago: University of Chicago Press, 1987).

17. Emile Durkheim, "L'individualisme et les intellectuels," *Reveu bleue,* 4e série, 10 (1898):7–13; reprinted in *Emile Durkheim on Morality and Sociology,* 231.

18. Durkheim, *Morality and Sociology,* 46.

19. Ibid., 49.

20. Ibid., 35.

21. Emile Durkheim, *Professional Ethics and Civic Morals,* trans. Cornelia Brookfield (Westport, Conn.: Greenwood Press, 1958).

22. See G.W.F. Hegel, *Natural Law,* ed. T.M. Knox (Philadelphia: University of Pennsylvania Press, 1975) for an early and relatively brief account of what would later become his dialectical theory. In this book he distinguishes *Sittlichkeit* and *Moralität.* The former refers to a conscientious abiding by the laws and customs of one's nation, while the latter refers to conscientiousness simply. Hence a wrongdoer may be moral, but not *sittlich.* Having made the distinction, Hegel demonstrates that this separation is ultimately a distortion of the real moral unity.

23. Indeed, the real experiences of Paul MacLean, Alan Fine, and Terrence Deacon in the context of the conference demonstrated this. In the heat of discussion sessions (and of the unusually sultry conditions of a popular summer retreat for marine biologists on this August 4–6), each of these three accomplished neuroscientists was asked to explain his motives, the driving force for why he does neuroscience. Rational language failed them in relating their lives as really lived in their home labs. As surely as each was failed by logos to communicate his personal ethos, so the ethos of neuroscience ultimately is the real, concrete human activity of the laboratory, known and understood by those who live it. This, I take it, is what Durkheim means when he says the ethos emerges from the activity itself. It also illustrates what Hegel means when he says that any verbal account is itself an empirical distortion of the real moral unity.

7

The Gendered Brain:
Some Historical Perspectives

LONDA L. SCHIEBINGER

Women have historically been excluded from science. The world's major scientific academies were founded in the 17th century—the Royal Society of London (the oldest continuous society of science) in 1660, the Parisian Académie royale des Sciences (perhaps the most prestigious academy of science) in 1666, the Akademie der Wissenschaften in Berlin in 1700. These academies first admitted women in the 20th century—the Royal Society admitted Kathleen Lonsdale and Marjory Stephenson in 1945, the Berlin Academy admitted Lise Meitner in 1949 but only as a corresponding member. The Académie des Sciences in Paris did not admit a woman until 1979 and even she was a "safe" woman, the daughter of a prominent mathematician and the wife of an academy member.[1]

But what about the substance of science? What effect does this exclusion have on the concepts of science? For more than 2000 years the brain has been seen as highly gendered. For Aristotle, woman was a man manqué; in regard to her sex, she was a monster or error of nature. In Aristotelian science, as in the later Galenic tradition, the inferiority of women—as odd as it may seem—depended on their lesser heat. Aristotelians held that women are cold and moist while men are warm and dry; men are active and women are indolent. Because women are colder and weaker than men, Aristotle argued, women do not have sufficient heat to cook the blood and purify the soul. Thus their power of reasoning is weaker than men's.

Scientific theory changed with the centuries; evaluations of women did not. In the late 18th and 19th centuries, craniologists tried to account for sexual differences in intellectual achievement by measuring the skull. Anatomists assumed that the larger male skull was loaded with a heavier and more powerful brain. By the mid-19th century, Social Darwinists invoked evolutionary biology to argue that woman was a man whose evolution—both physical and mental—had been arrested in a primitive stage. One of the most notorious arguments arose late in the century, when Harvard medical doctor Edward Clark (a man vehemently opposed to admitting women to university) attempted to show that women's intellectual development could proceed only at great cost to their reproductive development. If women exercise their brains, he argued, their ovaries shrivel. Clark sought to document his theory with cases from Vassar College. This kind of nonsense did not end in the 20th century. In the 1920s and 1930s, arguments for women's different (and inferior) nature were based on hormonal research. Today we are inundated with apologies for women's lowly status in the sciences based on studies of brain lateralization and sociobiology.

Historically, sexual science — the scientific study of sexual differences — has been used to keep women in their place.[2] The social inferiority of women has been (and still is) buttressed by scientific apologies. At the same time, these scientific studies were done by and large in the absence of women. Eliminating dissenting voices has been one factor insulating the scientific profession against correction of misreadings of female nature. Women have been excluded from the scientific community and that exclusion has been rendered "natural" through studies emanating from that community.

What does this mean for the study of sexual differences? Should scientists continue to study differences in anatomy, physiology, or brain function in men and women? Ignoring differences is certainly not the solution. The National Institutes of Health (NIH) have recently been taken to task for neglecting the study of sexual differences. A report by Congress's General Accounting Office revealed that a number of treatments — from cholesterol-lowering drugs to AIDS therapies — have been studied almost exclusively in men. Doctors have comparatively little information about how medications affect the female body. Therapies beneficial to men can be harmful to women. Studies have shown that diets that reduce cholesterols, for example, such as the one promoted by the American Heart Association, while useful to men, may actually harm women. Moreover, research on exclusively female ailments — menopause and breast cancer, for example — lags much behind research on distinctively male problems.[3] Medical technology has also privileged the male body. Though few mentioned it at the time, the artificial heart was designed *by* males *for* males: it was simply too large to fit most female bodies.

Scientists have long recognized that sexual differences are real and warrant study. Already in the 1780s, medical doctors realized that disease might take different courses in the male and in the female body and argued that in order to treat each sex properly, differences in body-build needed to be taken into account.[4]

There is a certain irony, then, in congressional concerns about our inadequate knowledge of the female body. Though scientists might have neglected research concerning women's health and well-being, they have never hesitated to make proclamations concerning the origins of human inequality. As we have seen, long before the 1986 NIH policy requiring grant applicants to consider women in their research, the study of sexual difference was a priority of scientific study; monies — both federal and private — have flowed freely into sexual science. This research, however, has been confined to particular questions and problematics. We have no dearth of information about women's supposed mathematical (in)ability or less specialized brains — information aimed at perpetuating divisions in power and privilege between the sexes. As the same time, there has been relatively little interest in soaring cancer rates among women or in new ways to make Pap smears and mammograms more reliable — an information gap robbing millions of women of life and well-being.

The scientific community has not operated in isolation in this regard. Reports that women's scientific inabilities are genetic or hardwired make big news in this country. When Jerre Levy of the University of Chicago reported in 1980, for example, that "males are good at maps and mazes and math" and females by contrast have "superior verbal skills" and a "sensitivity to context," her findings made front-page news across the country. *Discover* magazine highlighted similar research in a 1981 article on "The Brain: His and Hers," subtitled "Men and Women Think Differently. Science Is Finding Out Why." These reports, because they are scientific, are often thought to provide resolution to political struggles. *Quest Magazine* recently concluded that brain lateralization studies "threaten the axioms of militant feminists as well as homosexuals."[5]

Today we find ourselves in a curious situation. Congress reports that our knowledge of the female body in general is insufficient. We do not know basic things about whether taking an aspirin per day reduces heart disease in women. At the same time, certain scientists claim to have located in the female brain the reasons for women's underrepresentation in science. The solution to this complex social problem is imagined to be localizable in one aspect of female physiology.

Nature versus Nurture: 18th-Century Origins

The controversy over woman's sexual character has its origins in what has been called the nature versus nurture debate. Are the differences that we perceive between the sexes to be traced to *nature* or to *nurture,* are they rooted in the body (and more specifically, the brain), or are they the product of history and therefore of socialization? The distinction is significant: if gender roles or personality traits are rooted in the brain, traceable ultimately to congenital or genetic factors, it is easier to argue that the predominant sexual division of labor is natural, and that efforts to transform social relations between men and women are misconceived or foolhardy. This debate — still raging today — finds its origins in the 18th century.

The Enlightenment posed a challenge to existing orthodoxies of inequality: "All men are by nature equal." This rallying cry reverberated throughout Europe and America. Responses to Enlightenment calls for equality came primarily from two camps: the nurturists and the nativists. Taking the nurture side, feminists—both men and women—began performing a number of "experiments" designed to determine whether women and minorities, given the proper environment, could become the intellectual equals of European men. The issue centered on university education. Rather surprisingly, Germany was a pioneer in accepting the occasional woman or African man for admission to university study.[6] In each case the experiment was a success. Anton Amo, a black man from Guinea (Ghana), received a Ph.D. from Wittenberg in 1734; he went on to teach at that university and subsequently served as an advisor at the court of Frederick the Great. Dorothea Erxleben received her M.D. from the University of Halle in 1754 and had a flourishing medical practice; Dorothea Schlözer took her Ph.D. at the University of Göttingen in 1787. In England Francis Williams, a Jamaican boy, attended the University of Cambridge where he excelled in mathematics and the classics. Back in Jamaica Williams opened a school for black boys.

Despite the success of these few experiments, long-standing inequalities were not overturned in the 18th century. European universities that had been closed to women since their founding in the 12th century remained closed to women until late in the 19th and sometimes even into the 20th centuries. Nor were women to become the equals of men in the church, state, or professions. Now, however, unlike earlier, Enlightenment ideals required that these injustices be justified. The French 1791 "Declaration of the Rights of Woman" required resolution; the 1802 reintroduction of slavery (in French colonies) required explanation. The way that these inequalities were justified within prevailing egalitarian ideologies gave rise to both the perplexing impasses we encounter in current nature/nurture debates and the power of science in those debates. It is essential that we understand something about political history in order to understand the role of science in the study of female nature today.

When revolutionary legislatures decided that women would not become citizens in the new democratic orders, justifications for these decisions were taken in some cases directly from anatomy books—women were weaker and disposed to a deadly overexcitation. Within these new orders the claim of women to equality was increasingly considered a matter not of *ethics* but of *anatomy*. The belief that science could be called upon to arbitrate essentially political debates was encouraged by four factors.

The first was the new authority of nature. Nature and its laws played a pivotal role in the rise of liberal political theory. Natural law philosophers such as John Locke and Immanuel Kant attempted to set social convention on a natural basis by identifying the natural order underlying the well-ordered polis. Natural law (as distinct from the positive law of nations) was held to be immutable, given either by God or inherent in the material universe. Within this framework, an appeal to natural rights could be countered only by proof of natural inequalities. The marquis de Condorcet wrote in reference to the equality of women that if women are to be excluded from the polis, one must demonstrate a "natural difference" between men and women in order to legitimate that exclusion. In other words, if social inequalities were to be justified within the framework of Enlightenment thought, scientific evidence would have to show that human nature is not uniform, but differs in socially significant ways according to age, race, and sex.

This first factor is closely allied with the second and third: the growing authority given to science as a privileged source of knowledge about nature, and the belief that science is impartial (*unparteyisch,* as the German anatomist Samuel von Soemmerring and his colleagues called it). Since the Enlightenment, science has stirred hearts and minds with its promise of a "neutral" and privileged viewpoint, above and beyond the rough and tumble of political life. Perhaps science could provide objective evidence in debates over women's intellectual and physical abilities. Perhaps the knife of the anatomist could find and define sexual difference once and for all; perhaps sexual differences — even in the mind — could be weighed and measured.

Enlightenment enthusiasm for nature and its laws spawned already in the 18th century a vigorous and protracted search for new scientific definitions of sexual (and racial) differences. Beginning in the 1750s, discovering, describing, and defining sexual differences in every bone, muscle, nerve, and vein of the human body became a research priority for science. This broader search for sex differences brought forth the first representations of the female skeleton in Western anatomy. It was part of the materialism of the age that the skeleton as the hardest part of the body was thought to provide a "groundwork" for the body. If sex differences could be found in the skeleton, sexual identity would no longer be a matter primarily of sex organs, but would penetrate every muscle, vein, and organ attached to and molded by the skeleton.

Although drawn from nature with painstaking exactitude, great debate erupted over the features of the female skeleton. Here already we see the outlines of current debates. Political circumstances drew immediate attention to depictions of the skull as a measure of intelligence and the pelvis as a measure of womanliness. In her influential drawing of the female skeleton, the French anatomist Marie Thiroux d'Arconville (one of the few women to practice the art) portrayed the female skull as absolutely and proportionally smaller than the male; she also drew attention to the pelvis — remarkable, in her view, for its breadth and width. In contrast to Thiroux d'Arconville, the German anatomist Soemmerring portrayed the skull of the female (correctly) as larger in proportion to the body than that of the male. Soemmerring drew the ribs smaller in proportion to the hips, but not remarkably so. As one of Soemmerring's students pointed out, the width of women's hips should not be overemphasized; they only appear larger than men's because women's upper bodies are narrower, which by comparison makes the hips seem to protrude on both sides.

Despite (or perhaps because of) its exaggerations, Thiroux d'Arconville's skeleton became the favored drawing, especially in England. Soemmerring's skeleton, by contrast, was attacked for its "inaccuracies." Edinburgh physician John Barclay wrote, "although it be more graceful and elegant and suggested by men of eminence in modeling, sculpture, and painting, it contributes nothing to the comparison [between male and female skeletons] which is intended."

In subsequent years, when anatomists had to concede the truth of Soemmerring's depiction of the female skull as larger than the male skull in proportion to the rest of the body, they did not conclude that women's large skulls were loaded with heavy and high-powered brains. Rather than a mark of intelligence, the size of women's skulls signaled their incomplete growth. Anatomists in the 19th century used this as evidence that physiologically women resembled children and primitives more than they did white men. In 1829, Barclay, using Thiroux d'Arconville's plates, presented a skeleton family in order to "shew," as he said, "that many of those characteristics described as peculiar to the female, are more obviously discernible in the foetal skeleton."

What was the significance of the depute about skeletons? This brings us to the fourth factor in Enlightenment justifications of inequalities. Enlightenment thinkers privileged the physical over the intellectual and the moral. For many, physical differences underlay and determined the intellectual and moral differences one observed in men and women's character and daily lives; in other words, sexual differences determined gender differences. Rousseau, for example, held that differences in minds and morals were connected to sex, though (he admitted) "not by relations which we are in a position to perceive." In *Emile,* Rousseau opened his chapter "Sophie, or the Woman"—perhaps the single most influential piece written about women in the 18th century—with a discussion of comparative anatomy. One important aspect of his portrait of woman was her inability to do science: women, mired as they were in the immediate and practical, were incapable of discerning abstract and universal truths. Others agreed with Rousseau. Hegel compared the mind of woman to a plant because, in his view, both were essentially placid. In this way modern science began to establish the notion of a gendered brain.

These elements taken together—the authority given to nature and to science as the knower of nature, the belief that science is value-neutral, and the grounding of gender in sexual difference—rendered *invisible* (at least to some) the injustices in the system by sealing an already self-reinforcing system. The self-proclaimed neutrality of anatomists was premised on the absence of dissenting points of view. Those who might have criticized new scientific views were barred from the outset, and the findings of science (crafted in their absence) were used to justify their continued exclusion. The image of women developed in this context had the character of a self-fulfilling prophecy: women did not excel in science — but, then, they seldom had an opportunity to work in science. As we saw in the example of the female skeleton, an uncanny consensus developed within the medical community on the natural adeptness of women to bring forth children but not science. The paradoxical nature of this situation was captured by an anonymous 18th-century writer: "Men have not only excluded the Women from partaking of the Sciences and Employ by long Prescription, but also pretend that this Exclusion is founded in their natural Inability. There is, however, nothing more chimerical."

Beyond Nature and Nurture

Science that tries to show that "biology is destiny" remains entrenched in this old way of thinking. The issues, however, are more complex than the simple nature versus nurture dualism implies. One of the fallacies of sexual science has been the belief that nature — bodies and brains — is less mutable than culture. After the 1750s, the anatomy and physiology of sexual difference seemed to provide a foundation upon which to build natural relations between the sexes. Nativists saw the body as a static bedrock of organic life. The seemingly superior build of the male body (and mind) was cited more and more often in political documents to justify men's social dominance. In recent years we have come to recognize that nature responds sometimes rapidly to culture. The environment warms as we release increasing volumes of hydrocarbons. Turkeys stumble as we breed them with breasts too large for their legs to carry. Women grow taller and heavier in proportion to men as nutrition improves. Women's times in sports are faster as training improves. Nature is plastic: it can and does change.

Sexual science has also suffered from assuming rigid distinctions between the sexes, by failing to recognize that there is often as much variation *within* groups as *across* groups. Any individual woman may be physically and mentally as distinct from any other woman as from any one man — even in relation to *primary* sexual characteristics. It is important to keep in mind that in our culture sexually ambiguous babies are "fixed" at birth. (Endocrinologists make these babies mostly into girls; urologists make them mostly into boys.)[7] The ambiguities inherent in sexual typing were appreciated by early students of sexual science. Joseph Wenzel (one of Soemmerring's students) argued that a sharp physiological delineation between the sexes was impossible, stressing that individual variation was as important as group variation: "one can find male bodies with a feminine build, just as one can find female bodies with a masculine build." In fact, he wrote, one can find skulls, brains, and breastbones of the "feminine type" in men. The physician Johann von Döllinger similarly claimed that certain parts of the male genitalia (such as the prostate) were feminine and parts of the female genitalia (such as the uterus) were masculine. Though one might ask how Wenzel and Döllinger distinguished so finely "the masculine" and "the feminine," they rightly questioned those anatomists who made artificially sharp contrasts between the sexes. It is important to keep in mind that the desire to emphasize the distinctiveness of men and women as groups comes from culture, not nature.

One should also keep in mind that difference is multifactorial: class divides race and gender, race cuts through class and gender, gender bisects race and class.[8] Johann Blumenbach, the father of physical anthropology, set little store by the sexual distinctions much celebrated by his colleagues. Indeed, he found sexual and racial differences too often obscured by class and cultural situation. In regard to racial difference, Blumenbach taught that occupation or social class can determine skin color at times as much as race. "The face of the working man or the artisan," he wrote, "exposed to the force of the sun and the weather, differs as much from the cheeks of a delicate [European] female, as the man himself does from the dark American, and he again from the Ethiopian." Blumenbach also pointed to problems created by the fact that anatomists procured corpses from what he described as the "lowest sort of men." These (European) men, he remarked, have skin around the nipples and on the testicles that come nearer to the "blackness of the Ethiopians than to the brilliancy of the higher class of Europeans." Though with this example Blumenbach affirmed the meanness of both dark-skinned peoples and the lower classes of his own countrymen, one of his primary goals was to question emerging consensus on racial and sexual difference.[9]

But fault does not rest with the nativist alone. Nurturists—those who claim that nurture, not nature, explains women's poor showing in the sciences—have too often tried to explain away difference. For them, sexual science is simply "bad science," overheated with simplistic and exaggerated notions of sexual difference. Fighting science with science, these feminists (both men and women) begin unraveling the scientific myths surrounding men's and women's bodies and minds. In so doing, they often tend to deemphasize—almost to the point of denying—sex differences. Liberal egalitarians have tended to see men and women as interchangeable parts in the machine of science—for all practical purposes, women think and act in the same way as do men. Much in the nurturist argument has assumed *sameness* as the only ground for equality. This all too often *requires* that women be like men—culturally or even biologically (as when expectations are that working women need not take time off to have children). A simple call for equality ignores the complexities of gender in modern life. Why should women have to assimilate to the dominant culture in order to succeed? Why can't peoples be genuinely different in their skills and outlooks, and still be equal?

Should Sexual Differences Be Studied?

Science—its methods, priorities, and institutions—must be recast before it can be used as an effective tool for understanding difference. Sexual science needs to become aware of its history. We need to analyze how sexual differences have been studied, by whom, and to what ends.[10] We need to understand the ways that subtle (and not-so-subtle) forms of discrimination built into scientific institutions distort the findings of science. We need to disabuse ourselves of the notion that science is value-neutral, and study instead the realities of political and social forces molding the goals and priorities of science.[11]

The goal is not to suppress knowledge but to design socially responsible science projects. The design of any research project should include research into its history, politics, social costs, and consequences. The Human Genome Project is the first science project with funding set aside for the study of ethical consequences. It is obvious that we need to know about differences of all sorts—sexual, racial, differences due to age and stage of development, differences caused by stress, levels of nutrition, and so forth. The health and well-being of millions of women requires that NIH invest money in neglected areas of women's health.

It is not incidental to our problem that the double-edged sword of sexual science—the simultaneous exaggeration of sexual differences especially in the realm of intellect and neglect of research beneficial to women—has been wrought in institutions dominated by men. In regard to women, science is not a neutral culture but one shaped by its inhabitants. In the modern sexual division of labor that crystallized around the Industrial Revolution in the 18th century, science was part of the terrain that fell to the male sex. In the process, scientists sought to distance themselves from things defined as feminine, including women. Moreover, male scientists established the male as the measure of all things.

It is in this context that I believe it makes a difference who does science today. If women do, indeed, sometimes see things differently, if (as even newspaper polls tell us) they as a group hold significantly different opinions and have different values, might they not bring an enriching perspective to science? For hundreds of years the institutions of Western science have systematically excluded women. If these outsiders are now embraced and allowed to cultivate their own interests (that is to say, make their way into scientific institutions without having to assimilate completely to the dominant scientific culture), might they not make a difference? Might they not have different priorities, use different methods, relate to colleagues and their research subjects differently?

Let me put this more concretely through a historical example. Has there been a time in history when women developed a science, where the majority of its practitioners were women, and was that science significantly different than when men practiced it? Midwifery is the prime example of a science (or medical art) developed and practiced by women most often for the benefit of other women. Since ancient times women dominated the art of birthing and the whole field of women's health care. In the 18th century, midwifery was taken over by university-trained obstetricians — almost exclusively men. By the 19th century midwives had been banished to the countryside and the treatment of the poor.

I am not interested here in the institutional battles between midwives and medical men. What interests me is what difference the removal of women made in gynecological and obstetrical practice. Where the midwife saw her role as one of assisting the mother in birthing, the new man-midwife — trained as a surgeon — tended to set to work with his surgical instruments, treating parturition more like a disease than a natural process. Because surgeons had traditionally been called only in cases of emergency, few had ever seen a normal birth. At the same time, midwives and other women of the village or neighborhood had assisted the mother not only with the technical aspects of birthing but with other aspects of her daily regime, such as cooking and caring for the children, while the mother recuperated. Man-midwives, by contrast, attended the mother only during the hours of labor and eventually required women to give birth in hospitals — a process that further undermined women's support systems. It was only in the 1960s and 1970s that the women's movement was able to begin to reverse these trends and return to women some control over their health care.

The sex of participants in science should not necessarily be important to the results of science. But we must recognize that society is highly gendered. It is unfortunate that even at the conference where this paper was first delivered — purporting to study human values, knowledge, and power in brain research — more than 80 percent of the participants were male (and all the participants were of European descent). *Human* in Western society should not continue to mean white and male. As we have seen, systematic patterns of discrimination distort science. Objectivity in science cannot be proclaimed, it must be achieved. In order to edge toward greater objectivity, science must — as a first step — embrace a wider variety of perspectives among its investigators. We need to know about sexual and racial differences, but that knowledge should not aim at reaffirming social inequalities as has been the tendency in the past. It seems basic to human values that knowledge should be designed to improve the life and well-being of humans — of all races and both sexes.

Endnotes

1. Much of the material in this chapter is derived from my *The Mind Has No Sex? Women in the Origins of Modern Science* (Cambridge, Mass.: Harvard University Press, 1989). The reader is referred to that work for fuller arguments and specific references.
2. The term "sexual science" I have adopted from Cynthia Russett, *Sexual Science:*

The Victorian Construction of Womanhood (Cambridge, Mass.: Harvard University Press, 1989).

3. Andrew Purvis, "A Perilous Gap," *Women: The Road Ahead,* special issue, *Time* (Fall, 1990): pp. 66–67.

4. Jakob Ackermann, *Über die körperliche Verschiedenheit des Mannes vom Weibe ausser Geschlechtstheilen,* trans. Joseph Wenzel (Koblenz: 1788).

5. See Robert Proctor, *Value-Free Science? Purity and Power in Modern Knowledge* (Cambridge, Mass.: Harvard University Press 1991), chap. 17.

6. The question of education for African women residing in Europe was not even raised.

7. Suzanne Kessler, "The Medical Construction of Gender: Case Management of Intersexed Infants," in *From Hard Drive to Software: Gender, Computers, and Difference,* special issue, *Signs* 16 (Autumn 1990): 3–26.

8. Nancy Leys Stepan and Sander Gilman, "Appropriating the Idioms of Science: Some Strategies of Resistance to Biological Determinism," unpublished manuscript, Columbia University and Cornell University. See also Henry Louis Gates, Jr., ed., *"Race," Writing, and Difference* (Chicago: The University of Chicago Press, 1986).

9. See Londa Schiebinger, "The Anatomy of Difference: Race and Gender in Eighteenth-Century Science," in *The Politics of Difference,* ed. Felicity Nussbaum, special issue, *Eighteenth-Century Studies* 23 (1990): 387–406.

10. There is much valuable literature on this topic; see, for example, Ruth Hubbard, Mary Sue Henifin, and Barbara Fried, eds., *Biological Woman — The Convenient Myth* (Cambridge, Mass.: Schenkman Publishing Co., 1982); Janet Sayers, *Biological Politics* (London: Tavistock Publications, 1982); Ruth Bleier, *Science and Gender: A Critique of Biology and Its Theories on Women* (Elmsford, N.Y.: Pergamon Press, 1984); Steven Rose, Leon Kamin, and Richard Lewontin, *Not in Our Genes: Biology, Ideology, and Human Nature* (London: 1984); Anne Fausto-Sterling, *Myths of Gender: Biological Theories about Women and Men* (New York: Basic Books, Inc., 1986); Emily Martin, *The Woman in the Body: A Cultural Analysis of Reproduction* (Boston: Beacon Press, 1987); Ludmilla Jordanova, *Sexual Visions: Images of Gender in Science and Medicine between the Eighteenth and Twentieth Centuries* (Madison: University of Wisconsin Press, 1989); and Ruth Hubbard, *The Politics of Women's Biology* (New Brunswick, N.J.: Rutgers University Press, 1990).

11. Evelyn Fox Keller, *Reflections on Gender and Science* (New Haven, Conn.: Yale University Press, 1985); Ruth Bleier, ed., *Feminist Approaches to Science* (Elmsford, N.Y.: Pergamon Press, 1986); Sandra Harding, *The Science Question in Feminism* (Ithaca, N.Y.: Cornell University Press, 1986); Sandra Harding and Jean O'Barr, eds., *Sex and Scientific Inquiry* (Chicago: The University of Chicago Press, 1987); Donna Haraway, *Primate Visions: Gender, Race, and Nature in the World of Modern Science* (New York: Routledge, 1989); Helen Longino, *Science as Social Knowledge* (Princeton: Princeton University Press, 1990); and Proctor, *Value-Free Science?*

8

Walker Percy: Language, Neuropsychology, and Moral Tradition

EDWARD MANIER

Walker Percy and Sinclair Lewis: Two Authors in Search of a Character

Walker Percy's essay "Notes for a Novel about the End of the World" (1967), anticipating the novel *Love in the Ruins* (1971), posed several questions for John Steinbeck and Sinclair Lewis. Percy thought of Steinbeck and Lewis as social critics and cultural satirists who had failed to examine the radical bond connecting human beings with reality, conferring or failing to confer meaning upon human life. Percy (1967) asked, "What happens to the thousand Midwesterners who settle on the Riviera? What happens to the Okie who succeeds in Pomona and now spends his time watching Art Linkletter? Is all well with them or are they in deeper trouble than they were on Main Street and in the dust bowl? If so, what is the nature of the trouble?" (p. 103). Critic Harold Bloom (1988), whose characterization of the final scenes of Lewis's *Arrowsmith* (1924)—in which two dedicated seekers after the ultimate truths of experimental biology work alone in an isolated laboratory in the north woods—is cruel but on the mark, raises similar questions: "*Arrowsmith* is . . . a romance, with allegorical overtones, but a romance in which everything is literalized, a romance of science. . . . In the romance's pastoral conclusion, Arrowsmith retreats to the woods, a Thoreau pursuing the exact mechanism of the action of quinine derivatives" (p. 3). Bloom considers *Arrowsmith* a monument to "another American lost illusion, the idealism of pure science, or the search for a truth that could transcend the pragmatics of American existence" (p. 4). In Percy's terms, the transcendent truths of science cannot connect human beings with reality, cannot place them in the world, cannot illuminate, chart, or motivate the journey of a human life.

Percy's essay "A Novel about the End of the World" and his novels *Love in the Ruins* (1971) and *The Thanatos Syndrome* (1987) provide models of the place of biomedical research in the illumination of the human situation that contrast sharply with the model found in Lewis's *Arrowsmith*. I propose a brief comparison of those models.

Arrowsmith pans the vacuous Babbittry, the hucksterism, and the thoughtless political and economic opportunism of much of American medical education, practice, and research in the first quarter of the present century, finding gold only in the heroic but isolated monk of science, Martin Arrowsmith's mentor, Max Gottlieb. Lewis, with the help of Paul de Kruif, patterned Gottlieb after those aspects of Jacques Loeb's philosophy of science that came to the fore during the last stage of his career at the Rockefeller Institute.

As if to invite comparison with the perspectives of *Arrowsmith,* Gottlieb also appears in the Dr Tom More novels: *Love in the Ruins* and *The Thanatos Syndrome.* But Percy's Gottlieb is a marginal figure, the hero's occasionally effective friend and protector. He most emphatically is not the saint-savant after whom Tom More will pattern his own career. In *Arrowsmith,* in contrast, the thinly veiled ideology of Jacques Loeb/Max Gottlieb is both valorized and romanticized.

Historian Philip J. Pauly (1987) provides a useful analysis of some of the underlying ambiguities in Loeb's views on experimental biology. Pauly argues that Loeb, after his move to the Rockefeller Institute, shifted the focus of his work from "action and power to discourse and understanding" (p. 138). Theoretical interests in the subsumption of microbiology and eventually all biology under the unifying principles of biophysics gradually replaced a simpler, cruder image of the biologist as engineer. Earlier in his career, Loeb had "rejected the view that some 'complete' analysis of biological organization was the fundamental problem of the life sciences" (p. 54), had rejected the analytic perspectives of both cerebral localizers and the adherents of *Entwicklungsmechanik* in favor of an "engineering ideal" placing emphasis upon control of development and behavior. From this perspective, the concepts of instinct, will, and even ganglion cells were mystical or metaphysical because inaccessible. The value of the tropism concept was that under properly identified and controlled circumstances, the concept made it "possible to control the 'voluntary' movements of a living animal just as securely and unequivocally as the engineer has been able to control the movements in inanimate nature" (p. 47). The difficulties inherent in Loeb's morally ambiguous position, oscillating as he does between the ideals of engineering control, on the one hand, and of the pure truth of a comprehensively unified theory of biology, on the other, never come to the fore in the fictional world of *Arrowsmith* (Lewis, 1924).

Gottlieb (Loeb) insists that true scientific competence requires a knowledge of higher mathematics and physical chemistry. "All living things are physiochemical," he says to Arrowsmith; "How can you expect to make progress if you do not know physical chemistry, and how can you know physical chemistry without much mathematics?" (p. 285). This statement sets the stage for Gottlieb's profession of the pure religion of science: "The scientist is so religious he will not accept quarter-truths, because they are an insult to his faith. He wants . . . everything . . . subject to inexorable laws.

He is not . . . kindly to anthropologists and historians who can only make guesses . . ., he hates pseudo-scientists, guess-scientists — like these psycho-analysts . . . those comic dream-scientists!" (pp. 267–268). Gottlieb's scientist is at once heartless, living in a cold clear light, and the only true, lucid philanthropist (lover of humanity), whose opportunity to guide the world may arrive through default, through the complete failure of all competing worldviews. Gottlieb invokes Goethe, the archetypal romantic ideal of the investigator as scientist/philosopher/artist, while urging Arrowsmith to eschew "Success" in order to remain pure in his search for the Truth itself, as if the Truth for man were a matter of physical chemistry, something to be brought under the mass action law.

According to Charles Rosenberg's analysis, "Martin Arrowsmith: The Scientist as Hero" (Bloom, 1988), Lewis shared the view of his informant Paul de Kruif that "the effort to bridge the gap between the practice of medicine and pure science, the attempt to train each practitioner as a scientist was simply delusive" (p. 25) because the social necessity that had created the medical profession tied it to the shifting sands of popular favor and everyday life and the bedrock of bill collecting and commercialism. The odyssey of Martin Arrowsmith, through medical school, residency, and village practice, led to a temporary haven as assistant director of public health in a small city where his path crossed that of the splendid villain, Almus Pickerbaugh, the Babbitt of *Arrowsmith,* who, in the words of H. L. Mencken,

exists everywhere, in almost every American town. He is the quack who flings himself melodramatically upon measles, chicken pox, whooping cough — the orga-nizer of Health Weeks and author of prophylactic, Kiwanian slogans — the hero of clean-up campaigns — the scientific *beau ideal* of newspaper reporters, Y.M.C.A. secretaries, and the pastors of suburban churches. (cited in Bloom, 1988, p. 2)

When Pickerbaugh moves on to a career in Congress, Arrowsmith's stint as his successor is abruptly terminated by his inability to translate his laboratory skill into manipulation of the press, the lecture platform, and public opinion as effective as Pickerbaugh's. His stay in a high-fashion private clinic in Chicago is equally brief and unpleasant since his employers there are interested in his research only as a means of securing free advertising.

Nevertheless, his duties at the Rouncefield Clinic enable him to publish his first paper and to move on to the novel's penultimate setting, a research professorship at the McGurk Institute. There he encounters the centraliza-tion and bureaucratization of science in its more sophisticated forms, which, however inevitable they may be as the concomitant of the increasing complexity of both scientific knowledge and the society that supports its pursuit, are nevertheless inimical to the impulse of spontaneous creativity. A. de Witt Tubbs, director of McGurk, gushes, "If men like Koch and Pasteur only had such a system, how much more scope their work might have had! Efficient universal cooperation—that's the thing in science today—the time of this silly jealous, fumbling individual research has gone by" (Lewis, 1924, p. 308). But Martin finds that he cannot maintain his own methodological purity in an environment that demands that he accept international acclaim for creating and testing a vaccine against plague, when he knows that his emotions have overrun the need for rigorously maintained controls in a field test. He flees McGurk, leaving his wife and young son behind, to join the perfect iconoclast–biochemist Terry Wickett in a well-stocked wilderness laboratory in the north woods.

He began, incredulously, to comprehend his freedom. He would yet determine the essential nature of phage; and as he became stronger and surer—and no doubt less human—he saw ahead of him innumerous inquiries into chemotherapy and immunity; enough adventures to keep him busy for decades. It seemed to him that this was the first spring he had ever seen and tasted. . . . when they had worked all night, they came out to find serene dawn lifting across the sleeping lake. Martin felt sun-soaked and deep of chest, and always he hummed. (p. 428)

". . . and always he hummed." Oh well. Harold Bloom remembers that *Arrowsmith* had a kind of biblical status in the Bronx High School of Science in 1945, but his own mature judgment is merciless. "Though sadly dated, *Arrowsmith* is too eccentric a work to be judged a period piece. . . . Its hero, much battered, does not learn much; he simply becomes increasingly more abrupt and stubborn, and votes with his feet whenever marriages, institutions, and other societal forms begin to menace his pure quest for scientific research" (p. 3).

More than half a century after the publication of *Arrowsmith* and the other works of Lewis's great decade (the 1920s), there is general agreement that Lewis's talents were those of a journalist with a camera eye, limited by an inability to probe much below the surface of his characters, by a loathing of his roots that made him incapable of engaging his own tradition in significant dialogue, and by a glaring flaw in his satirical art and ironic perspective on American life: his romantic idealization of the Martin Arrowsmiths of the world.

In stark contrast, Percy's novels offer stinging parodies of Dr. Tom More, stuck someplace on life's way. In *Love in the Ruins* More describes himself as one who loves "women best, music and science next, whiskey next, God fourth, and my fellowman hardly at all. Generally I do as I please. A man, wrote John, who says he believes in God and does not keep his commandments is a liar. If John is right, then I am a liar. Nevertheless, I still believe" (p. 6). Dr. Tom More is susceptible to seductive claims that medical technology has the power to conquer every terror, from flatulence to the apocalypse. These novels also burlesque the "qualitarian" ethos, the "scientific humanism," of Dr. More's liberal utilitarian antagonists as sexually perverse, homocidal, and potentially genocidal.

So much for the ideals of Jacques Loeb. The targets of satire and irony change with time. The half-century of American life represented in the distance between Lewis's great decade and the 1970s of Walker Percy and Harold Bloom witnessed a remarkable transformation of the standard interpretive options for analyzing the relationship of science and society; the relationship of culture as it might be interpreted by the historian, the anthropologist, the artist, or the critic; and nature, as it presents itself to scientific analysis and explanation.

Walker Percy's Dr. Tom More: Neuropsychology Runs Amuck

Nietzsche saw the artist as the Promethean creator of the bond between the human mind and nature, (1870–1871/1886). But Percy denies that either science or art can accomplish this heroic project. His less-than-titanic familiar is the miner's canary, taken down the shaft so that its early distress will warn the miners "to surface and think things over." The artist is the "suffering servant" of the age, not its savior. The canary collapses for lack of oxygen. Percy's distress as a novelist has an equally tangible cause: the lack of meaning or significance in a prevalent view of life. The view that Percy subjects to merciless parody is exactly the one romantically valorized in *Arrowsmith*.

Percy (1967) argued that the language needed for discussing the radical bond that connects human beings with reality has been devalued — overused and misused — until it is no more effective than the most trite advertisement. "So it may be useful to write a novel about the end of the world. Perhaps it is only through the conjuring up of catastrophe . . . that the novelist can make vicarious use of catastrophe in order that he and his reader may come to themselves" (p. 118). In this effort, "the fictional use of violence, shock, comedy, insult, and the bizarre" are the novelist's everyday tools.

All these fictional tools are frantically deployed in the opening scenes of *Love in the Ruins: The Adventures of a Bad Catholic at a Time Near the End of the World*. In this novel, the basic institutions of society have decayed: children are at war with their parents; racial conflict has taken the form of open rebellion; churchs are trivialized or schism-ridden; the political process has collapsed; no one is repairing anything — vines are growing on the interstates, through patio blocks, and in motel pools, and the trademark Howard Johnson orange roof tiles are cracked and scattered on the pool deck.

In the midst of this chaos, Dr. Tom More is a fabulous composite of Don Juan and Faust (Lawson, 1989). He is the armed protector of three beautiful, talented, and acquiescent younger women ingeniously seques-tered in a Rube Goldberg/HoJo bower, and he is the inventor of a CORTscan device utilizing a dangerous isotope capable of saving or precipitously destroying the Earth or the local golf course. More has patient-staff status, "not yet on an open ward," in the psychiatric wing of Fedville, a gigantic hospital and medical school complex. Nevertheless, everyone is convinced that he has contributed and will continue to contribute to a genuine Copernican revolution in neuropsychology.

He is on the verge of bridging the "dread Cartesian chasm" between mind and body by measuring the quantitative correlation of ionic activity levels of the various Brodmann areas of the cerebral cortex with the "manifold woes of the Western world, its terrors and rages and murderous impulses." He has invented a machine for "measuring and treating the deep perturbations of the soul," rage, lust, otherwordly abstraction, and alienated detachment. With this machine, More's Qualitative Quantitative Ontological Lapsometer (MOQUOL), he expects to win the Nobel prize.

Before he can submit the relevant publication to *Brain* and apply to the N.I.M.H. for the funding needed to take his project into the next phase of development, he must convince his psychiatric supervisor, (the very one, but not the only) Dr. Max Gottlieb, that he has overcome his suicidal depression, addiction to alcohol, and delusions of grandeur. This Gottlieb is a neobehaviorist who traces More's symptoms to groundless feelings of guilt associated with sexual activity. Gottlieb prescribes a program of conditioning to eliminate the unwanted guilt. More rejects this solution, insisting that he does not feel guilty about activities that he sees as "great pleasures," albeit sinful. It is the *absence,* not the *presence* of guilt, that depresses him. Unless he can experience contrition for his sins, they cannot be forgiven. More is a bad, but believing Catholic.

Before things get better, they get worse. Lucifer appears as Art Immelman, a shadowy coordinator of private and public grant support, who offers to assist the development phase of MOQUOL in return for patent rights permitting his organization to "facilitate interaction of the opposite ends of the political spectrum" (still another code name for still another world war). Immelmann then adds the requisite differential stereotactic emission ionizer to the machine, rendering it fully if dangerously operational, enabling More to score a smashing therapeutic victory, gain his personal freedom, and ensure his professional reputation.

The victory consists in More's use of the adapted MOQUOL to assist Mr. Ives, an elderly patient at Fedville, enraged to the point of speechlessness and partial paralysis by the suffocating social structure of the Golden Years Retirement Center in Tampa, Fla. MOQUOL and More enable Ives to quiet his rage, speak, walk, and display his true self. It turns out that Ives is really a creative and inventive archaeologist and master linguist, author of soon-to-be-published articles deciphering the Ocala frieze and illuminating the links between the early Spanish explorers of Florida and the proto-Creek culture.

This remarkable medical breakthrough takes place in the context of a dramatic struggle between More and Dr. Buddy Brown, whose diagnosis of advanced atherosclerosis, senile psychosis, psychopathic and antisocial behavior, hemiplegia and aphasia following a cerebrovascular accident had qualified Ives as a prime candidate for the Happy Isles euthanasia facility. The More-Brown debate takes place in the Pit, a student-pleasing, semester-ending display of faculty wit and talent; and More's stunning reversal of Brown's apparently easy victory throws the assembled crowd into an uproar, exploited by Immelman to distribute several hundred MOQUOLs to untrained personnel. Its frivolous use ignites the salt deposits in all the sandtraps on the Paradise Estates Country Club, threatening a Christian Pro-Am Golf Tournament ("Jesus, the greatest Pro of them all") and the end of the world. Dr. Tom More prays to Saint Thomas More to drive the devil hence, and the danger subsides.

Five years later at Christmastime, More has married the only woman left to his name, started a small family, and abandoned his lust for the Nobel prize for the quieter joys of "just figuring out what I've hit on. Some day a man will walk into my office as ghost or beast or ghost-beast and walk out as a man, which is to say sovereign wanderer, lordly exile, worker and waiter and watcher" (1971, p. 326). At the novel's end, More has returned to communion after a conversation-confession with Father Simon Smith helps him discover that he is *sorry he cannot be sorry* about past sins of drunkeness, fornication, and delight in the misfortunes of others. Father Smith then leads Tom to true contrition concerning the small, pervasive flaws in his domestic and professional life, flaws that "seem more important than dwelling on a few middle-aged daydreams" (1971, p. 340) — Tom's former life as Faust/Don Juan.

Love in the Ruins moves strongly and concurrently in two realms: an inner struggle moves Dr. Tom More himself from angelism/beastialism, Faust/Don Juan, to wanderer, exile, worker, waiter, watcher; and an outer struggle causes him to do battle with the qualitarian ethos claiming to share the same basic goals: to enhance quality of life; protect the right of the individual to control his own destiny, realize his own potential; minimize human suffering; in general, to place a supreme value on human values. The sporadic and incomplete conversations between More and Buddy Brown gradually establish that the qualitarian ethos espoused by Brown is radically incommensurable with the values pursued by Dr. More. Percy, as always, regards fiction as the pursuit of philosophy by other means. Here he uses the tension between Brown and More to ask if utilitarian moral philosophy (the greater good of the greater number) has the conceptual resources required to protect, or even to conceptualize, the pricelessness of every human individual.

The Thanatos Syndrome, written 15 years later (1987), finds Dr. More marginally more secure in his professional identity as a follower of psychiatrist Harry Stack Sullivan, and of Freud, seeking that pearl of great price, the patient's truest unique self; venturing into the heart of darkness, listening to another troubled human for months; enabling her to render the unspeakable speakable.

The Thanatos Syndrome heightens the dramatic intensity of the external struggle with the qualitarian ethos, based on the false view of human nature as god/goat, angel/beast. Dr. More's earlier discoveries of the neuropsychological significance of radioactive metallic ions has been put to use in a project ("Blue Boy") modeled after clandestine tests of the effects of fluoride in the general water supply. The molecular and cellular-architectural mechanisms underlying this project are left shadowy, but its behavioral consequences are spelled out in impressively quantitative detail: crime in the streets is reduced overnight by 85%, child abuse by 87%, teenage suicide by 95%, wife battering by 73%, teenage pregnancy by 85%; hospital admissions for depression, chemical dependence, and anxiety drop by 79%; AIDS drops by 76%; the incidence of murder, knifings, and homosexual rape in prisons declines to zero, and the prison population is happy and content; IQ, SAT, and comparable measures of concentration and graphic and computational abilities dramatically increase, with comparable improvement in the work ethic of minority teenagers.

The general character of the project, which "cools" the human neocortex so that aspects of it function at a pongid level, is a spoof of Kurt Vonnegut's conceit that a segment of the human neocortex and consciousness is an aberration of evolution; that the "human brain is too fucking big," the scourge and the curse of life on earth: the cause of war, insanity, perversion (Vonnegut, 1985).

As More challenges the "brains" behind this scheme to control human behavior with standard libertarian arguments, they answer that "it's war . . . plague; crime; drugs; suicide; AIDs. . . . [that cause] decay of the social fabric. Every society has a right to defend itself." This gambit stymies More's effective opposition until Father Smith's confession, the turning point of the novel (1987, pp. 191–202, 265, 325–334).

The horror of this confession requires careful consideration of its context in Percy's fictional world. His scathing depiction of the "autonomous self" in that world has shocked some reviewers *(New York Times, Commonwealth)* of *Lost in the Cosmos* (1983). In that work, the chapter entitled "The Demoniac Self: Why It Is the Autonomous Self Becomes Possessed by the Spirit of the Erotic and the Secret Love of Violence, and How Unlucky It Is that This Should Have Happened in the Nuclear Age" offers Percy's most insulting *reductio* of the secular ethos of the old modern world.

A nauseating vignette pictures a Nobel laureate scientist–advisor on a scrambler phone from a hotel room advising the U.S. joint chiefs to implement the Hiroshima-like use of a powerful neurotoxin "for the ultimate good of man . . . in the interests of peace . . . why don't we call it project peace" (1983, p. 194), while not missing a beat in the pornographically assisted collection of his own semen for contribution to an elite sperm bank. Mercilessly hyperbolic parody of the moral implications of both the romantic-idealist and the positivist approaches to the ethos of science is continued in *The Thanatos Syndrome* where a prime mover of a project to add a psychotropic isotope to the drinking water of an entire Louisiana parish turns out to be also the ringleader of a gang of pedophiles.

The Dr. Tom More character is most sharply focused by its opposition to the demon–savant of German medicine, Dr. Alfred Hoche, mentioned briefly in the central scene of *The Thanatos Syndrome* (Percy's source for this allusion was "The Geranium in the Window," in Wertham, 1966). Father Simon Smith, the half-crazy Roman Catholic priest who has abandoned his flock and mounted a fire tower, confesses to Dr. More that as a youth of 17, visiting Germany in the early Hitler years, nothing but luck stood between his choosing to join the German *Schutzstaffel* rather than the Dekes at Tulane. His youthful impulses favored the S.S. He had not been repelled when a private gathering of the most enlightened members of the German medical profession calmly discussed the pros and cons of the concept of *lebensunwerten Lebens,* life unworthy of life.

The widely influential *Die Freigabe der Vernichtung lebensunwerten Lebens,* coauthored by Hoche and the eminent historian of jurisprudence Karl Binding, argued that "life unworthy of life" could be identified with 100% medical certainty and that in a new age a higher morality would undermine the current overestimation of the value of life as such (Binding and Hoche, 1920). In the new age, "reverence for nation" would precede "reverence for life." Alfred Hoche was a leading psychiatric critic of the Freudian "psychical epidemic among doctors." He judged the Freudian movement "pathological" precisely because its principle of empathy required the physician to enter into the mind of the patient (Lifton, 1986).

Although Percy places much emphasis upon "intersubjectivity"—the achievement of true mutual understanding in the acquisition and development of human language—the scene in which Father Smith confesses to Dr. Tom More is one of the few in these two novels portraying a sustained effort of self-revelation. The insertion of Father Smith's confession in the midst of the stunted and partial exchanges that occur in lieu of conversation elsewhere in the novel demands the reader's active collaboration. Smith offers the confession in response to Dr. More's request for help in deciding how to deal with the "qualitarian" (liberal-utilitarian) project, Blue Boy. The tone of the confession is quiet and carefully lucid, the awful violence of its subject matter requiring no rhetorical amplification. It is of a piece with Percy's continuing critique of the Enlightenment/post-Enlightenment ideal of the autonomous self. The story of Alfred Hoche and the concept of *lebensunwerten Lebens* evokes a sense of the failure of a culture, a nation, and a philosophy—all fundamentally identical with our own.

More's understanding of the implications of the confession are reflected in the heightened resolve and purpose of his action through the remainder of the novel. The confession drives home the ease with which the qualitarian ethos might be extended to acceptance of the Hoche-Binding concept of *lebensunwerten Lebens*. Once More assimilates this point, he masterfully exploits the accumulated evidence to terminate the Blue Boy pilot project, close the qualitarian centers for pedeuthanasia and gereuthanasia, and divert the resulting pool of unspent grant money to repair the loss of Medicaid funds and reopen Father Smith's hospice. For Father Smith, and perhaps Dr. Tom More as well, anything less than complete care for the dying, the afflicted, the useless, born or unborn, will culminate with a distressed majority scapegoating an unsubsumable minority. Dr. Tom tells one of the authors of Blue Boy: "he thinks you'll end by killing Jews." In Father Smith's own aphorism, qualitarian "tenderness leads to the gas chambers."

Analysis: Language and Alienation in Percy's Nonfiction

On May 3, 1989, a year before his death, novelist and essayist Walker Percy read the 18th Jefferson Lecture, "The Fateful Rift, the San Andreas Fault in the Modern Mind" at the National Endowment for the Humanities in Washington, D. C. (published as "The Divided Creature," Percy, 1989). The central theme of that lecture had been developed in Percy's fiction and nonfiction for almost 40 years. Human beings are strangers, lost in the cosmos, marked by two "singularities" or distinctive features — language and a deeply rooted alienation — often expressed as a penchant for "upside down feelings and behavior" (feeling bad in a good environment and good in a bad environment). These two features mark the conceptual limits of humans as organisms in an environment. Percy's essays develop these themes through abstract, philosophical argument.

The Jefferson Lecture goes beyond Percy's earlier work by explicitly charting an agenda for scientific research and evaluating the methodology of scientific research programs. It argues that the modern scientific worldview, *on its own terms,* is incoherent and incomplete. Neither Darwin nor Freud could account for his own scientific activity using his own scientific theory. Scientific theories of human nature, as Percy sees them, generally fail to comprehend those points about linguistic denotation, intersubjectivity, and existential authenticity that rest on rock-hard evidence of the kind no adequate science should ignore or deny. This does more than prick the bubble of scientific humanism. With a wink and a nudge, it seeks to set science straight in its own house.

The fundamentally Cartesian framework of Paul Churchland's menu of ontological and epistemological positions in *Matter and Consciousness* (1988) ignores the problem of historical and narrative knowledge (see Danto, 1985), assuming that our knowledge of ourselves and others has the status of a more-or-less verifiable theory, a progressive or a degenerating program of scientific research (Churchland, 1988, 1989). When the distinctive capacities of the human species are seen as either "transcendent" (scientific and artistic) or "immanent" (survival, consumption, reproduction), the existential plight of a people "in trouble" and "needing help" cannot be cogently articulated. Percy seeks to undermine what he takes to be the prevalent epistemic dichotomy of scientific knowledge and artistic fiction as leaving no room for simple "news," the message in the bottle announcing our plight, *placing* us in the world, identifying our particular, embodied historical predicament. Although Percy clearly refers to the news of the Gospel, his distinction of knowledge and news carries the epistemic import of a distinction between scientific theory and historical narrative: implying a basic distinction between such works as Darwin's *The Origin of Species* and Tocqueville's *L'ancien régime et la Revolution* and *Democracy in America*.

Percy sees language and alienation as topics for which an appropriately radical biopsychology might provide a prolegomena, even an explanation. Such a *biology* would be post-Darwinian, renouncing the continuity hypothesis that led Darwin to insist that all human "higher powers" differ in degree, and not in kind, and that animal communication and human language are closely analogous. It would also be post-Darwinian in its claim that humans are *castaways*, and its corresponding *denial* that the human species is completely *at home* on Earth, that its evolution has been driven by nothing but the natural selection of more adaptive traits. Percy's reformulated biopsychology would ground an anthropology charged with the task of making sense of "man as pilgrim . . . falling prey to the worldliness of the world, . . . seeking salvation."

Percy's 1989 Jefferson Lecture, in other words, goes well beyond "A Theory of Language," from *The Message in the Bottle* (1975). "A Theory of Language" can be read as an expression of preference for certain research projects in neurolinguistics and psycholinguistics, for example, those of Norman Geschwind and Roger Brown, while simultaneously denying its competence to speak about the ontological basis of the linguistic act of affirmation or denial: "The apex of the triangle, the coupler, is a complete mystery. What it is, an 'I,' a 'self,' or some neurophysiological correlate thereof, I could not begin to say." But the Jefferson Lecture insists upon the point "A Theory of Language" cannot begin to formulate. The latter essay offers to impose a new direction upon scientific activity. A scientist who would seriously read it this way would have to *learn* to operate with an ontology as baffling as anything Percy finds in the unreformed science of nature: (1) a material substance cannot name or assert; and (2) naming and asserting are natural phenomenon which must be dealt with by natural science. No self-respecting neuroscientist should think *that* pair of sentences identifies the path *away* from incoherence.

Since Percy thinks that Cartesian dualism is incoherent, the point might be more clearly expressed:

A human . . . biology that mistook neurobiology and evolutionary theory for a theory of psychology and culture would be wrong; so would a psychology that could never have evolved or a social science that posited a biologically impossible psychology. These different levels of theory must be mutually compatible, but no one level can be reduced to another. (Barkow, 1989, pp. 3–5)

Language is not merely a biological phenomenon. Nevertheless, the evolution of a naming and asserting animal must, in Percy's view, be a central topic for biology.

Through language human beings inhabit a *world,* not an environment, a world where there is a name or a question for every conceivable and imaginable object: past, present, or future, real, coherent, or phantasmagorial. Language *refers.* Terms such as "balloon" denote objects in a fashion irreducible to the response evoking a Pavlovian link between a ringing bell and meat powder in the dog's mouth. Percy argues that language can only be understood by concurrent resolution of the problems associated with the denotative relation of word and object and the intersubjective understanding achieved through the social matrix of speech. Language locates the speaker in a world, and it both creates and is created by a meeting of the minds of speaker and listener.

Percy also developed themes from Kierkegaard, Sartre, and Marcel probing language as an index of the ontological basis of human alienation and despair. As he put it in the "semiotic" essay at the center of *Lost in the Cosmos* (1983),

The self of the sign user can never be grasped, because, once the self locates itself at the dead center of its world, there is no signified to which a signifier can be joined to make a sign. The self has no sign of itself. . . . For me certain signifiers fit you, not others. For me, all signifiers fit me, one as well as another. . . . I am both the secret hero and the asshole of the Cosmos. (pp. 107–108)

Percy's account of the distinctive features of human language is linked with his account of the ethical axis of authenticity and inauthenticity, with the essential human capacity "to be able not to be what I am." Human alienation has deep ontological and epistemological roots in the human ability to name everything except the naming self.

The bifurcation of man, in the "old modern age," into angelic and bestial components (Percy's translation of Nietzsche's apotheosis of the divine duality represented by Apollo and Dionysos) — an "all-transcending 'objective' consciousness" and a consumer-self with a list of needs to be satisfied — is monstrous. Such "tempestuous restructuring" of consciousness leaves no room for questions and stories that must be plumbed by anyone with an explicit and ultimate concern with the nature of man and the nature of reality where man finds himself.

Science and Historical Narrative

Alasdair MacIntyre (1977) illustrates the notion of epistemological crisis by comparing the situation of Shakespeare's Hamlet with the Cartesian exercise of hyperbolic doubt. Hamlet suffers an embarrassing plethora of traditional schemata for interpreting and acting in response to his father's murder (the revenge theme of the Norse sagas, the insouciant, devious sophistication of Rosenkrantz and Guildenstern, the Machiavellian struggle for control of the state). Descartes would eliminate all traditions and leap to the solid truth of physics as a novel Archimedean point.

It is Descartes who misdescribes his own situation, and Galileo who successfully rewrites the narrative constituting the scientific tradition by enabling us to understand his predecessors in a new way. Enabling us to understand why the original beliefs were held and why the original beliefs were misleading. Enabling us to understand how criteria of truth and understanding were reformulated, new problems defined, and new evidence recognized and interpreted. For MacIntyre, the best account of the superiority of some scientific theories presupposes an intelligible dramatic narrative, it presupposes the historical truth of a narrative in which such theories are the subjects of successive episodes. We can compare theories rationally to the extent that we can rationally compare such histories. The philosophy of science presupposes a philosophy of history. Only when we view demands for justification in highly particular historical contexts are we freed from dogmatism or skepticism.

The borders between the biological sciences and the sciences of the mind have been disputed since the time of Darwin, Helmholtz, and Wundt (Hatfield, 1991). Currently, the issue has an aspect evocative of MacIntyre's model of an epistemological crisis. Some disputants suggest that "folk psychology," the traditional framework of literature and of any and all narrative accounts in the history of science, as well as standard discussions of rational theory appraisal in the philosophy of science, must be replaced by the categories of a mature cognitive neuroscience. This suggests that no traditional framework is available for the adjudication of theses dividing the opposed positions.

A plausible critique of the program known as eliminative materialism in neuropsychology is that it leaves the transition to the language of neuroscience in the limbo of a Kuhnian "gestalt switch." The putative incommensurability of cognitive science and folk psychology makes historical conversation unthinkable. Eliminative materialism places neuropsychology outside the linguistic realm that might enable it to find an audience capable of hearing and understanding the claims that the brave new categories represent a real advance in scope and explanatory power. Darwin has failed to convince us that evolutionary biology might eliminate the need for cogent historical criticism of the dimensions of the human life-world. How can computational neuroscience succeed in this same effort when its methodology seems even less historical, even more detached from the conditions of life on a particular earth?

Summary

Percy's *The Thanatos Syndrome* provides an effective illustration of his sharp distinction of *world* and *environment,* and of his view of the role of language in placing human beings in a *world.* As that novel begins, several of Dr. Tom More's patients, later found to be dosed with psychotropic sodium-24, present remarkably modified language patterns: two-word sentences, automatic verbal responses in graphic or computational form, reduced and stereotypically pongid displays of affect and sexual response, and little or no capacity for or interest in real conversational exchange, the many mininarratives of everyday life. In this fictional world, human beings have been drugged back to the level of talking chimps who read off environmental features with devastating speed and accuracy, but respond to nothing but the routine stimulus-response ("Slab!") aspects of language. Recent articles by Edward Hardy (1989) and John Desmond (1989) on place and time in Percy's novels demonstrate that fiction translates and transforms these dimensions of the lived world in ways that heighten its significance and personal import, at the same time widening the gap between *Welt* and *Umwelt,* world and environment. The elaborate complexity of Percy's fictional world is placed in sharp contrast with the relative poverty of a pongid environment.

MOQUOL, More's Quantitative Qualitative Ontological Lapsometer, correlating the electronic aspects of the cortical Brodmann areas with the major ills of Western life, imaginatively mocks the pretensions of neuropsychology to bridge the old dread Cartesian chasm. This fantastic device, with no basis in fact or logic in the world outside the novel, displays the power of fiction to heighten and to undermine the pretensions of medical technology in the everyday world dominated by that technology. The devices of irony, parody, satire, and comic burlesque sensitize human discourse to levels of questioning unknown to the strategies of animal communication. The ironic *style* of these representations of animal models of human speech is itself the strongest evidence of the inadequacy of such models.

The epiphany of Mr. Ives, the enraged archaeologist–linguist, and the conversational confessions of Dr. More to Father Smith, and then of Father Smith to Dr. More, illustrate the sense in which language creates and is created by relations of intersubjectivity permitting human persons to share deeply significant, otherwise tacit aspects of their lives. These relations are not "in the head," not in any particular neurobiological circuit, for the simple reason that they are relations, polyadic not monadic properties. The reason for this is that such predicates may be and often do refer to the uniquely determinative relationship of two or more particular people, idiosyncratic interpersonal relations, as in the case of Dr. More and Father Smith, with widely spreading roots in nonlinguistic aspects of two shared lives and worlds. In a sense, Percy bridges the dread Cartesian chasm by an effective fictional display of the poverty of the Leibnizian image of windowless monads (substances all of whose relations are "internal") and of the inability of that ontology to capture the reality of interpersonal relations or the resultant complexity of the lived world.

All the novelist's ingenuity in the display of epiphanic interpersonal relations (Ives–More, Ives–Brown, More–Smith, Smith–More) display the fragile balance, the radical tension connecting alienation and authenticity in human life. Mr. Ives is seen by Dr. Buddy Brown as afflicted with a plethora of untreatable maladies, almost a paradigm of *lebensunwerten Lebens.* More engages Ives's personal history, his search for a particular kind of meaning. More makes the dumb Ives speak by dissolving the raging solitude that silenced and paralyzed the old man. This tension is also apparent in the confessions of Dr. Tom More and Father Simon Smith.

Each confession clarifies the sense in which it is *open to a person not to be what he is:* to More to be Faust/Don Juan and to orbit the earth with no sense of the place of his inventions in the quest for the meaning of everyday life; to Father Smith to be a member of the *Schutzstaffel,* all but indifferent to the terror of *lebensunwerten Lebens,* caught up in the youthful exuberance of the pursuit of glory and as much awed by the rigors of science as repulsed by the shallow vacuities of mere romanticism.

Percy's account of alienation locates it in a human ability not to hear, not to recognize, not to differentiate the *news* that enables a person to place himself or herself in the world, in a true, real, mundane situation. This news does not come in the form of scientific knowledge. Such knowledge must be articulated from a perspective independent of the concrete life world of the investigator.

It is not within the scope of this chapter to follow Percy's exegesis of Kierkegaard's account of the religious dimensions of these ideas. Instead, it must suffice here to illustrate the distinction of *knowledge* and *news* by pointing to the sense in which Darwin's theory of evolution provides us with few or none of the resources needed to understand the differences Tocqueville found in postrevolutionary France and in Jacksonian America. How could one describe these radically distinct social and cultural situations in biological language, the language of the theory of natural selection?

So it follows, for Percy, that humans are not just animals, not just machines, although there are senses in which they are both. They are also more. Science, however, finds itself unable to go beyond the controlling imagery, the dominant metaphors of organism and machine. Darwin's appropriation of the imagery of Adam Smith's analysis of the British economy and his expression of that machinery in terms of the romantic metaphors of Wordsworth's quest of the sublime in nature offer a thin veneer concealing the chasm between the environments of organisms and the worlds of human beings. Percy rediscovers a familiar paradox in all this: "the more science has progressed . . ., and benefitted man, the less it has said about what it is like to be living in the world" (1983, p. 116).

All this culminates in Percy's project to expand the boundaries of science itself to include reflection on those features of human life that are distinctive to it: language and alienation. Language is more than a set of syntactic algorithms or an array of semantic rules, it is the origin-product of the shared worlds of persons whose complex relations create a frame within which significant processes of translation and criticism can go forward. Alienation is the continuing condition of being "in trouble," "needing help," of being able to turn away and "not to be what one is."

If my reading of Percy's work is correct, these views are intended to challenge such rigorous reductionist perspectives as those of Norman Geschwind to take account of the deeply significant relational aspects of the human life-world. It is not reductionism itself that it is challenged; Percy only insists that the reductionist project should not be trivialized by preshrinking the human world. The intersubjectivity of human language and the deep sense in which human beings seem "lost in the cosmos" are data that must be addressed by the human sciences and their biological foundations on pain of continuing to tell us "less what it is like to be living in the world."

Walker Percy invites cautious reconsideration of the claim that folk psychology is nothing but a stagnant theory. His distinction of knowledge and news, and his illustration of the illuminating power of narrative may help us avoid the implausibility of a new Cartesianism in epistemology, and to prefer more historically nuanced and realistic views of the great moral traditions. His reading of the moral bankruptcy of enlightenment liberalism and its mirrored antithesis, Nietzschean perspectivism, has been articulated more systematically by other conservatives, including MacIntyre and Robert Bellah (MacIntyre, 1988, 1990; Bellah et al., 1985).

As Dostoyevsky uses Smerdyakov's crude violence to reduce Ivan Karamazov's philosophy to absurdity, Walker Percy uses the tools of violence, shock, comedy, insult, and the bizarre to give us a nauseating sense of the limits of the ethos of the autonomous self, the ethos of modern liberalism. His rhetoric is invidious, insulting, and question begging. He demands our attention. Walker Percy is an unhappy canary whose plaintive cry is as easy to ignore as the roar of a neuropsychologically well-informed biblical prophet.

Acknowledgments

The author is most grateful for the support of National Science Foundation Grant 8706817 during the time this chapter was written.

References

Barkow J (1989): *Darwin, Sex, and Status: Biological Approaches to Mind and Culture.* Toronto: University of Toronto Press

Bellah R, Madsen R, Sullivan, WM, Swidler A, and Tipton S (1985): *Habits of the Heart: Individualism and Commitment in American Life.* Berkeley and Los Angeles: University of California Press

Binding K Hoche A (1920): *Die Freigabe der Vernichtung lebensunwerten Lebens.* Leipzig: F. Meiner

Bloom H, ed. (1988): *Sinclair Lewis's Arrowsmith.* New York: Chelsea House

Churchland PM (1989): *A Neurocomputational Perspective: The Nature of Mind and the Structure of Science.* Cambridge, Mass.: Bradford Press MIT

Churchland PM (1988): *Matter and Consciousness.* Cambridge, Mass.: Bradford Press MIT

Desmond JF (1989): Disjunctions of Time: Myth and History in *The Thanatos Syndrome. New Orleans Review* 16(4): 63–71

Desmond JF (1990): Walker Percy's "new science" of fiction. Unpublished manuscript. Walla Walla, Wash: Department of English, Whitman College

Hardy JE (1989): No place like home: The world of Walker Percy's fiction. *New Orleans Review* 16(4): 73–81

Hardy JE (1987): *The Fiction of Walker Percy.* Urbana: University of Illinois Press

Hardy JE (1991): Man, beast and others in Walker Percy. In *Walker Percy, Novelist and Philosopher,* Gretlund J Westarp K. (eds.) University of Mississippi Press

Hatfield G (1991): *The Natural and the Normative: Theories of Spatial Perception from Kant to Helmholtz.* Cambridge, Mass.: Bradford MIT

Lewis S (1924): *Arrowsmith.* New York: Harcourt, Brace & World, Inc.

Lifton RJ (1986): *The Nazi Doctors: Medical Killing and the Psychology of Genocide.* New York: Basic Books

MacIntyre, A (1990): *Three Rival Versions of Moral Enquiry: Encyclopedia, Genealogy, and Tradition.* Notre Dame, Ind.: University of Notre Dame Press

Nietzsche F (1870–1871, 1886): *The Birth of Tragedy,* Golffing F (trans). New York: Doubleday/Anchor

Pauly P (1987): *Controlling Life: Jacques Loeb and the Engineering Ideal in Biology.* Oxford: Oxford University Press

Percy W (1967): Notes for a novel about the end of the world. Republished in *The Message in the Bottle* (1975), New York: Farrar, Straus and Giroux

Percy W (1971): *Love in the Ruins.* New York: Farrar, Straus and Giroux

Percy W (1975): *The Message in the Bottle.* New York: Farrar, Straus and Giroux

Percy W (1983): *Lost in the Cosmos.* New York: Farrar, Straus and Giroux

Percy W (1987): *The Thanatos Syndrome.* New York: Farrar, Straus and Giroux

Percy W (1989): The divided creature. *The Wilson Quarterly* 14: 77–87

Stich S (1983): *From Folk Psychology to Cognitive Science: The Case Against Belief.* Cambridge, Mass.: Bradford Press MIT

Wertham F (1966): *A Sign for Cain: An Exploration of Human Violence.* New York: Macmillan

9

Walker Percy: Neuroscience and the Common Understanding

JAMES H. SCHWARTZ

> He backed away and went to the baby's room, where it was dark except for the tiny night-light, a small bulb shining in a plastic lamp, with Mickey Mouse on the lampshade. The sight of it angered him. That was the new icon, that cartoon. It was not wicked but it was very stupid, and something so stupid was dangerous. Mickey Mouse was God—the fatuous smiling creature consoled people, because it was not human and not an animal and did not threaten anyone. It was a toy and it was not associated with any particular age or region or country. Mickey was a universal symbol of acceptance, and so people worshipped it—this item of colorful vermin. The worst of it was that people turned away from reality to venerate it. They pretended not to take it seriously, but they held on.
> —Theroux (1990, pp. 149–150)

Percy's views of mind and behavior really do not differ greatly from those current in modern neuroscience. Further, the purposes for which Percy and neuroscientists would have their work used are not so different either: both wish to impose a view of the world that, if not exactly comforting, at least makes sense of it in a generally acceptable way—Percy and the modern neurobiologist do not want to be left with the problem of evil unexplained. Percy (1973) wants to know why

> after three hundred years of the scientific revolution and in the emergence of rational ethics in European Christendom, Western man in the twentieth century elected instead of an era of peace and freedom an orgy of wars, tortures, genocide, suicide, murder, and rapine unparalleled in history? (p. 27)

Three aspects of Percy's writings bear on neuroscience in particular and science in general. The first is Percy's postulate that personal or societal values cannot be obtained through science. The second is that a different discourse must be used to arrive at reasons for living. The third is the unusual formal features of his books.

Clearly, Percy and the neuroscientist approach the problems of "wars, tortures, genocide, suicide, murder, and rapine" somewhat differently. But first, by his assertion that science cannot set values and provide a guide to moral behavior (an assertion with which almost any scientist would reflexively agree), Percy implicitly and optimistically asserts that values *are* obtainable, morality *is* achievable, and meaning and consciousness *are* understandable by *some* form of discourse (the second aspect). Third, Percy is an artist. We therefore must pay attention to the form in which his ideas are cast. Most critics divide his work into essays and fiction. But his essays are not in the style usual for professional philosophical journals; rather, they are serious homilies delivered in the tradition of the great radio comedians. I believe reviewers make the mistake of calling his fiction novels (if they are favorably inclined) or soap operas and flat television plays (if they are not).

A severe example of critical impatience with Percy's form is found in Robert Tower's (1987) review of *The Thanatos Syndrome:*

Much of *The Thanatos Syndrome* treats Percy's serious concerns in a manner more playful than impassioned. . . . While many of the characters are too briskly skeletal to be memorable, they keep things moving. . . . Care about form . . ., a concern to concentrate rather than to broadcast his effects, a degree of subtlety and surprise in the rendering of characters and their speech—these have all been sacrificed . . . in the author's headlong rush to fictionalize his diagnostic observations and his tragicomic vision of the horrors to come. *The Thanatos Syndrome* . . . lacks the shapeliness and substance of achieved literary art.

For want of a better formulation, Percy's work is a kind of comedy that is not all that funny. We might call it satyric perhaps, reflecting the ancient origin of comedy. Most of all, it is diagrammatic, almost animated, in the sense of a movie cartoon, it does not portray depth or singularity of character (as, say, in Tolstoy or Proust) or intricate formalism (as in Henry James or James Joyce). While I disagree with Tower's assessments that Percy's fiction lacks the parameters necessary for "achieved literary art," its form is difficult to name, for Percy borrows elements from old comedy (*The Birds* of Aristophanes), moralistic allegory (Bunyan's *Pilgrim's Progress*), bitter satire (Swift's *Gulliver's Travels*), and modern hyperrealistic burlesque (Robert Coover's *Gerald's Party,* David Lynch's movie *Blue Velvet,* or Andrew Bergman's *The Freshman*).

To restate my premises: Percy's view is first, that science provides no values. Therefore, second, a different sort of discourse is needed to help us, for we are all children of crisis, to borrow the phrase of his friend Robert Coles (Coles, 1978, 1990), and lost in the cosmos. Third, Percy's art is not exactly what it seems. How do these three aspects of Percy's work fit together, and how do they bear on contemporary biology in general, and neuroscience in particular?

Science and Values

First, let us examine the idea that science provides no values. From the end of World War II we all have become accustomed to the cliché that knowledge and discoveries can be used either for good or evil. Thinkers and writers loosely characterized as "post modern" — members of the Frankfort school, Michel Foucault — and most recent trendy historians of science concur that, like every other human activity, science has a social and an economic agenda (e.g., see Latour, 1987, 1988). Hardly new, similar opinions were expressed during the 19th century, and certainly as soon as Darwin's theories pervaded popular thinking (Himmelfarb, 1986). A succinct but obtuse statement of this sort of thinking is to be found in the quirky writings of Eric Gill, an artist, a convert to Catholicism, and a man who approximates Percy's (1971) description of Dr. Tom More in *Love in the Ruins:* "I believe in God and the whole business but I love women best, music and science next, whiskey next, God fourth, and my fellowman hardly at all." Gill, writing in 1918, said:

Renaissance, Reformation, Industrialism — these three are exactly antithetic to Poverty, Obedience and Chastity — the new to the old. The Renaissance is essentially the acclamation of man, his power and riches. The Reformation is the refusal, the denial of authority. Industrialism, to which the Renaissance and Reformation inevitably proceed, is the whoredom in which man is exploited by man.

The period of decay . . . began about four hundred and fifty years ago. It is now past repair. We are in a sinking ship and each man must save himself. There is no question even of "women and children first" for in this matter all are equal and no man can save another's soul.

Despite the superficial similarity of sentiment, it would be a mistake to see Percy's ideas as being congruent with Gill's reactionary views of technology. Percy was a medical doctor, if not a scientist; he did not advocate returning to the technology of the Middle Ages, as did Gill, nor to the way medicine and psychiatry were practiced then. Moreover, as far as I can judge from his writings that bear on neuroscience, Percy was completely comfortable with the principles of neural science, neurology, and psychiatry as he understood them.

The History and Principles of Neuroscience

What are those principles? Briefly, during the first quarter of the 19th century, most people came to believe that the mind is in the head, not in the heart. The German neuroanatomist Franz Joseph Gall initiated the modern period of neuroscience:

The body of man consists of matter: hence it is subject to all the laws of matter. . . . Man unites all physical, chemical, vegetable, and animals laws; and it is only by particular faculties that he acquires the character of humanity. The brain is exclusively the organ of the feelings and intellectual faculties. . . . Thus all concurs to prove that the brain must be considered as the organ of the moral sentiments and intellectual faculties. (Spurzheim, 1815)

Thus, in the words of Thomas Sewall (1837), an American professor of anatomy and physiology:

1. The brain [is] the material organ of the mind.
2. Just in proportion to the volume of the organ . . . [is] . . . the power of its mental manifestations.
3. Exercise of the mind promotes the development of the brain.
4. The character of the mind is . . . determined by the configuration of the brain.
5. The brain is a multiplex organ, composed of a definite number of compartments, each of which is the appropriate seat of a propensity, sentiment, or intellectual faculty.

Gall's reputation has suffered because his persuasive writings led to the enormous popularity of the pseudoscience phrenology (Cooter, 1984; Stern, 1971). Nevertheless, he did much to convince both scientists and the general public that the brain is the organ of the mind and that mental function is located in specific parts of the brain. The modern view that has developed from Gall's ideas is the anatomical localization of function. This view derived support from the clinical observations of the great 19th-century neurologists, Pierre Paul Broca, Carl Wernicke, and John Hughlings Jackson (Harrington, 1987; Young, 1970). At the turn of the century, convincing experimental evidence for the localization of brain function came from neuroanatomists and histologists, most prominently Santiago Ramón y Cajal, and from neurophysiologists, notably Charles Sherrington. From later clinical, neuroanatomical, and neurophysiological work done during the course of this century, we now know that specific nerve cells in the brain carry out specific aspects of behavior. Many of these functions are mediated by identified nerve cells in the brain. Since 1950, work with microelectrodes, biochemical and pharmacological techniques, and with recombinant DNA technology has refined and further identified the functions of specific parts of the brain. Most recently, the enormously powerful but noninvasive techniques of imaging, like positron emission tomography (PET) scanning, permit us to see regional changes in function as they happen in the living brain.

Brain functions are localized, but how are those functions performed? Neuroscientists agree that the functional unit of the brain is the nerve cell. Nerve cells communicate with one another by a process called synaptic transmission. A presynaptic cell releases a chemical messenger from its terminal that acts on another neuron or a target cell (like muscle). At all chemical synapses, signaling is achieved by the interaction of specific chemical messengers with specific receptors on the postsynaptic target cell.

In summary, during the past century, the functions of many brain areas have been identified; within the past 40 years, the mechanisms of neuronal signaling, electrical as well as biochemical, have been elucidated, and within the past decade, many of the molecules important to nerve function — for example, receptors and ion channels — have been isolated or cloned, and their structures determined. This information provides prodigious insights into how the signaling components of the brain work (Kandel et al., 1991).

Percy and Cognitive Neuroscience

I cannot see Percy being troubled by any of this information about brain function – from its basic materialistic assumptions (that the mind is in the brain), to the reductionist explanation of how nerve cells function. But I do see two philosophical frontiers that Percy would wish to defend.

The first frontier is the uniqueness of a particular event (Percy, 1973): "Science cannot utter a single word about an individual molecule, thing, or creature in so far as it is an individual but only in so far as it is like other individuals" (p. 22). With this existential restatement of the uncertainty principle, Percy implies that the scientist cannot speak to the individual or vice versa because the individual thinks or behaves in an indeterminate manner. Yet as a physician, who received his M.D. degree in 1941 from Columbia University's College of Physicians and Surgeons, and who then "interned at Bellevue Hospital [and] performed autopsies on more than 125 corpses" (Mitgang, 1977), Percy certainly knew, that for most experiences, individuals think, behave, and die in ways predictable from studying populations of other individuals. Personally important to him, they fall ill in much the same way. He himself caught pulmonary tuberculosis from one of the many bodies on which he had performed autopsies, catching it from "the same scarlet tubercle bacillus I used to see lying crisscrossed like Chinese characters in the sputum and lymphoid tissue of the patients at Bellevue" (Percy in Kazin, 1971). Particularly important to the physician, infections can be cured by antibiotics, pains relieved by analgesics and narcotics, and depressions lifted by antidepressive drugs or electroshock therapy. Dr. Percy knew that a broken leg is a broken leg.

The second frontier Percy wished to defend is the difference between humans and other animals, especially with regard to signs and symbols, consciousness and self-awareness. Other neuroscientists are better qualified than I to judge how similar are the inner representations of stimuli, images, and motivations in other animals to signs or symbols in humans. Again, I believe that most would either agree with him or reserve judgment, regarding this question as uncharted territory. But neuroscientists would be hesitant to agree with Percy that the difference is not biological. The patrolling of this second frontier preoccupied Percy from at least the mid-1950s, when he wrote the philosophical essay "Symbol as Need" (reprinted in Percy, 1973), to 1989, when he delivered the Jefferson Lecture in the Humanities (Percy, 1989).

Percy uses Helen Keller's story (Keller, 1903; Brooks, 1956) of how she became aware of the meaning of the word *water,* how the sign (the sensation of water flowing over her hand) became associated with the symbol *water,* to explain the linguistic difference between man and beast, a characteristic he called the *Delta factor* (Percy, 1973).

But it is striking that Percy fails to say how formally parallel Helen Keller's wonderful story is to the account of the Fall of Man in Genesis. She acquires knowledge in her garden's well-house; by her own description, she then becomes insatiable for more knowledge. From then on, her life is changed. Despite the conventional sweetness and charming innocence of this autobiography, we can be sure that she now thought about evil and anticipated separation and death (Keller, 1908).

According to Genesis, all human troubles follow as the result of acquiring this linguistic knowledge. God says to Eve: "in sorrow shalt thou bring forth thy children" (Gen. 3:16). God says to Adam: "Cursed is the earth in thy work. . . . Thorns and thistles shall it bring forth to thee. For dust thou art, and into dust thou shalt return" (Gen. 3:17–19). God says to the Angels: "Behold Adam is become one of us, knowing good and evil: now, therefore, he [might] take also of the tree of life, and eat, and live forever" (Gen. 3:22). To prevent man from becoming immortal, God casts Adam out from Eden to till the earth from which he was taken. (Translations of the Bible are adapted from the Douay version.)

Both modern science and Walker Percy appear to trivialize the wisdom inherent in the story of the Fall. Knowledge, or the *Delta factor,* comes at a price. By *knowing,* Helen Keller could appreciate her disabilities, not just as signs, not as a wounded, dumb animal with no *symbolic* idea of cosmic justice. By *knowing,* humans can make war systematically, not in response to *signs* of hunger or immediate danger, but over *symbolic* slogans. By *knowing,* Hitler can systematically slaughter six million Jews.

I have asserted that both Percy and modern, or what is perhaps better termed "institutionalized," science—at least as practiced in the United States—make the common mistake of believing that injustice, evil, and death can be explained in a comfortable way. Percy and institutionalized science make this error for different reasons or, to use up-to-date language, because of different social or economic agendas.

Form

Now I wish to return to Percy's form of presentation. It is striking how similar is the manner in which science and Percy present their points of view. Percy's art is diagrammatic, characteristically bordering on caricature or animated film. A glance at the most influential textbooks and journals in the biological sciences, for example, *The Molecular Biology of the Cell* (Alberts et al., 1989), Stryer's (1988) *Biochemistry,* and even Kandel, Schwartz, and Jessell's (1991) *Principles of Neural Science,* read like just-so stories illustrated by complicated, but animated and colorful, cartoons. Similarly, to an ever-increasing extent, the prestigious journals in the field, *Science, Nature,* and *Cell* (which people in my laboratory sometimes refer to as *Sell*), bear marked resemblances to the *Esquires* and *Playboys* of the 1960s. They present important scientific results in a style influenced greatly by television commercials, billboard advertisements, and ultimately by animated cartoons: Mickey Mouse, Bugs Bunny, Tom and Jerry and their friends, and Looney Tunes (Maltin, 1987; Solomon, 1987). Most important for my—perhaps hyperbolic—assertion are two aspects of this sort of presentation: first, as in the cartoons, in these texts, diagrams of sick mice do not die, cancer cells do not appear to be the savage monstrosities they actually are, and all the molecules, even, for example, digestive enzymes, do not appear destructive, only absurdly cute. Just as in cartoons, when Bugs Bunny is run over by a steamroller, he is briefly flattened but rapidly springs back to his lagomorphic self.

Historically, this kind of diagrammatic representation runs parallel to Walt Disney's seriously sentimental animated films, which first retold stark moralistic tales of the 19th century, *Snow White* (1937), *Pinocchio* (1940), and then saccharine animal stories *Bambi* (1941) and *Dumbo* (1941). After *Victory through Air Power,* produced at the request of the Air Force in 1949, Disney started to release his series of popular nature "documentaries" in which anthropomorphic penguins behave like spectators at Ascot, wolves like Robin Hood's band in Sherwood Forest, and ostriches like Bolshoi prima ballerinas (Schickel, 1968).

The Common Understanding

Why this immensely popular sentimentality? Some might say to sanitize the sexual behavior of animals. I believe *that* theory is as accurate as the idea that the story of the Garden of Eden is a Victorian lesson in proper table manners. The purpose is more likely to disguise the essential *amorality* of nature and animals, the vital necessity of the food chain, and the fact, still not grasped by civilized nature lovers, that you can't have your forest and your condo, too: knowledge or the Delta factor, in the sense of the story of the Fall, makes it implicit that the cosmos simply is not fair, nor is there any reason to think that it was meant to be just.

Despite the enormous progress made in understanding the brain and neurons as cells, understanding the molecular means by which information is transferred, how it is stored in the nervous system and how animals behave, and despite the increasing availability of successful treatments for psychiatric and neurological disorders, a deep popular pessimism pervades our society. In part, this is so because scientific knowledge has been used at least since the middle of the 19th century to control markets and regulate the behavior of people in modern industrial society. The sense of disappointment is all the greater because of the enormous optimism of the past. The mottos of the 1933 Chicago World's Fair, "A Century of Progress," and of the 1939 New York World's Fair, "The World of Tomorrow," now look rather threadbare. Despite the unquestionable progress in developing new and effective drugs, it should be evident from the history of this century that the promise of better life through chemistry, industry, and technology has a much hollower ring now in 1991 than it did in 1933 or 1939.

I have said that both science and Percy make the same mistake, but for different reasons. Science errs in its belief in the idea of infinite progress, whatever its presumed social and economic agendas. But why does Percy make the error that we can comprehend injustice, evil and death within some coherent system of analysis? Because of Percy's satirical *form,* I am not sure whether Percy makes the mistake sincerely or as a deliberate cosmic joke: in *Lost in the Cosmos: The Last Self-Help Book* (Percy, 1983), he asks a characteristic question:

Why is it possible to learn more in ten minutes about the Crab Nebula in Taurus, which is 6,000 light-years away, than you presently know about yourself, even though you've been stuck with yourself all your life.

Because of its black humor and emblematic absurdity, I am uncertain whether Percy (1983) is serious when he suggests that the answer to evil can emerge from a "scientific" study of semiotics. After all, he writes:

The pleasure of such transcendence derives not from the recovery of self but from the loss of self. Scientific and artistic transcendence is a partial recovery of Eden, the semiotic Eden, when the self explored the world through signs before falling into self-consciousness.

I say "partial recovery of Eden" because even the best scientist and artist must reenter the world he has transcended and there's the rub: the spectacular miseries of reentry—especially when the transcendence is so exalted as to be not merely Adam-like but godlike.

So, after all, he ends up with Augustine's formulation of the Fall, so succinctly put by Willie Stark in Robert Penn Warren's (1946) *All the King's Men:* "Man is conceived in sin and born in corruption and he passeth from the stink of the didie to the stench of the shroud."

By conversion to Catholicism, Percy appears to have embraced this simple formulation, which can only be appreciated by creatures endowed with language and the ability to distinguish symbols from signs. Even though Percy's formulation attempts to discredit the Panglossian but threadbare optimism of popular modern science, nonetheless, it is still fundamentally optimistic because it, too, implicitly asserts that injustice, evil, and death can be coherently explained. Twentieth-century neuroscience, philosophy, and history do not provide much reason for this optimism. Nor does traditional religion, Catholic or any other faith, which has had its chance for over two millennia to prove itself successful.

Nature, dumb animals, and humans simply are not what Walt Disney, the surgeon general, and the members of the National Academy of Science would have us believe, even though man can now "draw out the Leviathan with a hook . . . and put a ring in his nose" (Job 40:20–21).

The Lord's answer out of the whirlwind appears to be essentially correct, despite Girard's (1987) brilliant formulation of Job's sufferings. We cannot hope to make coherent a cosmos that is ethically incoherent, or perhaps, a cosmos to which ethics simply do not apply. The best we can do is follow God's stoic advice, again from the whirlwind: to gird up our loins, clothe ourselves with beauty, and try to be glorious.

References

Alberts B, Bray D, Lewis J, Raff M, Roberts K, Watson JD (1989): *Molecular Biology of the Cell,* 2nd ed. New York: Garland

Brooks, VW (1956): *Helen Keller: Sketch for a Portrait.* New York: E. P. Dutton & Co.

Coles R (1978): *Walker Percy: An American Search.* Boston: Little, Brown and Company

Coles R (1990): *The Spiritual Life of Children.* Boston: Houghton, Mifflin

Cooter R (1984): *The Cultural Meaning of Popular Science: Phrenology and the Organization of Consent in Nineteenth Century Britain.* Cambridge: Cambridge University Press

Gill E (1918): *Sculpture: An Essay in Stone Cutting, with a Preface about God.* Ditchling, England: St. Dominic's Press

Girard R (1987): *Job: The Victim of His People,* Freccero Y (trans). Stanford, Calif.: Stanford University Press

Harrington A (1987): *Medicine, Mind, and the Double Brain: A Study in Nineteenth Century Thought.* Princeton, N.J.: Princeton University Press

Himmelfarb G (1986): The Victorian trinity: Religion, science, morality. *In: Marriage and Morals among the Victorians and Other Essays.* New York: Knopf

Kandel E, Schwartz JH, Jessell T, eds. (1991): *Principles of Neural Science.* New York: Elsevier

Kazin A (1971): The pilgrimage of Walker Percy. *Harper's Magazine,* June, p. 84

Keller H (1903): *The Story of My Life.* Reprinted, Garden City, N.Y.: Doubleday and Company, 1954

Keller H (1908): *The World I Live In.* New York: The Century Co.

Latour B (1987): *Science in Action.* Cambridge, Mass.: Harvard University Press

Latour B (1988): *The Pasteurization of France,* Sheridan A, Law J (trans). Cambridge, Mass.: Harvard University Press

Maltin L (1987): *Of Mice and Magic — A History of American Animated Cartoons,* Rev. ed. New York: New American Library

Mitgang H (1977): A talk with Walker Percy. *New York Times Book Review,* February 20

Percy W (1971): *Love in the Ruins.* New York: Farrar, Straus and Giroux

Percy W (1973): *The Message in the Bottle.* New York: Farrar, Straus and Giroux

Percy W (1983): *Lost in the Cosmos: The Last Self-Help Book.* New York: Farrar, Straus and Giroux

Percy W (1989): "The fateful rift: the San Andreas Fault in the modern mind," 18th

Jefferson Lecture in the Humanities. Delivered May 3, 1989. Excerpted as "The Divided Creature," *The Wilson Quarterly* 13, 77–87; now available in Percy W (1991): *Signposts in a Strange Land,* Samway P (ed.) New York: Farrar, Straus and Giroux, pp 271–291

Schickel R (1968): *The Disney Version.* New York: Simon and Schuster

Sewall T (1837): *An Examination of Phrenology: In Two Lectures.* Washington City: B. Homans (printer)

Solomon C (1987): *The Art of the Animated Image: An Anthology.* Los Angeles: American Film Institute

Spurzheim JG (1815): *The Physiognomical System of Drs. Gall and Spurzheim, founded on an anatomical and physiological examination of the nervous system in general and of the brain in particular and indicating the dispositions and manifestations of the mind,* 2nd and rev ed. (1st ed. in 1814). London: Baldwin, Cradock and Joy; William Blackwood, Edinburgh

Stern MB (1971): *Heads and Headlines: The Phrenological Fowlers.* Norman: University of Oklahoma Press

Stryer L (1988): *Biochemistry,* 3rd ed. New York: W. H. Freeman

Theroux P (1990): *Chicago Loop.* New York: Random House

Tower R (1987): *New York Review of Books* 34:45

Warren RP (1946): *All the King's Men.* New York: Harcourt, Brace and Company

Young RM (1970): *Mind, Brain, and Adaptation in the Nineteenth Century: Cerebral Localization and the Biological Context from Gall to Ferrier.* Oxford: Oxford University Press. (Reprinted 1990. New York and Oxford: Oxford University Press)

PART 3

Neuroscientific Knowledge
and Social Accountability

10

Reconstructing the Brain: Justifications and Dilemmas in Fetal Neural Transplant Research

ALAN FINE

My purpose in this chapter is to present some issues of controversy that have emerged in the course of efforts to establish clinical trials of human fetal neural transplantation for the experimental therapy of Parkinson's disease. We were asked for the workshop upon which this volume is based to consider what constraints neuroscience might place on the study of meaning and value. Here, however, I am in part posing the opposite question: What sort of constraints do values place on neuroscience? While this question arises in many contexts in neuroscience, research on fetal neural transplantation confronts a range of value issues that is unusually broad; for this reason, it may be a useful case to consider.

Although widespread interest in neural transplantation is a recent development, the idea of such transplantation is not new. As early as 1917, Elizabeth Hopkins Dunn reported survival of intracerebral grafts of mammalian brain tissue (Dunn, 1917). Additional such reports appeared infrequently over the subsequent decades, and transplantation of embryonic central nervous system tissue in *lower* vertebrates became an established and important tool of developmental biologists. Only in the 1970s, however, did the ability of embryonic neurons to survive transplantation in the mammalian central nervous system receive widespread attention. The pioneering work of the 1970s was done by Das and collaborators (Das and Altman, 1971), who first demonstrated the survival of such grafts in an unambiguous way. However, widespread attention accrued largely in response to the results of groups in Sweden (Björklund and Stenevi, 1979) and in Washington, D.C. (Perlow et al., 1979) who demonstrated that transplanted neurons could contribute to the *function* of the recipient animal — in particular, that they could partially restore normal movement in animals with an experimentally induced form of Parkinson's disease. This finding led at once to a recognition that neural transplantation might serve not only as a tool for studying development and function in the central nervous system, but also as a potential means of treating certain currently incurable degenerative or traumatic neurological disorders.

Parkinson's disease, for example, is currently treated with L-dopa, the precursor to dopamine—a transmitter substance whose deficiency causes the symptoms of the disease. Daily intake of exogenous precursor brings substantial symptomatic improvement to most Parkinson's disease patients. Unfortunately, the relentless progression of the disease, with continuing reduction in numbers of the nerve cells that convert L-dopa to dopamine, usually leads after 5 to 10 years to diminished effectiveness or unacceptable side effects of L-dopa. At present, then, there is no cure or definitive treatment for Parkinson's disease, nor indeed for many other important neurological disorders such as Alzheimer's disease and stroke. The brain suffering from many of these disorders may, however, be amenable to reconstruction, by putting back into it cells of the type that have degenerated or been injured. An awareness of these therapeutic possibilities has, over the past decade, led numerous medical researchers to enter the field of neural transplantation.

Problems soon emerged that influenced the course of this research. For example, people who were contemplating the clinical application of embryonic neural transplantation had at once to consider the possible use of human fetal tissue. In the United States, where abortion has for decades been an issue of great political controversy, this was seen to be a problem best avoided. Largely for this reason, neural transplantation researchers at the National Institute of Mental Health in Washington, D.C., shifted their efforts to the use of adrenal tissue. It had previously been found that adrenal medullary cells—the cells that normally make adrenaline—have the capacity to "transdifferentiate," that is, to change their shape and biochemistry from that of adrenal gland cells to that of dopamine-secreting nerve cells (dopamine is a metabolic precursor of adrenaline) (Unsicker and Chamley, 1977). On this basis, Freed and coworkers explored the possibility that political controversy could be avoided by using adult adrenal tissue as an alternative source of dopaminergic cells for transplantation in Parkinson's disease. Their animal experiments suggested that such adrenal medullary grafts into the brain could substantially restore normal locomotion in an animal model of Parkinson's disease (Freed et al., 1981). On the basis of those studies, clinical trials were initiated in Sweden in 1984, grafting Parkinson's disease patients' adrenal medullary tissue into their own brains (Backlund et al., 1985). Although these "autologous" grafts (or "autografts") were without significant deleterious effects, the therapeutic results were disappointing (Lindvall et al., 1987), and adrenal medullary autografts were discontinued in Sweden.

Similar clinical trials performed in Mexico, however, were reported in 1987 to produce striking clinical improvements (Madrazo et al., 1987). This report led in the following 2 years to a veritable epidemic of clinical intracerebral adrenal transplantation, so that by mid-1990 more than 200 such procedures had been performed around the world. The results of most of these were summarized at an international symposium (Lindvall and Björklund, 1991), with a general consensus that the outcome of these adrenal autografts for Parkinson's disease has been disappointing. As a result, most groups are discontinuing these procedures (although it remains possible that modifications such as the use of nerve growth factor or other "trophic" substances may enhance survival or transdifferentiation of these grafts).

This episode raises difficult questions. For example, at what point in the development of an experimental therapy are clinical trials justified? To what extent are animal experiments a suitable basis for this decision? For how long are clinical trials warranted in the face of negative results? This last question is especially problematic for surgical procedures, where it is often quite difficult to compare results of different groups where there can be subtle but crucial variations in technique. Questions of conflict of interest may arise, particularly where investigators charge their patients large fees for the experimental procedures (as was reportedly the case in some of the adrenal medulla transplant programs). It is unclear whether such investigators could then be relied upon to report results objectively, or even to obtain proper informed consent from participating patients. Indeed, the very notion of informed consent is problematic in the context of doctor-patient relations, particularly for a patient faced with a progressive and terminal degenerative disease. The adrenal medulla transplantation trials also raise interesting questions about the ethical responsibilities of the medical journals. For example, the rash of clinical trials following publication of the Mexican results in 1987 was, I suspect, in part due to the publication of those results in the very prestigious *New England Journal of Medicine*.

In the course of preclinical development and evaluation of neural transplantation therapies, the question of what circumstances, if any, can justify animal experimentation has presented itself in stark terms. The anatomy and physiology of the brain areas involved in Parkinson's disease differ in significant ways between rodents, on the one hand, and primates and carnivores, on the other. Moreover, monkeys—like humans, but unlike most other animals—are extremely sensitive to certain toxins that cause parkinsonian symptoms. For these reasons, such parkinsonian monkeys have greatly aided efforts to understand the causes and possible treatments of Parkinson's disease. While such experimentally induced parkinsonism is unlikely to cause physical pain, and despite the best efforts of the researchers (and rigorous institutional safeguards) to guarantee hygienic and humane veterinary care, it can scarcely be denied that the sudden onset of parkinsonian symptoms is highly stressful to these intelligent animals. Scientists (including myself) who carry out such experiments in the face of this suffering do so, I hope, because they foresee a clear and substantial human good—improved medical therapy—resulting from them. It is less clear whether "mere" increase in human knowledge, without obvious immediate medical consequences, might justify similar animal experimentation. The debate is complicated by the difficulty, or even impossibility, of predicting the outcome of any particular experiment. In such circumstances, the theoretical utilitarian calculus may be a practical impossibility.

Even while the adrenal medullary transplant programs were underway, work continued in a number of laboratories on fetal neural transplants. Investigators continuing this work believed that the likelihood of reconstructing a damaged nervous system was greatest if they replaced the very elements that had degenerated in the patient, using fetal brain cells that were destined to become those elements in the developing fetus. In laboratory studies, transplantation of fetal dopamine-secreting neural transplants into dopamine-depleted brain regions of rats with experimentally induced Parkinson's disease led to complete recovery from certain movement impairments (e.g., Dunnett et al., 1983), and substantial recovery has been observed in experimentally parkinsonian monkeys with fetal dopamine-secreting neural grafts (e.g., Fine et al., 1988). In aggregate, these results have led a number of investigators, including myself, to establish clinical trials of fetal neural transplantation for treatment of Parkinson's disease patients who no longer benefit from available medical therapy.

The first reported clinical trials of fetal neural transplantation for treatment of Parkinson's disease were performed in Sweden in 1987 (Lindvall et al., 1989), leading to modest and transient improvements, and, shortly thereafter, in Mexico (Madrazo et al., 1988) where substantial patient improvement was claimed. These reports again raise the issue of when clinical trials are justified: the Swedish and Mexican procedures were begun before any successes had been reported in primate studies. In any event, by late 1987 several groups had come to the conclusion that clinical trials were warranted. A request in 1987 to NIH for permission to conduct such a trial led the United States Department of Health and Human Services (DHHS) to impose a moratorium on such research—or, more precisely, on the use of federal funds for such research—until the department was able to act upon the recommendations of a specially convened panel. In Sweden, in contrast, approval for clinical trials was granted by the necessary agencies uneventfully, with modest public debate. The process in Mexico was similar, with the significant difference that the researchers claimed to be using tissue from spontaneous, rather than induced, abortions—a choice they felt removed them from the controversy of abortion.

(The use of tissue from spontaneous abortions has often been proposed in the public debate about fetal neural transplantations. Unfortunately, it is an ill-considered proposal. Spontaneous abortions most often result from fetal pathology; that, together with an indeterminate interval between fetal death and expulsion and the possibility of contamination as the fetus passes through the vaginal canal, renders spontaneous abortion an unsatisfactory source of tissue for neural transplantation. There is thus reason to doubt the veracity of the Mexican report of successful use of tissue from spontaneous abortions, which doubt in turn raises disturbing questions about veracity in the medical literature.)

In China, Cuba, Czechoslovakia — that is to say, in countries where easy access to therapeutic abortions is official policy — there have been no governmental obstacles to clinical trials of fetal neural transplantation, even where the investigators' prior research experience in this field has been limited or nonexistent. In the United Kingdom, there has been substantial public debate but no legislation; as a result, one group has already transplanted fetal dopamine-secreting brain tissue in approximately 20 patients. In the United States, the moratorium on use of federal funds for this research has been widely interpreted as a call for cessation of all such work. This situation leads to the interesting question of whether voluntary restraint is actually possible in a field of intensive biomedical research; to my knowledge, the Asilomar agreement is the only (approximate) precedent. In the event, one group at the University of Colorado proceeded with a clinical trial in defiance of the moratorium (on the disputable basis that no federal funds were used), announcing preliminary results before the DHHS was able to respond to its panel's recommendations; shortly thereafter, a group at Yale University also began trials. The DHHS has since announced that the moratorium will be continued indefinitely (it is still in force at the end of 1991), in what has been interpreted by some as a response to this failure of voluntary restraint.

In Canada, my collaborators and I first sought approval for such clinical trials in 1988. The established process that we followed for approval of new experimental medical procedures involves initial submission of a research protocol to an institutional ethical review board. Protocols deemed acceptable on ethical grounds go to a medical review committee, which investigates the medical and scientific basis for the new procedures. If passed, applications go to the hospital administration for final approval. In addition, our particular protocol was sent first to a university scientific review board because of our intent to perform ancillary laboratory experiments. In the course of this review process, and in the concurrent public debate in local and national newspapers, radio, and television, questions were raised such as whether the proposed procedures were likely to work, and whether they were likely to benefit the particular patients under study. Our positive replies to these questions were supported by results of animal experiments, and by explicit restriction of the study to patients whose response to available medication was failing. The issue of informed consent was dealt with by requiring extensive prior counseling of potential participants and by excluding patients with detectable dementia. Some questions evidently arose from a lack of familiarity with the underlying science: for example, it was more than once asked whether patients undergoing transplant procedures might not acquire the personality of the donor (an impossibility).

Faced with questions about compliance with guidelines for fetal research, we have had to point out the fundamental difference between *fetal research* and *research involving fetal tissue*. It has been our contention that we are using tissue from fetuses — already dead, for reasons independent of tissue transplantation — that otherwise would have been disposed of. We have been asked whether this distinction is warranted. Parallels have been drawn between our assertions and the claims of Nazi scientists that without their studies of "scientific material" from concentration camps valuable specimens would have been wasted. This criticism is disturbing, for there are crucial differences between the two cases. Rigorous precautions are taken in the case of fetal neural transplantation, to minimize any influence of subsequent use of aborted tissue upon the woman's decision to abort; in contrast, as pointed out by Müller-Hill (1984), there were requests by Nazi scientists for tissue from specific types of individuals, who were presumably killed for these "scientific" purposes. Certainly the motive of treating the incurably ill is fundamentally and importantly different from the motive of investigating notions of racial difference. The implications of these parallels — that a developing fetus is ethically equivalent to a person with history, relations, feelings, and consciousness, and that abortion is ethically equivalent to genocide — are unjust. Similar considerations may be relevant to the previously raised issue of animal experimentation.

Questions of complicity require us also to address certain issues from the perspective of the woman undergoing the abortion. Informed consent for use of fetal tissue may be requested of the woman at some time after she has given informed consent for the abortion. While this is the procedure followed in Canada and Sweden, legal, ethical, and practical objections have been raised: it is unclear, for example, whether a woman undergoing abortion can be considered to be acting as an ethical guardian of the fetus, serving its best interests. The legal status of the use of fetal tissue for research and the requirement for consent of the woman undergoing abortion vary among, and even within, countries. In Germany, the Clinical Code reform of 1987 makes it illegal to procure a dead fetus without consent, although it makes no reference to research on fetal material (Gunning, 1990, p. 28). In the United States, state laws in Arkansas, Arizona, Illinois, Indiana, Ohio, Louisiana, New Mexico, and Oklahoma prohibit experimental use of fetal tissue from induced abortions (Gunning, 1990, p. 44). Specific legislation on transplantation of fetal tissue from induced abortions is nonexistent or under study in many other countries, leaving such procedures to be regulated by professional guidelines or regional review committees; as might be imagined, there is a wide spectrum of policies. In Norway, for example, such procedures will not be approved by these committees, whereas in Canada separated fetal tissue is regarded as routine pathological tissue suitable for research (Gunning, 1990, pp. 36 and 19).

I will conclude by raising the question of whether modifying the abortion procedure to improve the recovery of useful fetal tissue would be ethical. I have argued (Fine, 1988) that separation of the abortion procedure from subsequent uses of the aborted tissue provides a "fence" against certain possible abuses. However, it can be imagined that the chance of a successful outcome of the transplant might be much enhanced by slight modification of the abortion procedure, so as (for example) to recover less-fragmented tissue; such change might add no additional risk or inconvenience to the procedure. In fact, modifications for this purpose have been made by one of the two groups (University of Lund) participating in the Swedish fetal neural transplantation program. It thus remains to be seen whether effective moral "fences" can indeed be maintained. While we have taken pains to ensure that the existence of fetal tissue transplantation procedures will not influence the decision to obtain an abortion, we cannot exclude the possibility, however unlikely, that widespread institution of these procedures might nevertheless increase (or decrease) the incidence of abortion: the motives of the woman can be known only to her. What the public *can* expect soon to know is whether reconstruction of the damaged central nervous system by fetal neural transplantation will provide a useful new treatment for devastating neurological illnesses.

References

Backlund E-O, Granberg P-O, Hamberger B, Knutsson E, Martensson A, Sedvall G, Seiger A, Olson L (1985): Transplantation of adrenal medullary tissue to striatum in parkinsonism. First clinical trials. *J Neurosurg* 62:169–173

Björklund A, Stenevi U (1979): Reconstruction of the nigrostriatal dopamine pathway by intracerebral nigral transplants. *Brain Res* 177:555–560

Das GD, Altman J (1971): Transplanted precursors of nerve cells: Their fate in the cerebellums of young rats. *Science* 173:637–638

Dunn EH (1917): Primary and secondary findings in a series of attempts to transplant cerebral cortex in the albino rat. *J Comp Neurol* 27:565–582

Dunnett SB, et al. (1983): Intracerebral grafting of neuronal cell suspensions: 4. Behavioural recovery in rats with unilateral implants of nigral cell suspensions in different forebrain sites. *Acta Physiol Scand Suppl* 522:29–37

Fine A, (1988): The ethics of fetal tissue transplantation. *Hastings Center Rep* 18/3:5–8

Fine A, et al. (1988): Transplantation of embryonic marmoset dopaminergic neurons to the corpus striatum of marmosets rendered parkinsonian by 1-methyl-4-phenyl-1,2,3,6-tetrahydropyridine. In: *Transplantation into the Mammalian CNS,* Gash DM, Sladek JR Jr, eds. Amsterdam: Elsevier

Freed WJ, et al. (1981): Transplanted adrenal chromaffin cells in rat brain reduce lesion-induced rotational behavior. *Nature* 292:351–352

Gunning J (1990): *Human IVF, Embryo Research, Fetal Tissue for Research and Treatment, and Abortion: International Information* London: Her Majesty's Stationary Office

Lindvall O, et al. (1987): Transplantation in Parkinson's disease: Two cases of adrenal medullary grafts to the putamen. *Ann Neurol* 22:457–468

Lindvall O, et al. (1989): Human fetal dopamine neurons grafted into the striatum in two patients with severe Parkinson's disease. *Arch Neurol* 46:615-631

Lindvall O, Björklund A, eds. (1991): *Intracerebral Transplantation in Movement Disorders — Experimental Basis and Clinical Experiences,* Fernstrom Symposium, Vol. 16. Amsterdam: Elsevier

Madrazo I, Drucker-Colin R, Diaz V, Martinez-Mata J, Torres C, Becerril JJ et al. (1987): Open microsurgical autograft of adrenal medulla to the right caudate nucleus in two patients with intractable Parkinson's disease. *N Engl J Med* 316:831-834

Madrazo I, et al. (1988): Transplantation of fetal substantia nigra and adrenal medulla to the caudate nucleus in two patients with Parkinson's disease. *N Engl J Med* 318:51

Müller-Hill B (1984): *Tödliche Wissenschaft.* Reibeck bei Hamburg, pp. 23-24. (Cited in Proctor RN (1988): *Racial Hygiene: Medicine under the Nazis.* Cambridge, Mass.: Harvard University Press, p. 44)

Perlow MJ, et al. (1979): Brain grafts reduce motor abnormalities produced by destruction of the nigrostriatal dopamine system. *Science* 204:643-647

Unsicker K, Chamley J (1977): Growth characteristics of postnatal rat adrenal medulla in culture. *Cell Tissue Res* 177:247-268

11

Therapeutic Exuberance:
A Double-Edged Sword

Elliot S. Valenstein

The 1990s have been designated "The Decade of the Brain" primarily because of the belief that advances in the neurosciences will soon make it possible to treat hitherto incurable brain diseases. Louis Sullivan, secretary of the Department of Health and Human Services, observed in his remarks at a National Institute of Mental Health symposium launching this decade that the knowledge gained in the neurosciences will help "find solutions to the major disorders that ravage the human brain." Other speakers at the symposium noted that neuroscience research would lead to effective treatments for many of the 30 million people now widely assumed to be suffering from mental illness because of some brain dysfunction.

Certainly, the knowledge gained in the neurosciences will eventually help many desperately ill patients, but along the way there will surely be many unwarranted reports of successful treatments that later prove to be ineffective and, in some instances, even harmful. Judging from the past, when the conditions are right some of these treatments will be widely accepted and vigorously pursued despite flawed evidence that they are safe and effective. While it must be recognized that it is impossible to avoid all mistakes and that there can be little progress without risk, uncontrolled therapeutic exuberance may cause great harm to patients and actually hinder progress by diverting energy and resources into blind alleys. It is necessary that ways be explored for reducing the magnitude of therapeutic mistakes without unnecessarily hindering progress. In this context, I offer a brief account of the history of prefrontal lobotomy (Valenstein, 1986) in order to emphasize the factors that have in the past promoted unjustifiable therapeutic exuberance and with the hope that it may stimulate more discussion of the safeguards that should be instituted.

The History of Prefrontal Lobotomy:
A Historical Cautionary Tale

During another decade, the one following the Second World War, tens of thousands of people were subjected to prefrontal lobotomy and other so-called psychosurgical procedures as a treatment for their mental illness.

Even allowing for what was known at the time, prefrontal lobotomy was a crude procedure that often mutilated the brain. Yet, it was widely promoted and adopted even though it was justified by flimsy theory and little evidence. How did this happen, and what can be learned from this history?

Many argue that we cannot generalize from the history of lobotomy because those events arose from conditions that could exist only before the major modern advances in psychiatry. Others simply dismiss the history of lobotomy by considering it an instance of isolated "therapeutic zealotry," never a part of mainstream medicine. Such arguments, however, are not valid. Lobotomy was very much in the mainstream of medicine. It was promoted from the outset by the *New England Journal of Medicine,* which claimed in an editorial (December 3, 1936) that it was based "on sound physiological observations." Support for prefrontal lobotomy was quite widespread within the medical community. Although many today believe that lobotomy was promoted primarily by psychiatrists and neurosurgeons, in fact, neurologists also played major roles in its promotion. The two principal figures in the history of lobotomy were Egas Moniz, professor of neurology at the University of Lisbon, who was awarded the Nobel prize for introducing this treatment, and Walter Freeman, professor of neurology at George Washington University, who did more than anyone else to promote this operation around the world. Presidents of the American Neurological Association as well as presidents of the American Psychiatric Association endorsed lobotomy with enthusiasm. At the time, prefrontal lobotomy received substantial support from many prominent and influential physicians and scientists, among them John Fulton, Donald Hebb, Ashley Montagu, Adolf Meyer, Edward Strecker, and scores of others (see Valenstein, 1986, for quotations and references) who let it be known that they were convinced that prefrontal lobotomy was ushering in a new era of scientific psychiatry.

In short, prefrontal lobotomy was widely accepted. By 1950, lobotomies were being performed in 41 percent of the nation's inpatient mental hospitals. Even this figure is an underestimation, as many patients were sent to a central facility for a lobotomy if the hospital in which they were housed lacked the facilities or personnel to perform these operations. Thus, patients from many psychiatric institutions in Massachusetts were sent to the Boston Psychopathic Hospital (now the Massachusetts Mental Health Center) for a lobotomy. Lobotomies were performed at all 37 of the Veterans Administration hospitals that existed at the time. In New York, 16 of the 18 state mental health institutions used lobotomy. Governor Thomas Dewey observed at the opening of the huge Pilgrim State Hospital on Long Island that lobotomy had made it possible for hospitals to start curing patients rather than merely providing custodial care for them. Lobotomy became a showpiece in mental hospitals, and superintendents referred to its use in their annual reports to illustrate the alertness of their institution to medical progress. The factors that were responsible for the widespread acceptance and use of prefrontal lobotomy are discussed below. It is the thesis of this chapter that, even though much has changed since the peak period of prefrontal lobotomy, all the major classes of influence that fostered its growth in the past are still important today.

A procedure as radical and dangerous as lobotomy would not have been so widely adopted had there not been large numbers of desperate patients who were not being helped by any treatments available. In the early 1930s, the main treatments for serious mental illness consisted of such nonspecific drugs as bromides, paraldehydes, chloral hydrate, and opium, or various forms of "hydrotherapies," including "wet packs," baths, showers, and douches. These "treatments" were at best only temporary sedatives and palliatives, and, as a result, the number of institutionalized patients steadily increased.

Between 1900 and 1940, the number of institutionalized mental patients in the United States increased from 150,000 to nearly 500,000, a rate of increase more than twice that of the population at large. Understaffed and underfunded, most public institutions were abominable. They were justly exposed in a series of articles and books published in the 1940s. Typical was the article in *Life* magazine entitled "Bedlam 1946: Most U.S. Mental Hospitals Are a Shame and a Disgrace," which described abused mental patients who sat naked all day on cold concrete steps in rooms reeking from excrement. Superintendents of most state hospitals were willing to try any proposed treatment as long as it was not expensive or labor-intensive. Physical treatments were attractive because they usually made patients more manageable and involved relatively little expense. Even the liberal reformers, like Albert Deutsch (1948), who did so much to expose these horrendous conditions in his influential book *The Shame of the States,* concluded by recommending more extensive use of the new physical treatments in psychiatry, namely insulin, metrazol, electric shock, and prefrontal lobotomy. When no effective treatment exists, the medical community is under great pressure to accept, often uncritically, new treatments that are claimed to have had success, especially if these claims are made by (or supported by) well-known figures in medicine.

Economic factors in many guises also influence the acceptance of new procedures in medicine. Over and over again, the money that would be saved if inmates could be discharged from psychiatric institutions to the care of their families was used as an argument to justify lobotomy. Early in its acceptance, the superintendent of the Delaware Mental Hospital, the first state hospital to adopt lobotomy, observed that if a small state like Delaware has already been able to save more than a million dollars, much more would be saved by the nation as a whole if the procedure were more widely adopted. John Fulton claimed that "if only 10 percent of the patients occupying neuropsychiatric beds could be sent home, it would mean a savings to the American taxpayer of nearly a million dollars a day."

Economic factors influenced the acceptance of lobotomy in other ways, too, particularly as a by-product initially of the jurisdictional disputes between neurologists and psychiatrists, and later, between psychiatrists and psychologists. Before 1930, a significant amount of the practice of most neurologists was involved with treating mentally ill patients. By the mid-1930s, psychiatrists, who had become board-certified, gained political clout and as a result they started to take over much of the treatment of the mentally ill. Leading neurologists like Henry Alsop Riley—in his presidential address to the New York Neurological Society—warned that the viability of neurology was jeopardized if the treatment of the mentally ill was completely lost to psychiatrists. Similarly, Derek Denny-Brown, Harvard University's professor of neurology, observed that "the mounting aggressiveness of psychiatry encroached on what had been considered the neurologist's field, and by the 1930's, many predicted the extinction of the genus neurologist."

On the other hand, psychiatrists did not have it all their own way, for psychologists and other nonmedical psychotherapists began in turn to compete with them. In the mid-1930s, successive presidents of the American Psychiatric Association (for example, William Russell in 1932 and James May in 1933) alerted the membership to the growing number of nonmedical people who were using psychotherapy to treat the mentally ill and to the dangers that would ensue if psychology departments continued to offer courses in abnormal psychology.

The competition between these professional groups made physical treatments of mental illness more attractive. Neurologists benefited because their knowledge of the brain was opening up opportunities for them as part of therapeutic and research teams. Psychiatrists benefited because their use of "somatic treatments" such as prefrontal lobotomy not only answered those physicians who accused them of not being part of scientific medicine, but it also provided a treatment realm in which nonmedical psychotherapists could not compete. It is not coincidental that at this time many of the first separate departments of psychiatry in medical schools were established while other somatic treatments such as insulin, metrazol, and electric shock were also widely adopted.

Although there had been early attempts to treat mental illness by brain surgery, it is generally agreed that prefrontal lobotomy was introduced by Egas Moniz, a Portuguese neurologist. Born in 1874 into a landed, aristocratic family, Moniz traced his roots back to the 15th-century wars against the Moors. After graduating from Coimbra University where he had been active in politics, Moniz studied neurology in France. His interests, however, ranged widely into Portuguese history and art, but especially politics. Although he was appointed professor of neurology at the newly established University of Lisbon, Moniz served, without relinquishing his professorship, in Portugal's parliament, as ambassador to Spain, and as foreign minister at the Paris Peace Conference following World War I.

Enormously ambitious and fiercely competitive, he had hoped to become prime minister, but this proved impossible when the political party he helped to form lost out in a power struggle during the economic crisis that followed the First World War. This was 1925, the year Salazar took over the reins in Portugal.

Moniz, then 51, turned to neurology for fulfillment. At the time, several investigators in Europe and North America were developing techniques for using X-rays to obtain pictures of blood vessels in the arms, legs, and spinal cord. Although almost everyone thought that it was too dangerous to inject foreign substances into cerebral blood vessels, Moniz was in a position to take the risk. One patient died and several were seriously injured by the irritating substances he injected to make the blood vessels of the brain visible under X-rays. Moniz succeeded in obtaining "cerebral angiograms," which eventually proved to be extremely useful in localizing brain damage through their ability to detect abnormal blood vessel configurations.

Moniz published his results rapidly and resented anyone else working in what he now regarded as his exclusive domain. As a result, he became embroiled in an unpleasant and wholly unnecessary priority dispute with Japanese and German neurologists. Moniz had his sights set on the Nobel prize; indeed, there is evidence that he solicited nominations for the award. The Nobel Committee, however, decided initially that the usefulness of the technique was not yet proven, and on a second round, that there were too many people who had contributed to the field to single out one person for an award. Although Moniz was not awarded the Nobel prize, this work established him as a major figure in neurology, one who is still generally regarded today as the "father of cerebral arteriography and angiography."

Not someone to rest on his laurels, Moniz began prefrontal lobotomy in 1935 at the age of 61. While attending the 1935 Congress of Neurology in London, Moniz had heard Carlyle Jacobsen, a psychologist working in John Fulton's laboratory at Yale, report that chimpanzees could not solve certain problems after the frontal lobes of their brains had been destroyed. What most interested Moniz, however, was an incidental observation described by Jacobsen. Before the brain surgery, Becky, a female chimpanzee, had had temper tantrums whenever she made an error that deprived her of a food treat. Eventually, she became untestable, refusing to even enter the test chamber. After destruction of her frontal lobes, however, she happily participated in the experiment, even though unable to solve any of the problems. Moniz who, like most neurologists at the time, often treated mentally ill patients, was clearly interested in this anecdotal report of Becky's behavior. He has been reported to have stood up after Jacobsen finished his report and asked, "If frontal lobe removal eliminates frustrational behavior, why would it not be feasible to relieve anxiety states in man by surgical means?"

Although few in the audience thought Moniz was seriously contemplating such a brain operation on a human, Moniz was serious. As soon as he returned to Lisbon he instructed the neurosurgeon Almeida Lima to operate on 20 mental patients obtained from local asylums. Lima, although a neurosurgeon, was a much younger man who was beholden to Moniz and very much his assistant. Typically, Moniz was in a great hurry to obtain results. Twenty prefrontal "leucotomies," as Moniz called the operation, were performed in rapid succession, averaging an operation every 4 days. A monograph was produced, claiming that 7 of the 20 psychiatric patients, whose condition was described as hopeless, were "cured," 7 were significantly improved, and the remaining 6 unchanged after the prefrontal leucotomy. If true, this would have been a remarkable achievement at a time when no other effective treatment existed, but the evidence for the claim was totally inadequate. Moniz had presented only "thumbnail" sketches of patients who had been followed for only brief periods after the operation, in most instances less than 2 weeks. Virtually no evidence was provided to convince any critical reader that the patients claimed to be cured had been able to return to a normal life. Only much later was it revealed that many of the "cured" patients had either relapsed or suffered some permanent impairment. The monograph, however, was published less than 9 months after the first operation and was written before adequate time had elapsed to make adequate judgments about the lasting effects of the operation.

Having been embroiled in an earlier priority dispute, Moniz took no chances that he would have to share any of the credit for prefrontal lobotomy. Before the end of 1936 Moniz reported these and similar results in articles published in six different countries. Moniz had reason to be concerned that others might try similar operations on the mentally ill. Indeed, several other physicians at the time had already thought about or had actually tried various frontal lobe interventions to treat mental illness. The mood of the time is well characterized by a remark made by the surgeon William Mayo, one of the founders of the famous clinic in Minnesota. In a speech delivered before Moniz had performed his first lobotomy, Mayo commented on the lack of treatment for the mentally ill: "Are we not now in the same position in the treatment of the mentally afflicted that we were in surgery of the abdomen when I began fifty years ago, or as we were in the surgery of the chest? . . . Day by day, I can see the extension of remedial measures to the brain at earlier stages" (cited in Woltman, 1941).

Moniz's rationale for prefrontal leucotomy was simplistic at best, although there were some who thought it was "inspired." The rationale was based on two major assumptions. First, because the prefrontal area of the brain was not known to play a sensory or motor function, it had been assumed for some time to be a brain area important for integrating "higher intellectual functions," such as thinking, problem solving, and planning. It was argued that the conclusion that the prefrontal area was involved in these "higher functions" was supported by the evidence that it was obviously most highly developed in humans. Although the argument was not in any way tightly reasoned, Moniz apparently believed that it followed that the prefrontal area was also the area underlying the abnormal thought processes seen in the mentally ill. The second assumption in Moniz's rationale for justifying prefrontal was his assertion that all mental illness resulted from ideas becoming pathologically "fixed" as a result of abnormally stabilized neural activity in the prefrontal area of the brain.

For Moniz, it apparently was a straightforward conclusion that to alleviate mental illness it was necessary to destroy the pathways responsible for the abnormally stabilized neural activity that he assumed was what sustained the "fixed ideas." Moniz had no way of testing the merits of his speculations about the presence of abnormal neural activity and he never provided a justification for destroying any particular fiber tract. In fact, he frequently changed the location and number of brain areas destroyed, on what can only be concluded was a trial-and-error process. The operation itself was almost unbelievably crude. After holes were drilled in the skull, a device called a leucotome was inserted "free hand" into the brain. When the leucotome was believed to be in place it was rotated, leaving behind cores of dead brain tissue. As many as six to nine such cores were made on each side of the brain. Because the leucotome worked much like the kitchen utensil used to remove the core of an apple, Moniz's prefrontal leucotomy procedure was called the "core operation."

Despite all the shortcomings in the theory, the procedure, and the evidence, Moniz's claims were accepted uncritically by many physicians around the world. Within a short time, Moniz's surgical procedure was being tried on patients in many other countries. During the ensuing years, a great number of different surgical procedures for destroying the prefrontal brain area and other brain regions replaced the Moniz procedure. By 1948, physicians from 28 countries attended the First International Congress of Psychosurgery in Lisbon. Plans to nominate Moniz for the Nobel prize were orchestrated at the Lisbon Congress and the next year the prize was awarded to him. As Portugal's only Nobel prize winner to this date, Moniz has been honored in that country by four commemorative stamps and by several statues.

How ready the psychiatric world was to accept psychosurgery can be appreciated from the fact that less than 3 months after Moniz's monograph was published in June 1936, prefrontal leucotomy was being performed in Italy, Romania, Brazil, Cuba, and the United States. The man responsible for the first operation in the United States and who eventually promoted psychosurgery throughout the world more effectively than anyone else was Walter Jackson Freeman. Like Moniz, he also had a family tradition to live up to. Freeman was the grandson of William Keen, president of the American Medical Association, an internationally respected surgeon, author of important medical texts, and the man called upon to operate on President Grover Cleveland's cancerous jaw. Keen's reputation opened doors and provided opportunities for the young Walter Freeman, but the young Freeman did not simply take a free ride on his grandfather's coattails. Bright and ambitious, Freeman's capacity for work impressed everyone who knew him. Following postgraduate work at the Salpêtrière in Paris and at prestigious medical research centers in Rome and Vienna, his knowledge of neurology and neuropathology became truly impressive. Freeman's first position was as chief of pathology at St. Elizabeths Hospital in Washington, D.C. At the almost unprecedented age of 30, Freeman became chairman and professor of neurology at George Washington University and while still in his 30s he completed an important neuropathology text and was appointed secretary of the AMA's Section on Nervous and Mental Disorders. Freeman was the youngest member of the committee that founded the American Board of Psychiatry and Neurology and during the several decades that followed he was elected president of several regional medical societies.

Freeman was a charismatic teacher and students regarded his classes as "the best show in town." Later, many former students and residents helped him introduce psychosurgery in institutions around the country. Freeman loved the limelight and he knew how to manipulate the media. He called in science reporters for an "important story" before his lectures at professional meetings, and his work was often described in *Time,* the *Saturday Evening Post,* and other widely circulated magazines and newspapers. This publicity played a major role in promoting lobotomy.

In treating mental patients, Freeman preferred biological explanations and physical treatments. He is believed to have been the first person in Washington, D.C., to have used insulin, metrazol, and electric shock. After reading Moniz's monograph, Freeman and the neurosurgeon James Watts performed the first psychosurgery in the United States. Freeman preferred calling the operation a "lobotomy" rather than a "leucotomy" because he recognized that nerve cells bodies as well as the so-called "white matter" (*leuco* = white) axon fibers were destroyed. The initial surgery was done at George Washington University Hospital during September 1936. Two months later they reported that 6 lobotomized patients were improving remarkably. This report was given wide coverage in the national media.

Despite Freeman's enthusiasm and all the publicity, it became evident after several more months had elapsed that the outcome of lobotomy was not all that it seemed at first. Most of the "remarkably" improved patients relapsed and second and even third more radical lobotomies were subsequently performed on the same patients with the result that seizures and other neurological and psychological complications ensued. Attributing these "complications" to the shortcomings of the Moniz method, Freeman and Watts developed their own lobotomy procedure.

In 1942, Freeman and Watts summarized the experience with their new lobotomy procedure as well as the state of the field throughout the world in their book, *Psychosurgery: Intelligence, Emotions, and Social Behavior.* The book had an enormous impact, increasing interest in lobotomy among the public and physicians throughout much of the world. The *New York Times* reviewer claimed that "no novelist ever had more fascinating material" and the English neurosurgeon Eric Cunningham Dax wrote that when the book arrived in Great Britain in 1942, "It hit us like a bomb! It seemed to answer all our questions."

The book did answer many questions, describing which patients were appropriate candidates, how to perform the operation, and the postoperative care that was required. Based on the latest neuroanatomical information, Freeman, who wrote most of the book, argued that lobotomies work because they sever the nerve fibers connecting the frontal lobes to the dorsomedial nucleus of the thalamus, a region believed to regulate emotions. In the popular language Freeman used when speaking to journalists, the operation separated the "thinking brain" from the "feeling brain" and prevented thought processes from being overwhelmed by inappropriate emotional states. This rationale was much more acceptable than anything Moniz had offered. The prevailing biological theories of emotions at the time were "thalamic theories," and the dorsomedial nucleus was considered the prime candidate for an "emotional center." The theory was certainly logical and Freeman was able to cite considerable experimental evident that was consistent with his rationale for performing lobotomies. It was not long before neurosurgeons began arriving in large numbers at George Washington Hospital to learn what was to be called the "Freeman-Watts Standard Lobotomy."

During the 1940s, the popular media promoted lobotomy in article after article, calling it a medical "breakthrough" and describing how "hopeless" patients had returned to productive lives. A 1947 article in *Life* magazine was entitled "Psychosurgery: Operation to Cure Sick Minds Turns Surgeon's Blade into an Instrument of Mental Therapy." Front-page articles in small-town newspapers were often headed by such memorable hyperboles as "Wizardry of Surgery Restores Sanity to 50 Raving Maniacs." A *New York Times* editorial stated that Moniz had shown "it was time to cut worry, phobias, and delusions out of the brain," adding that "surgeons now think no more of operations on the brain than they do of removing an appendix," while the *New England Journal of Medicine* concluded that "a new psychiatry was born . . . when Moniz took his first bold steps in the field of psychosurgery."

On the other hand, there were psychiatrists and other physicians who were strongly opposed to psychosurgery from the outset. Psychiatrists were divided between "functionalists" who believed that mental illness arose from life experiences, and should be treated by psychotherapy, and "somaticists" who preferred biological explanations and physical treatments. In general, the "functionalists" opposed lobotomy, but their criticism was ineffective because they had little to offer the seriously disturbed patients deteriorating in large numbers in mental institutions across the country.

The increased interest in lobotomy led to a proliferation of new psycho-surgical procedures. By the 1940s new techniques introduced among many others included suborbital undercutting, cingulotomies, gyrectomies, topec-tomies, and amygdalectomies, each touted by a different neurosurgeon. Preferred brain targets varied within the prefrontal area, but some of the targets were located within the limbic system, a neural circuit that many were arguing was especially concerned with regulating emotionality. Some of these procedures were based on the claim that the destruction of specific targets made it possible to achieve symptom relief without impairing judgment or intellect. Encouraged by some preliminary results from animal experiments that were unfortunately often misunderstood, and by the introduction of stereotaxic surgical techniques, the new psychosurgical operations were designed to destroy circumscribed regions within the frontal lobes and other brain areas.

Several of the new psychosurgical procedures, however, were based solely on purported technical advantages to the neurosurgeon. An article pub-lished in the *American Journal of Psychiatry,* for example, described "a more perfect surgical procedure" that could be performed under conditions enabling the surgeon to see the "field" more clearly, to stop bleeding with more certainty, and to be able to weigh the brain tissue removed and examine it under the microscope. This procedure, called a prefrontal lobectomy, actually involved the removal of a much larger part of the brain then was generally destroyed during most lobotomies (Peyton et al., 1948). By the end of the 1940s, psychosurgery was so accepted as a treatment modality that it was no longer necessary to justify destroying part of the brain: it was sufficient only to describe how to destroy brain more elegantly.

Although lobotomies were always justified as a procedure of "last resort," once the personnel and facilities were in place it actually became a standard part of the psychiatric armamentarium, which was considered almost routinely in many institutions if electric shock and insulin coma treatments proved to be ineffective. More and more applications were sought, and in time lobotomies were performed on alcoholics, drug addicts, psychopaths, pedophiles, habitual criminals, the feeble-minded, and even psychoneurotics.

Despite the many abuses in practice, it would be wrong to conclude that no one appeared to improve after a lobotomy. Without doubt, many of the symptoms most disturbing to the patient as well as to the staff were less severe after the surgery. Many agitated and violent patients, who had formerly been confined to locked rooms and even restrained, could afterward be safely permitted more freedom within the institution, and a number were discharged to the care of their families. It was much easier, however, to describe the alleviation of symptoms following a lobotomy than it was to describe the intellectual impairment that may have been produced. This was true for many reasons, not the least of which was the lack of sensitive and standardized testing procedures and especially the impossibility of assessing the preoperative baseline of intellectual capability in psychotic patients.

It is not easy to summarize the outcome of all the psychosurgery performed. With so many different operations, the parts and the amounts of brain tissue destroyed varied enormously from case to case. The patients, even when they had the same diagnostic label, also differed greatly because during this period the labels were applied idiosyncratically, varying enormously between institutions. Furthermore, the evaluation of patients was pitifully inadequate. Patients were often counted as "cured" if they could be sent home, even if they still required constant supervision. Intelligence tests were commonly used even though they were known to be insensitive to frontal lobe damage. Not uncommonly, IQ was reported to be higher after a lobotomy, but this was only due to the fact that the patients, being less agitated, could be encouraged to answer a few more questions. There were no controlled studies to compare the outcome of lobotomy with a truly comparable group of unoperated patients. It was not true, as was often assumed, that without lobotomy all the patients selected would have inevitably deteriorated and died in the institutions. Many seriously ill mental patients, particularly those with affective disorders, did get better and were discharged without a lobotomy.

Although behavior often improved following a lobotomy, there was always a price. In the best cases, the patients were able to go home and even be employed, but usually at a job lower than might otherwise be expected judging from their education and experience. There were a few "show cases" of lobotomized patients who held responsible positions, but these were not representative. In the worst cases, lobotomized patients became impulsive, childish, and so distractable that they could not pursue even short-term goals. Sometimes just the opposite resulted — they became inert, practically mute, justifying the label "zombies." But to assert, as some critics do, that all lobotomized patients became intellectual and emotional "vegetables" implies that the physicians involved were all stupid, callous, or malevolent. This was clearly not the case. Such a charge is a disservice and trivializes the problem. One of the lessons to be learned from this history is that well-meaning and highly trained physicians can often be convinced of the effectiveness of a treatment by a few dramatic, but unrepresentative, instances of success. It is easy to understand how such physicians would attribute the success to their treatment even though equally plausible alternative explanations may exist.

By 1946, after having promoted prefrontal lobotomy for a decade, Walter Freeman decided that lobotomies as normally performed could not solve the massive problem of the mental patients housed in state mental hospitals around the country. Few state hospitals could afford either neurosurgeons or the extensive postoperative nursing care required following a major lobotomy. What was needed, Freeman reasoned, was a "simple" operation that did not require neurosurgeons or protracted postoperative care and for that reason might be considered at an earlier stage of mental illness. Freeman had by this time become convinced that there was a danger in waiting too long because lobotomies did not help chronic, deteriorated patients.

Freeman was aware of an easy-to-perform lobotomy procedure that had originally been introduced under the name "transorbital lobotomy" by Amarro Fiamberti, an Italian psychiatrist. The procedure had never attracted much interest, but after Freeman modified it and began to promote it, it was widely adopted in the United States. As Freeman performed the operation, patients were first made comatose with two or three electroconvulsive shocks — thereby eliminating the need for an anesthetist as well as a neurosurgeon — and then a pointed instrument resembling an ice pick was inserted over the eyeball and driven with a mallet through the bony eye socket into the frontal lobes. By moving the instrument's handle, sideways and up and down, a part of the frontal lobes could be destroyed. Freeman initially did 10 transorbital lobotomies in his office in downtown Washington, D.C., operating on private patients on an outpatient basis. Incredible as it may seem, in most instances patients could go home within an hour, usually wearing sunglasses to conceal their "black eyes." Although there were several tragic accidents, usually caused by the tearing of a cerebral blood vessel during the procedure, Freeman always insisted that the statistics indicated his procedure was much safer and more effective (because it would be undertaken at an earlier stage in the illness) than the major prefrontal lobotomies, which often produced seizures and had a mortality rate of around 5%.

Convinced that he could train psychiatrists to perform transorbital lobotomies, Freeman began to travel around the country demonstrating the procedure in as many state hospitals as possible. These trips evolved into a personal crusade, which James Watts (1972) characterized as Freeman's "magnificent obsession." On some trips he drove 15,000 miles before returning home. The trips were well planned in advance, enabling him to visit as many institutions as possible. Typically, he would arrive at midmorning and review the cases; by midafternoon he would have completed 20–30 transorbital lobotomies with young psychiatrists, often residents, observing him. Patients were lined up in an assembly line, receiving several electroconvulsive shocks just before the operation was scheduled to begin. After Freeman had completed the operations, the young psychiatrists who had observed him would descend to the morgue and practice on several cadavers saved for the occasion. The next day, with Freeman standing by, the psychiatrists would perform three transorbital lobotomies on patients. Before two full days had elapsed, Freeman was driving to the next hospital, leaving behind several psychiatrists who were now ready to do the procedure on their own. The operation itself took only 5 minutes; in one 12-day period in 1952, Freeman performed 225 transorbital lobotomies on patients assembled from the four West Virginia State Mental Hospitals.

Freeman died in 1972, convinced to the end that transorbital lobotomy had helped an enormous number of people and that it had played a major role in reducing the number of mental patients in state hospitals. He believed that without his operation most of these patients would have deteriorated beyond help. Even after the introduction of new drugs for treating mental illness, Freeman thought that transorbital lobotomy still had a place for those patients who did not respond to drugs. Toward the latter part of his life, despite repeated bouts with rectal cancer, he drove all over the country in a van tracking down former patients—to a North Dakota farm, a jail in Mobile, large cities where he arranged to have them come to his hotel, and remote areas in Appalachia. He kept track of all his former patients, through exchanges of letters and Christmas cards and by periodic visits. All of this was done at his own expense, for he never had a grant to support the follow-up studies. Freeman followed all his patients until they died and even then he made an effort to obtain a death certificate to verify the cause of death. His last manuscript, published posthumously, described a study on thousands of lobotomized patients who had been followed for more than 25 years.

To obtain some broader perspective, it is necessary to consider the growth rate and the extent of lobotomy and the institutions where these operations were performed. The number of lobotomies did not increase radically in the United States during the first years after the first operation performed by Freeman and Watts in the fall of 1936. It was several years before the demand began to build up. By that time the Second World War had diverted attention to other medical problems and many neurosurgeons were called into military service. Prior to the United States's official entry into the war in December 1941, no more than 200 lobotomies had been performed in any single year, but immediately afterward a large increase occurred. Approximately 500 lobotomies were performed in 1946 and the number doubled each successive year, leveling off at about 5000 annually in 1949. This rate was maintained for about 5 years.

Two clear influences account for the large increase in lobotomies that started in 1946. First, the transorbital lobotomy procedure was introduced in 1946, and, due to Freeman's promotion, one-third of all the lobotomies performed in the United States during the peak years and 56 percent of those done in state hospitals used this procedure. Second, and probably of greater importance, was the economic and political pressure to find a quick way of treating the many returning soldiers who were suffering from mental problems. Over 55 percent of all the Veterans Hospital beds were occupied by neuropsychiatric patients. This was not only expensive, but the families and the public conscience were demanding that something be done for these veterans. The Veterans Administration started training psychiatrists and neurosurgeons in "modern psychiatric treatment methods," namely electric shock, insulin therapy, and lobotomy. About 12 percent of all lobotomies performed during the peak period that began in 1946 was carried out in Veterans Hospitals. Of greater significance was the fact that those trained in the VA programs to perform lobotomies almost always used these skills in other hospitals as well.

It is not true, as often charged, that lobotomies were performed only on indigent patients in state hospitals. Although 56% of all lobotomies *were* performed in the state hospitals, patients from wealthy and prominent families, at expensive private sanitoriums, who had access to the "best" medical advice, were also lobotomized. In addition, the medical staff affiliated with many of the most prestigious university hospitals played leading roles in the practice of lobotomy at private sanitoria and in state mental hospitals. The often cited Columbia-Greystone Lobotomy Project, for example, was directed by a team of Columbia University neurosurgeons, psychiatrists, neurologists, neuroanatomists, and psychologists—but the patients were inmates at the New Jersey State Hospital in Greystone Park. Neurosurgeons at other major universities and leading medical research centers—Yale, Duke, Pennsylvania, Minnesota, the Mayo Clinic, and many others—all had similar arrangements with state hospitals in their sphere of influence. Following a visit and demonstration by Walter Freeman, the professor of psychiatry at the University of Cincinnati performed over 200 transorbital lobotomies at the Longview State Hospital in that city.

Although lobotomy had its critics from the outset, before 1949 there were almost no critical articles published in professional journals or in the popular media. Articles in the popular media had the effect of promoting the operations by exaggerating their success and minimizing any possible deleterious effects. Nothing in medical journals would have made physicians hesitate to refer patients who had a serious mental illness for a lobotomy; in fact, most would have been encouraged to do so. Starting in 1949, however, some criticism of lobotomy began to appear in print. As the number of lobotomies performed began to increase significantly, articles began to appear in which some serious reservations were expressed about the consequences of a procedure that seemed to make patients more manageable by exchanging "an agitated psychosis for the quiet dementia of brain damage." Because it was not possible to offer any alternative treatment, however, the criticism had relatively little influence on those regularly resorting to lobotomy. There was no substantial decrease in the amount of psychosurgery until 1954, the year chlorpromazine was introduced. By 1960, several other psychiatric drugs became available and the number of lobotomies performed had dropped precipitously to about 500 annually, only 10% of the peak rate. The continuation of lobotomy and other forms of psychosurgery used today is justified by the existence of desperate patients who do not respond to drugs.

Guarding against Therapeutic Exuberance

When I began to summarize the factors responsible for the proliferation of lobotomy, I realized that all of them still influence the practice of medicine. There are still desperate patients for whom no treatment exists and there often appear premature claims of new treatments for them that are based on uncontrolled experiments and inadequate or flawed data. The popular media still promotes these unproven treatments and often these "promising results" have been supplied by physicians and scientists, some of whom are surely motivated by a desire for "name and fame." Economic pressures in many guises still influences medical treatment. Competition between medical specialties is as fierce as ever and physicians and hospital administrators continue to discover new applications for new procedures once the staff and facilities are in place.

It is even possible to argue that some of the factors that shaped the early history of lobotomy are even more influential today than they were during the peak era of lobotomy. Communication is more rapid and new treatments, some of which may ultimately prove to be ineffective or dangerous, are more quickly spread around the world, amplifying their impact. Although science and medicine receive more coverage in the popular media today, it is not clear that the average level of reporting has improved. Jay Winston of the Harvard School of Public Health has reported that science reporters today frequently skirt the "boundary of truth" by exaggerating results and implications in order to prove to their editors that they have something "groundbreaking" and of "vital importance" (Winston, 1985).

The better-known science reporters receive over 500 letters a week, the majority of them either from individuals describing their own "breakthroughs" or from the public relations agencies hired by them or by the hospitals, universities, and research institutions where they are employed. Often the most preliminary results are widely circulated—as happened recently following a report of a new treatment for Alzheimer's disease. The treatment involved implanting a "chemical pump" for delivering to the brain a drug similar to the neurotransmitter acetylcholine. The report was really not much more than a feasibility study of the "pump." The only data available were subjective impressions obtained from the relatives of the few patients involved. Nevertheless, a press conference was arranged at the New England medical center where the work was done. As a result, the story was widely covered in newspapers and on the NBC "Nightly News" and the "Today" show, the "CBS Morning News," ABC's "Good Morning America," PBS's "MacNeil-Lehrer Report," and the Cable News Network. *Newsweek, McCall's, Family Circle, People, Forbes,* and other magazines also described the new treatment. The *Boston Globe* ran the story under the heading, "Researchers Describe Possible Alzheimer's Cure."

An examination of the phenomenal growth in the number of coronary artery bypass operations performed over the last 2 decades may be useful to consider in the present context. The procedure and its justification is much more straightforward for coronary artery bypass operations than it is likely to be for any treatment that may be used for neurological and mental disorders. A patient's saphenous vein (or a synthetic substitute) is used to bypass one or more occluded arteries that are not capable of supplying sufficient amounts of blood to the heart. The lack of oxygen to the heart may cause angina pectoris, severe "ischemic" pain, a reduced capacity to pump sufficient blood to the rest of the body, and permanent heart damage. The importance of increasing blood to deprived heart muscles is easy to understand and difficult to argue against. There may be questions raised, however, about the relative effectiveness (compared to other treatments) and safety of a particular procedure used to increase blood to the heart and about its appropriateness for different heart conditions.

The first coronary artery bypass operation was performed in 1964. Soon after the number rapidly soared: there were 1000 bypass operations performed in 1968, 2000 in 1970, 17,000 in 1972, 60,000 in 1975, 180,000 in 1982, 250,000 in 1988, and an estimated 340,000 in 1989. It is now the most common elective surgery in the United States. At a cost of $15,000–25,000 per operation, the economic consequences are enormous. The total cost of these operations (physician and hospital fees) is currently estimated to be between $5 and 10 billion annually and the economic consequences are much higher when the number of companies supplying the special supplies and equipment needed is factored into the equation. In those clinics that have become "coronary artery bypass centers," it is the major source of income.

From the outset, the enthusiasm for the coronary artery bypass operation was almost boundless, with the popular press playing the role of cheerleader. *Life* reported in 1971 that the operation "has saved two thousand lives" and "the odds of dying during the bypass surgery are most likely no higher than the chance of having yet another heart attack within the year." The same year *Newsweek* described it as "a major breakthrough in the conquest of the nation's No. 1 killer," and when *Time* estimated that nearly 250,000 Americans a year might eventually have a bypass operation, most people thought this was an exaggeration. At the time these statements were being made there was no convincing evidence that the operations were actually saving many lives. Nor was there much information concerning how long the new vessels would remain effective or the risks involved. Certainly, during the early stages of "bypass surgery" when the operation was being promoted unreservedly in the media and by some surgeons, the risks were not insignificant and the evidence that it prolonged lives was not available. Initially, the operative mortality was as high as 25%, but this was soon reduced to about 10% and at present it may be between 1 and 5%, depending on the patient population and the skill and experience of the surgeon. Moreover, the bypass vessels do not always remain patent and a significant number of patients must undergo a second operation. Patients undergoing coronary artery bypass operations may experience serious side effects, both psychologically and physically. According to some studies, approximately one-third of the patients are troubled postoperatively by depression and many patients who counted on the operation making it possible for them to return to work are disappointed (Halperin and Levine, 1985).

The coronary artery bypass operation was first performed on patients suffering from the severe "ischemic pain" that often occurs in cases of angina pectoris. There is no doubt that many of these patients experienced dramatic relief from their pain. However, today bypass operations are being performed almost prophylactically in a number of hospitals as a means of prolonging the life of patients whose symptoms are not crippling. This was done even though during much of the period of rapid growth of bypass surgery there were no reliable statistics on the life expectancy or future quality of life of patients undergoing the surgery or the more conservative alternative of medical (mostly drug) treatment (McIntosh and Garcia, 1978). Only during the last few years have some well-controlled studies been completed, and while several of these do report statistically significant increases in life expectancy, the results to date apply mainly to a subgroup of patients (those with isolated left main coronary artery conclusion), who constitute less than 10% of the total undergoing the bypass surgery.

Throughout most of the period of the phenomenal growth in popularity of the bypass operation, there were no controlled studies to back up the claims that it increased longevity and the quality of life. Few surgeons have the time to follow up patients once they are medically stable, and hospitals, which are structured for acute care, have rarely followed patients for any length of time. The enthusiasm for the bypass operation is said to be based solely on the desire to help suffering, desperate patients, but as two cardiologists commented, "as the numbers of procedures increase and performance of the procedure becomes more mechanized, it becomes difficult not to wonder if there are not other goals such as self-aggrandizement, keeping the beds full, or supporting further personal or institutional expansion" (McIntosh and Garcia, 1978).

The issue is not whether the coronary artery bypass procedure ultimately proves to be as effective as its strongest supporters claim. We should all hope it does. Rather, the issue derives from the fact that the growth in the number of operations performed occurred before there was any reliable evidence available to back those claims and the possibility that the phenomenal growth was driven by many factors besides patient welfare.

Over 150 years ago, Samuel Taylor Coleridge observed that history is "but a lantern on the stern" shining only on the waves behind us. Unquestionably, it is far easier to recognize past mistakes than to avoid them in the future. Moreover, it will probably prove impossible to formulate any universal ethical guidelines that could be applied in all cases where decisions have to be made about the advisability of undertaking medical risks. It is commonly argued, for example, that lobotomy should have been stopped on the grounds that it involved the irreversible destruction of healthy tissue, or at least tissue not demonstrated to be pathological. However, this is not as persuasive an argument as it may seem. Certainly, destroying healthy tissue should not be undertaken lightly, but it *can* be justified if there is convincing evidence that the progression of a serious disease is slowed by such an intervention. What is most important to the patient is not the irreversible loss of healthy tissue but whether there are irreversible losses in capacities that overbalance any benefits that are likely to accrue.

Psychosurgery has also been denounced because in tampering with the brain it can change personality and interfere with personal freedom. Here, too, the arguments are not very compelling on closer inspection. There are many noncontroversial operations on the brain that do not produce discernible changes in personality. The changes produced by brain damage depend on many factors such as the location and extent of the damage, the age of the patient, and so forth. It certainly was not the case that psychosurgery altered personality in the sense of manipulating basic values or political ideology, as has been implied by some critics. In any case, criticisms of psychosurgery because it may alter personality are specious. All psychiatric treatment, including psychotherapy, attempts to alter personality in the sense of reducing fears, anxiety, and other emotional states that interfere with the quality of life and the ability to perform adequately. How else can a patient improve? Nor is the argument that psychosurgery reduces personal freedom any more convincing, for it can be argued that these operations actually increase freedom by reducing those emotional problems that interfere with a patient's ability to achieve desired goals.

We hear so often today about cost-benefit analyses that we might assume this to be the obvious way of evaluating a proposed treatment. Unfortunately, it is often difficult to apply such an analysis in practice. In the first place, the costs and benefits have usually not been adequately established at the time it would be most desirable to make the decision. Moreover, the costs and benefits are impossible to quantify. How much loss of spontaneity following psychosurgery should be considered to be equal to how much relief of anxiety?

Nor is it possible to base the justifiability of a proposed treatment on the adequacy of the explanation of how it is supposed to work. Biological explanations are often controversial and they are constantly subject to change. Even the use of aspirin could not have been approved if such a criterion had to be met. In retrospect, the rationale used to justify past treatments (including equally those that later proved useful, ineffective, or harmful) were often so weak, flawed, lacking in supporting evidence, or even inherently untestable that we can only wonder how they could have been accepted. In contrast, present arguments always seem logical and supported by a body of convergent scientific evidence. Judging from history, however, there are good reasons for doubting whether the evidence and the arguments today are as qualitatively superior as usually assumed. Unless this fact is fully appreciated, we are likely to repeat the mistakes of the past.

It should not be thought that the history of psychosurgery only reflects the inevitable, if regrettable, trial-and-error course that every innovative therapy must undergo during its refinement. Much of the harm done to patients by lobotomy would not have been done if the whole process had been slowed down in order to establish the reliability and validity of the claims made for it. In 1939, when Oskar Diethelm, professor of psychiatry at Cornell University, reviewed the history of somatic treatments in psychiatry, he found the same pattern repeated over and over again. Unjustified claims of success were used to promote the adoption of innovative treatments by other physicians. Only too late was it realized that the benefits had been overestimated and the risks underestimated. Diethelm concluded: "The pioneer has his place in medical research. . . . On the other hand it is important in medicine to recognize fully the responsibility with regard to those who follow blindly, that is lay people; and to those who are forced to follow, that is patients" (Diethelm, 1939). While there have been instances of fraudulent reporting of results, this is a relatively infrequent event. In the great majority of instances where success is overestimated and failure underestimated, where there are biases in the selection of patients, and where other numerous influences introduce distortions, the innovator is genuinely convinced of the conclusions. Self-deception is always difficult to guard against and especially so when the desire to succeed is strong.

My own view is that we will never be adequately protected by establishing some abstract code of ethical behavior. While discussion of ethical codes may help to sensitize people to those areas of questionable behavior that deserve more thought, it is rare that codes of conduct apply unambiguously to the specific cases under dispute. Physicians who can be shown in retrospect to have caused great harm can always claim to have been motivated by the noblest of goals and the highest ethical principles. Where patients are desperately ill and no available treatment exists, it is not possible to know with any certainty what risks should be taken, and physicians who are equally ethical and knowledgeable may disagree.

It is unrealistic to believe that physicians will adequately regulate themselves. It has been argued that the major restraint in medicine today is derived from fear of malpractice suits. But such fear is not a rational way of controlling progress in medicine. What is needed is some way of regulating limited trials of innovative therapies to assure that they will be performed under conditions that increase the probability of producing reliable conclusions. Too often in the past, physicians have convinced themselves that an ineffective treatment has great value or that treatment truly beneficial to a small subgroup of patients can help a much larger group.

What is needed is some way of preventing uncritical enthusiasm and unbridled ambition from running rampant. Each field of medicine has its unique problems. The Food and Drug Administration has established steps for approving new drugs. Animal experiments to estimate the dangers and, where possible, to provide evidence of effectiveness, are followed by limited

204 Elliot S. Valenstein

clinical trials with an approved experimental procedure. If feasible and ethical, a randomized control group of patients is used. The system is certainly not perfect and it has been criticized both for preventing useful drugs from becoming available to desperate patients and for not adequately protecting patients from potentially dangerous drugs. Where innovative surgery is involved, evaluation is more complex because the question of skill can make it more difficult to obtain replicable results, but this is not an insurmountable problem.

Without doubt, restraints will be resisted by those on whom they are imposed. We will hear arguments that progress will be hampered and desperate patients deprived of help. My own view is that, in most instances, the undesirable effects of reasonable regulation and restraint have been exaggerated, certainly so when compared to the cost of the unbridled use of dangerous treatments. Ultimately, some effective regulation will have to be instituted. It is time to start thinking about what form it should take.

References

Diethelm O (1939): An historical view of somatic treatment in psychiatry. *Am J Psychiatry* 95:1165–1179

Halperin JL, Levine R (1985): *Bypass.* New York: Random House/Times Books

McIntosh HD, Garcia JA (1978): The first decade of aortocoronary bypass grafting, 1967–1977: A review. *Circulation* 57:405–431

Peyton WT, Noran HH, Miller, EW (1948): Prefrontal lobectomy (excision of the anterior areas of the cerebrum). *Am J Psychiatry* 104:513–523

Valenstein ES (1986): *Great and Desperate Cures: The Rise and Decline of Psychosurgery and Other Radical Treatments of Mental Illness.* New York: Basic Books

Watts JW (1972): Walter Freeman, M.D., 1895–1972. *Med Ann District Columbia* 41:553–554

Winston J (1985): Science and the media: The boundaries of truth. *Health Aff* 4:5–23

Woltman H (1941): *Proceedings of the Staff Meetings of the Mayo Clinic* 16 (26 March):200

12

Brain Research, Animal Awareness, and Human Sensibility: Scientific and Social Dislocations

JOHN DURANT

> if we choose to let conjecture run wild then <our> animals our fellow brethren in pain, disease death & suffering <<& famine>>; our slaves in the most laborious work, our companion in amusements. they may partake, from our origin in one common ancestor we may be all netted together.
>
> —Charles Darwin (1838)

Animal Liberation

In 1985, I attended the International Congress for the History of Science at the University of California at Berkeley. One of the particular pleasures of that meeting for me was the opportunity it provided to visit Frank Oppenheimer's famous Exploratorium in San Francisco. I was impressed by the quality of the interactive exhibits, and in particular by a "hands-on" neurophysiology experiment in which visitors were able to elicit electrical activity in a living nerve-muscle preparation. Having been trained as a biologist, I knew how difficult it was to make such an experiment work for undergraduates in a well-provided laboratory; to make it work for the general public in the relatively austere setting of the Exploratorium seemed to me an extraordinary achievement.

The Exploratorium has been enormously influential around the world. In Britain alone there are around 20 science centers scattered across the country, and many of them contain exhibits based on Oppenheimer's ideas. So far as I am aware, however, the Exploratorium's neurophysiology experiment is nowhere on display in Britain; and I doubt very much whether it will be tried in the near future. This is not because of the technical difficulty of the experiment, but rather because of the public sensitivity to the issue of animal experimentation. Any science center that gave the British public the opportunity to perform experiments on living animals would risk incurring the wrath of the animal liberation movement, and such wrath could literally close the institution in a matter of days.

Concern for animal welfare has a long history. In Britain, this history includes 19th-century developments such as the banning of cruel sports, the establishment of animal welfare agencies, and the passage of legislation controlling the use of animals in scientific research (Rupke, 1987). In the late Victorian and Edwardian periods, antivivisectionism and vegetarianism were important rallying-points for those who wished to defend other animals from human depredations. In recent years, however, concern for animals has taken a new and altogether more radical form. Increasingly, protestors have abandoned the paternalistic language of humane treatment and welfare in favor of the egalitarian language of political justice. In the 1970s, the notion of political rights was extended to embrace nonhuman animals; and to racism and sexism was added another form of human prejudice, "speciesism": the tendency to discriminate against other sentient beings on the grounds that they are not human.

The moral philosopher Peter Singer is one of the most influential writers in the animal liberation movement. Singer's (1974) position is clear in the following extract from an early position paper:

I am urging that we extend to other species the basic principle of equality that most of us recognize should be extended to all members of our own species.

If a being suffers, there can be no moral justification for refusing to take that suffering into consideration. No matter what the nature of the being, the principle of equality requires that its suffering be counted equally with the like suffering—in so far as rough comparisons can be made—of any other being. If a being is not capable of suffering, or of experiencing joy or happiness, there is nothing to be taken into account. This is why the limit of sentience . . . is the only defensible boundary of concern for the interests of others.

This is essentially the same argument that was used by the utilitarian moral philosopher Jeremy Bentham in the early 19th century; but restated powerfully by Singer in his book *Animal Liberation* (1976) it has become the focus of a powerful political movement. In Britain, animal liberationists have opposed a wide range of agricultural and industrial practices, from "factory farming" to the routine use of animals in the testing of cosmetics, and they have given particular prominence to the use of animals in scientific research. Over the past decade, scientists and scientific laboratories have been the targets of considerable and occasionally violent protest, to the point where individual researchers have been obliged to seek police protection and many institutions have been transformed into high security facilities.

A Shift in Public Opinion?

The animal liberation movement is the tip of a much larger political iceberg of public opinion. In recent years, the general public has turned against many traditional ways of using other animals. Big game hunting, whaling, the ivory trade, the farming of mink and other animals for fur, and the slaughter of seal pups have all been widely condemned. In part, public opposition to such practices has been rooted in concern for wildlife conservation; but in part, also, it has been rooted in aesthetic and moral repugnance at the apparently callous exploitation of other animals.

Evidence of a transformation in public attitudes toward other animals is all about us: in the obvious ambivalence felt by many toward the confinement and public exhibition of animals in circuses and zoos; in the success of "nonexploitative" cosmetics companies such as The Body Shop, which sells only products that have not been tested on other animals; in the decline of live animal work in school and university science courses; and in the steady increase in the popularity of vegetarianism. In the summer of 1988, random samples of adults in Britain and America were asked whether they agreed or disagreed with the following proposition: "Scientists should be allowed to do research that causes pain and injury to animals like dogs and chimpanzees if it produces new information about human health problems." In Britain 40% agreed and 60% disagreed with this proposition, while in the United States 56% agreed and 44% disagreed (Evans and Durant, 1989). Clearly, worries about the use of the higher mammals in medical research are greater in Britain than in the United States; but in both countries, a substantial fraction of the public is prepared to say no to such practices. It is sobering to contemplate what respondents might have said had they been asked about the acceptability of using animals like dogs and chimpanzees in nonmedical (i.e., "pure") research.

Two Voices of Science

The scientific community has great difficulty in dealing with the question of the moral status of other animals. In part, this is because science is generally ill-equipped to deal with moral issues of any sort; but in part, also, it is because science addresses human sensibilities concerning other animals in two contrasting and potentially conflicting voices. First, there is voice of the brain sciences (neuroanatomy, neurophysiology, neuropharmacology, etc.), and, second, there is the voice of the evolutionary sciences (evolutionary theory, ethology, sociobiology, etc.). Both of these areas of modern science have things to say about animal nature and human nature, but what each has to say is utterly different.

For all their modern sophistication, the brain sciences are rooted firmly in the Cartesian tradition. Within this tradition animals are seen as complex machines, and all specifically mental phenomena (including sentience) are downplayed or else completely ignored. Typically, the brain sciences employ a rigidly mechanistic vocabulary to describe nervous structures and functions. Sense organs are "stimulated," sensory nerves "send signals" into the central nervous system, spinal and/or brain centers are "stimulated" or "inhibited," and in due course motor nerves "issue commands" to muscles; but nowhere in all of these reflex operations do conscious agents actually feel, think, or decide anything. Technically, at least, the brain sciences are silent on these matters; but since some of their methods involve the "sacrifice" of animals in ways which, if the animals were aware, might be presumed to cause them a certain amount of suffering, such silence is easily mistaken for passive consent to the proposition that nonhuman animals are either nonsentient or else radically less sentient than human beings.

It is right to observe in passing that of course cognitive neuroscientists are now investigating the physiological basis of such things as knowledge acquisition, retention, and recall, and some of them are even beginning to take an interest in the physical basis of consciousness, but my point is that neuroscientific theory is singularly ill-equipped for these tasks, at least in so far as they concern the "inner life" of subjective experience. The reason for this is extremely simple: neuroscientific theory contains no terms by which even to refer meaningfully to subjective experience, let alone to explain it. In short, the unsolved mind-body problem stands as a barrier in the path of any substantial engagement of brain science with animals as sentient beings.

Having said this, it is important to note that the regulations by which brain scientists actually conduct their research are not couched in terms of what I have termed the Cartesian tradition. In Britain, for example, the 1986 Animals (Scientific Procedures) Act is based on the assumption that vertebrate animals are capable of experiencing a variety of feelings, including pain and distress. Scientists who wish to apply for a Project Licence to undertake research involving these animals must describe such pain or suffering as the proposed research may cause, together with the steps they intend to take to prevent or minimize it. Also, they must justify the proposed research in terms of the anticipated contributions that it will make to the advance of knowledge and the improvement of clinical practice. In dealing with other animals as sentient beings, however, these and similar regulations in other countries are drawing on a very different tradition of thought about other animals. For reasons that will become obvious, I shall refer to this very different way of dealing with other animals as the Darwinian tradition.

The Darwinian tradition stands in sharp contrast to the Cartesian on the question of the place of mind in nature. Within the Darwinian tradition, animal nature and human nature are seen as being fundamentally alike; that is, other animals and humans are presumed to share not only similar anatomies and physiologies but also similar psychologies. Of course, this way of viewing other animals long predates Darwin; indeed, it long predates René Descartes. For as long as people have looked after other animals, there has been a tendency to view them as sentient and even intelligent beings; and at a more formal level, ideas of "animal soul" may be traced back to classical times (Young, 1967). Nevertheless, within the world of science the mechanistic approach to animal nature has been very much in the ascendant since the 17th century, and it is principally to Charles Darwin and his followers that I believe we owe the renaissance of an alternative scientific approach.

Darwin himself presumed that other animals have mental lives composed of sensations, feelings, and even thoughts essentially like our own; and it was on this basis that he approached the question of the evolution of mental capabilities. His descriptions of behavior are based upon noninvasive observation of animals in natural or seminatural conditions, and they generally ignore the mechanistic strictures of the Cartesian tradition in favor of an altogether richer and (implicitly or explicitly) mentalistic explanatory vocabulary of feelings and moods, thoughts and desires, aims and intentions. Since at least 1838, Darwin's evolutionary vision that, as he put it, "we may be all netted together" has constituted a radical scientific alternative to the mechanistic view that other animals are so much flesh and blood in complicated, but nonetheless fundamentally Cartesian, motion (Darwin, 1838).

To some readers, it may seem inappropriate to oppose the Cartesian and the Darwinian traditions in this way. Surely, they will say, the two are fundamentally compatible with one another? Indeed, may not the Darwinian revolution of the 19th century be seen as the fulfillment in biology of the Cartesian program of the 17th century, according to which the study of the whole of nature was to be conducted in entirely nontheological terms? Did not Darwin extend to the living world Descartes's vision of nature as a law-bound system of matter in motion? This, for example, is the view of the historian John Greene (1981), who claims to find continuity of tradition where I claim to find discontinuity and even potential conflict.

In fact, these apparently divergent historical interpretations are not fully incompatible. Certainly, Darwin fulfilled one aspect of the Cartesian program by bringing organic adaptation and diversity under the aegis of natural law. The theory of evolution by natural selection provided a natural law governing the origin of species, and in this sense it advanced the cause of the mechanical philosophy. However, Darwin was interested in the origin of mental as well as physical characteristics, and here his approach was thoroughly un-Cartesian. Where Descartes had seen a fundamental divide between animal nature (the product of matter alone, hence automatic) and human nature (the product of the ineffable union of matter and mind, hence rational), Darwin saw a fundamental continuity of evolving mental life. Darwin detected remorse in apes, religious devotion in dogs, and loyal self-sacrifice in bees. He had no compunction about describing animal behavior in human terms, and this for the very simple reason that he believed that what went on in the heads of other animals was fundamentally like what went on in his own and other people's heads. Consistently, Darwin used human experience as a yardstick against which to measure the mental attributes of other animals; everywhere, he saw "our fellow brethren in pain, disease, death & suffering" (for a fuller discussion of Darwin's work in this area, see Durant, 1985).

Windows into the Mind

Darwin's method of exploring the relationship between animals and humans was fundamentally un-Cartesian, and so, too, have been the methods employed by many of those who have followed in his footsteps. The history of ethology—the biological study of behavior—is the history of the controlled use of anthropomorphism as a means of obtaining scientific insights into the nature and significance of animal behavior. Time and again, ethologists have uttered ritual denunciations of anthropomorphism, and time and again, they have promptly ignored their own strictures by providing richly anthropomorphic descriptions of animal behavior. These descriptions have been full of scientific utility precisely because they have also been full of constructive analogies and metaphors drawn from the world of human affairs (Durant, 1981).

Consider one very familiar example. Jane Goodall is a pioneer of the close study of the behavior of apes in the wild. For the past 30 years, she has been studying the chimpanzees in the Gombe National Park in Tanzania. During this time, Goodall has discovered a huge amount about chimpanzee behavior; and she has been followed by many others, including a number of other eminent women primatologists (e.g., Sarah Hrdy, Alison Jolly, and the late Dian Fossey). Thanks to the work of Goodall and her colleagues, we now have reasonably full accounts of the behavior of some of our closest living relatives.

How did Goodall set about her work? The answer is provided in her recent autobiography, *Through a Window: Thirty Years with the Chimpanzees of Gombe* (1990). The second chapter of this book, entitled "The Mind of The Chimpanzee," opens with the following words: "Often I have gazed into a chimpanzee's eyes and wondered what was going on behind them. . . . As long as one looks with gentleness, without arrogance, a chimpanzee will understand, and may even return the look. And then—or such is my fantasy—it is as though the eyes are windows into the mind." "Gentleness"? "arrogance"? "a chimpanzee will understand"? "windows into the mind"? What possible place is there in modern science for grossly anthropomorphic terms such as these? It seems that Goodall herself is rather doubtful about the legitimacy of her speculation, for she dismisses it in passing as "my fantasy," but here she entirely fails to do herself justice. For as her chapter title indicates, Goodall has a serious purpose in beginning as she does; and as the chapter unfolds, this purpose becomes clear. She describes the difficulty she faced in the early years in presenting her work at scientific meetings:

How naive I was. As I had not had an undergraduate science education, I didn't realize that animals were not supposed to have personalities, or to think, or to feel emotions or pain. . . . I didn't realize that it was not scientific to discuss behaviour in terms of motivation or purpose. And no one had told me that terms such as childhood and adolescence were uniquely human phases of the life cycle, culturally determined, not to be used when referring to young chimpanzees.

This may seem like a confession of sin, but we should not be deceived. As Goodall continues, it becomes clear on which side of the argument she still stands. After having won her first battle with a journal editor, who had demanded that impersonal be substituted for personal pronouns in all references to the subjects of her study, Goodall was advised by her doctoral supervisor, the Cambridge ethologist Robert Hinde, on methods of circumventing the conventions of scientific discourse: " 'You can't know that Fifi was jealous,' he admonished me on one occasion. We argued a little. And then: 'Why don't you just say *If Fifi were a human child we would say she was jealous.*' I did." This, of course, was no more than a political convenience. Goodall (1990) states her own view of the matter quite explicitly:

If we ascribe human emotions to non-human animals we are accused of being anthropomorphic—a cardinal sin in ethology. But is it so terrible? If we . . . accept that there are dramatic similarities in chimpanzee and human brain and nervous system, is it not logical to assume that there will be similarities also in at least the more basic feelings, emotions, moods of the two species?

The Popularization of the Darwinian Approach

It is not the purpose of this chapter to defend Goodall in particular, or the Darwinian tradition in general. For honesty's sake, however, perhaps I should confess that from the point of view of scientific methodology my own sympathies in this matter are with the Darwinians. Hinde advised Goodall to report her findings by stating how she would have described a particular scene if the actor had been a human child rather than a chimpanzee. Faced with the strict conventions of postwar behavioral biology, this may have been good politics, but it was bad philosophy and bad science. As Goodall herself points out, none of us has direct access to the emotions of other human beings. We know directly only our own feelings, and we infer or intuit the feelings of others chiefly by observing what they do. In this respect, the evidence for jealousy in a chimpanzee may be every bit as good as the evidence for jealousy in a (preverbal) human child. If, therefore, we are permitted to name the emotion in the one case, on what grounds shall we be barred from doing so in the other? It is only a half-baked Cartesian bias against the mental that persuades some scientists to impose the mindless and utterly unilluminating jargon of behaviorism upon what every field-worker with experience of chimpanzee behavior knows perfectly well: that these are beings very like us, with complex mental lives of their own.

Thankfully, there are signs of a growing acceptance of animal mind within the behavioral sciences. In this sense, things that Darwin and his followers have been asserting for more than a century are at last becoming scientifically respectable. For present purposes, however, this is not as relevant as the fact that the Darwinian approach has had a huge influence on the general public. There are at least two obvious reasons for this influence. First, as I have already pointed out, the Darwinian tradition operates with what may be termed a "commonsense" view of animal nature that is very close to the view of most pet owners and animal handlers or trainers. Such people routinely talk about their animals in ways that are consistent with the Darwinian tradition but that are utterly inconsistent with the Cartesian. This, I presume, is what underlies the animal trainer Vicki Hearne's observation that, "in the trainers' world different kinds of animals exist than the ones that I heard and read about in the university" (Hearne, 1987).

The second reason for the public influence of the Darwinian approach is the extraordinary popularity of natural history, especially among highly industrialized and largely urban communities. In Britain, there is a mass market for natural history books, films, and radio and TV program. David Attenborough, for example, has achieved huge audiences around the world for each of his major BBC TV series: "Life on Earth," "The Living Planet," and, most recently, "Trials of Life." Though not academically very ambitious, each of these TV series has brought a fundamentally Darwinian perspective into the homes of millions of people. Significantly, "Trials of Life" is about animal behavior. Stunning wildlife photography is accompanied by richly anthropomorphic descriptions that portray many of the more advanced mammals as complex mental and social beings.

It is hard to judge accurately the impact of this sort of popular exposure upon public perceptions of other animals. From the outset, Jane Goodall's work has had a very high public profile. Her chimpanzees have been portrayed as the individual personalities Goodall believes them to be in the pages of *National Geographic,* as well as in numerous films and TV documentaries. Similarly, the work of the late Dian Fossey has been presented in several popular books, and it has recently been the subject of a full-length feature film (e.g., Mowat, 1987). The public is fascinated by our closest living relatives, and by the rich insights into their behavior that have been provided by primatologists working firmly within the Darwinian tradition. As the English philosopher Mary Midgley (1983) has observed:

For the first time in civilized history, people who were interested in animals because they wanted to understand them, rather than just to eat or yoke or shoot or stuff them, have been able to advance that understanding by scientific means and to convey some of it to the inquisitive public. Animals have to some extent come off the page. With the bizarre assistance of TV, Darwin is at last getting through. Town-dwellers are beginning to notice the biosphere.

Significantly, Midgley makes this statement in her own contribution to the literature of animal liberation, entitled *Other Animals and Why They Matter*. I say significantly, because it seems to me that we have arrived at the point where the public impact of the Darwinian tradition intersects with the issue of animal awareness and animal suffering. The evolutionary vision of animal nature and human nature emphasizes the complexity and the richness of the mental lives of other animals, as well as the multiple continuities between them and us. This vision has been very widely popularized over the past 20 years, and at the same time there has been a growth of public concern about the exploitation of other animals for food, work, amusement, or scientific research. Is it too much to suppose that the one process has helped to cause the other? The following passage from Goodall's (1990) autobiography suggests that it is not:

The notion of an evolutionary continuity in physical structure from pre-human ape to modern man has long been morally acceptable to most scientists. That the same might hold good for mind was generally considered an absurd hypothesis— particularly by those who used, and often misused, animals in their laboratories. It is, after all, convenient to believe that the creature you are using, while it may react in disturbingly human-like ways, is, in fact, merely a mindless and, above all, unfeeling, "dumb" animal.

Here, then, we appear to have a paradoxical situation in which one kind of science (the Darwinian) may be contributing indirectly to the moral critique of another (the Cartesian). This point may be made with the help of a purely rhetorical question: how many field primatologists, I wonder, are entirely at ease in a primate laboratory?

Conclusion

I believe we may be witnessing a fundamental transformation in public attitudes regarding the treatment of other animals. In part, this transformation must be credited to the process of scientific discovery. For the evolutionary view of the animal world has been brought to mass audiences in the pastwar period, and by extending Darwin's vision of the interconnectedness of living things, and especially of the bonds between humans and other animals, this process of popularization may have helped to expand the circle of people's moral sympathy. Such expansion is, of course, a central element in the case for animal liberation.

All of this causes difficulties for scientists who continue to advance humankind's scientific knowledge and medical wisdom by means of experimental work on other animals. I have suggested that there is a growing sense of moral unease about this area of science today. This moral unease should not be dismissed by experimental physiologists as merely anti-intellectual and antiscientific. As I have tried to show, part of the concern may in fact be coming from within the scientific community itself. Unfortunately, animal liberation is an extremely charged issue. For some years, extremist elements within the animal liberation movement have been using violence or the threat of violence to intimidate scientists and their families. In response, some scientists have avoided publicity, while others have acted with considerable personal bravery by defending themselves and their colleagues against the charge that they are acting immorally. All forms of intimidation are, of course, to be roundly condemned; and at the same time, open dialogue between people of good will is to be greatly encouraged. I believe there should be more discussion between brain scientists and ethologists, with a view to clarifying questions concerning the mental attributes and moral status of other animals. Without such clarification, I see no prospect of our being able to resolve the scientific and social dislocations that lie at the heart of the current debate about animal suffering.

Acknowledgments

This chapter deals with issues that are exceptionally difficult and emotive. It has benefited in revision from the comments and criticisms of a number of colleagues, including Colin Blakemore, Richard Dawkins, Jane Gregory, Anne Harrington, and Michael Lockwood. I am grateful to these colleagues for taking the time and trouble to review my work. Needless to say, however, the views expressed in the article remain mine alone.

References

Darwin C (1838): "B" Notebook. Reprinted in: *Charles Darwin's Notebooks 1836–1844.* Ithaca, N.Y.: Cornell University Press and British Museum of Natural History, 1987, pp. 228–229

Durant J (1985): The ascent of nature in Darwin's *Descent of Man.* In: *The Darwinian Heritage,* Kohn D, ed. Princeton, N.J.: Princeton University Press, pp. 283–306

Durant J (1981): Innate character in animals and man: A perspective on the origins of ethology. In: *Biology, Medicine, and Society 1840–1940,* Webster C, ed. Cambridge: Cambridge University Press, pp. 157–192

Evans G, Durant J (1989): Understanding of science in Britain and the USA. In: *British Social Attitudes: Special International Report,* Jowell R, Witherspoon S, Brook L eds. London: Gower Press, pp. 105–119

Goodall J (1990): *Through a Window: Thirty Years with the Chimpanzees of Gombe.* London: Weidenfeld and Nicholson

Greene JC (1981): *Science, Ideology, and World View: Essays in the History of Evolutionary Ideas.* Berkeley and Los Angeles: University of California Press

Hearne V (1987): *Adam's Task: Calling the Animals by Name.* New York: Vintage Books

Midgley M (1983): *Other Animals and Why They Matter.* Harmondsworth, England: Penguin
Mowat F (1987): *Woman in the Mists.* New York: Warner Books
Rupke N (1987): *Vivisection in Historical Perspective.* London: Croom Helm
Singer P (1974): All animals are equal. Reprinted in: *Political Theory and Animal Rights,* Clarke P, Linzey A, eds. London: Pluto Press, pp. 162–167
Singer P (1976): *Animal Liberation.* London: Jonathan Cape
Young RM (1967): Animal soul. In *The Encyclopaedia of Philosophy,* Edwards P, ed. New York: Macmillan, 1:122–127

PART 4

Sociohistorical Perspectives
on Values and Knowledge
in the Brain Sciences

13

Securing a Brain:
The Contested Meanings of Kuru

WARWICK ANDERSON

I want to consider the practices of a knowledge community in order to understand how values are reproduced or nuanced through interaction. Values are the shared cultural standards that integrate as well as guide and channel the organized activities of the members of a subgroup. But what happens when the shared cultural standards of neuroscientists are in competition with those of another knowledge community?

My concern here is with a struggle for authority over the meaning of a phenomenon that anthropologists and medical scientists, each with a distinct grammar of practice, claimed to be competent to interpret.[1] What were the characteristic resistances of each practice, and which segments were most easily traded off, or symbolically appropriated? It is on the boundary between a conventional context and another, alien context, that one sees most clearly how each knowledge community struggles to attract phenomena into its own orbit and reinflect them with its own values. The situation is not uncommon. We operate in these "trading zones" of practices and values every day, whether we are trying to reconcile ethology with our knowledge of brain development, or psychophysics with neurophysiological recordings, or ethnography with reductionist models of behavior—or even just trying to explain the neurosciences to a lay audience.[2] But usually we are so confident in our own moral economy of competence that we do not even notice how our practices are nuanced to preserve or to gain a cultural authority in these interactions.

Entering the Kuru Region

Let me give you a picturesque, though ultimately rather grim, example of one of these trading zones. The story I want to tell may seem a little too distant from the conventional concerns of an audience of laboratory neuroscientists, but it has in its favor a contemporary feel to it, and definite heuristic potential. My story tells of the resistances and trade-offs in the interaction between neuropathology and anthropology that occurred in New Guinea during the 1950s.[3]

So imagine, for a moment, that you are in the Highlands of New Guinea, 3 days walk beyond the Okapa patrol post, in the Fore region. During a wet season in the 1950s, you have entered one of the more remote parts of the country, recently opened and still rarely patroled. You have come into an atmosphere charged with suspicion and distrust. Cannibalism is practiced. Spirit possession is regularly assumed. The tribes live in stockaded villages where the men always carry spears, unsure of even their family's loyalty, never knowing from which direction the next barrage of sorcery "poison" might come. Uncertain if they, too, will tomorrow become possessed by the spirit that causes the shaking and shivering they call "guria" or "kuru."

What might one make of this? Ronald and Catherine Berndt, anthropologists who entered the region in 1952, described the "spirit possession" in the following terms: "There were involuntary twitchings, a feeling of 'abnormal' coldness, dilatation of the eyes . . . and lack of control over the limbs. . . . A number of examples imply that death took place when maggots appeared on the surface of the body: death, it is said, 'always' takes place, and no known cures are mentioned."[4] They initially suspected a psychosomatic etiology. Perhaps it was a way in which vague feelings of resentment, focused on recent contact with the aliens and their wealth, could find expression. After all, anthropologists had previously observed similar paroxysms, also with an apparent contagious quality, in the Baigona and Taro cargo cults.

But within 5 years, a manifestation of culture contact and emotional insecurity could equally well be represented as a neuropathology. In early 1957, Carleton Gajdusek, a young American virologist who had been working at the Hall Institute in Melbourne, visited the Fore territory. As soon as he arrived, he began to translate "kuru sorcery" into a medical vocabulary:

Classical advancing "Parkinsonism" involving every age, overwhelming in females although many boys and a few men have it, is a mighty strange syndrome. To see whole groups of healthy young adults dancing about, with athetoid tremors which look far more hysterical than organic, is a real sight. And to see them, however, regularly progress to neurological degeneration in three to six months . . . and to death is another matter and cannot be shrugged off.[5]

Neuroscience fieldwork thus attempted to give a new meaning and context to a society, an environment, and a disease during the 1950s.

Kuru and Social Understanding

In early 1952, the Berndts, anthropologists from Sydney, established a base at a village in the Fore region. No anthropologists, and only a few patrol officers, missionaries, and prospectors, had ever been in that part of the

Highlands before. It was soon obvious that Europeans were regarded with a mixture of respect, distrust, curiosity, and fear. The field-worker was a stranger, a source of pride, a nuisance, and a confidant all at the same time. In this setting, the Berndts took as their task the description of as much of the whole culture as possible. From their own observations, and information gleaned from native informants, patrol officers, and missionaries, the Berndts described the local society. They stressed kinship and language groupings, the insecurity and the warfare, mythology, and commerce.

The problem they focused on was well recognized in anthropology. The suspicion, social disruption, and sorcery that they observed in the region could be explained, it seemed, by the Fore's effort to adjust socially and psychologically to European contact.[6] "Kuru" was thus a manifestation of a belief in spirit possession and magical performance, a result of sorcery. The Berndts, in an echo of Malinowski, recognized sorcery as an effort to resolve both internal and external conflicts at the same time as it perpetuated them.[7] They sought to view kuru sorcery "in its cultural perspective, against the background of indigenous life."[8] They looked for evidence of attitudes and behavior; they studied the transmission, development, and modification of a "contagious" belief.

But kuru sorcery was not the only phenomenon of interest: they described another feature of Fore life that later seized the public and scientific imagination (somewhat to the Berndts's annoyance). In their earliest papers they noted that cannibalism was ordinarily practiced, as a preferred method of disposal of the dead.[9] People did not, however, kill in order to obtain human meat. Generally it was a relative who "was often cooked and eaten almost immediately after death, [though] a favoured method was first to bury the corpse, and then to exhume it after a few days when the flesh was sufficiently decomposed to be tasty."[10] Cannibalism later intruded on all explanations of kuru.

So the anthropologists watched the "spasmodic individualized movements" of kuru, and then questioned informants in order to understand what was happening. The Berndts wrote: "Facts are obtained from informants, who when relating them may embroider and distort (giving a highly conventionalized or biased picture), but will also present what they consider relevant to the subject under discussion; and what they consider relevant is important for the clues they reveal."[11] One can witness many of the incidents, and "if the events took place within living memory, they are partially 'verifiable,' in so much a consensus . . . may be obtained from a number of individuals."[12] Verification is thus achieved through internal consistency.

In a reflection on her work, written in 1964, Catherine Berndt emphasized the importance of fieldwork experience in an anthropologist's training: she agreed it was, in effect, an initiation rite. The demand for participant observation required specialists in alien cultures to leave the university at an early stage in their careers and to translate themselves as much physically as intellectually into the world of their subjects. One was expected to live a "sufficient" length of time in a native village, learn the local language, and investigate "classic" themes, such as cargo movements and sorcery. Fieldwork called for a peculiar mixture of intense personal experience and dispassionate observation. Malinowski and other anthropologists wanted to convince their readers that the material had been objectively acquired, refracted through the eyes of a trained observer. Unlike the patrol officers and missionaries, against whom they defined themselves, the field-workers had to become detached cultural relativists. Their concern for internal consistency, dispassionate attention to what could readily be seen or heard, and precision in accounts of everyday life, gave the readers of anthropology texts a sense of reality, of having been present.[13] Whether one was convinced depended on how well the vividness of the experience and rapport of the participant scholar had been conveyed. When it was done well, the reader might assume that the empathetic field-worker had exerted little influence on the selection of information, but rather had acquiesced in the reality of the informant.

Kuru and the Field Laboratory

Gajdusek entered the Highlands in 1957. His career, his orientation to field practice, and his image of a significant text differed considerably from any anthropologist's. He had been drawn there through medical connections. Just after Christmas 1956, Vin Zigas, a visiting medical officer, had written to his superior in the Papua-New Guinea Public Health Department, reporting "a form of encephalitis among the Okapa people."[14] Gajdusek had just arrived in the Australian territory and when he heard of the report, decided to go through Kainantu and visit Dr. Zigas.[15]

The report of an epidemic of an encephalitis-like "disease" led him to think of "infection, and particularly virus infection," before he had even entered the field.[16] His preparations and his repertoire of investigations suggest above all a medical model. "We even delayed our departure to obtain buffered glycerine in which to store autopsy tissues for virus studies — the classical way of attempting to preserve a virus in tissues in the absence of refrigeration or of animals, chick embryos, and tissue cultures that could be immediately inoculated." And "when we entered the kuru region, we brought with us equipment to do further autopsies and to collect further specimens for extensive microbiological studies, especially serological and virological."[17] But the first object was a detailed clinical examination of everyone afflicted with the remarkable tremor that appeared more hysterical than organic. Following a long epidemiological tradition, Gajdusek and Zigas set out to identify the cases securely, and then to establish the limits of the disease.[18]

Like anthropology fieldwork, medical case identification initially relied on informants. Some knowledge of the Fore language was therefore helpful. But medical validation depended, too, on the performance of a standard neurological examination, including the use of plessors, ophthalmoscopes, and tuning forks. This gave the investigators knowledge that was invisible to the Fore. And the final confirmation of kuru always rested on autopsy findings, generally determined in metropolitan laboratories that were beyond the control even of the field-workers. Gajdusek's letters and journals describe the problems he encountered trying to transform the unruly, messy field into a standard laboratory. He writes: "our immediate need is a treatment hut . . . to study the disease in the home and village is hopeless."[19] Eventually he got his "mat-floor hospital . . . in which we have a microscope, hemocytometer, a host of reagents, and all the diagnosis instruments that such a 'bush' hospital would be expected to possess."[20]

But problems remained: "We have no appropriate cannulas," he wrote, "nor is the cold wind and rushed excitement of 'bush autopsies' conducive to careful and accurate perfusion."[21] For a medical significance to be assigned with confidence in this situation, much intervention was needed: it was necessary to reduce the bodies of the Fore to a standard physiology and pathology, and specimens had to be extracted in a regular and predictable way. All this, of course, goes against the grain of anthropology fieldwork: instead of observing and recording the distinctness of the natives, Gajdusek had altered conditions so that he could dissolve at least the sense of physical difference, and so incorporate the Fore into a generalizing medical discourse. The Berndts would later emphasize that their study, in contrast to the more contrived medical work, "was made under what could be called relatively traditional conditions, and not in the artificial situation of an organized clinic or hospital, which we understand was set up as part of the kuru project."[22]

Medical fieldwork thus implied intervention in the environment and the social life of the Fore. This could occur at the level of diagnosis or treatment, in the clinic or on patrol. It meant more than the taking of specimens and the performance of examinations. For a time, Gajdusek was also enthusiastic about

sending an ideal case to Brisbane, Sydney or Melbourne for study in a unit such as Dr Wood's Clinical Research Unit. This would yield, in the long run, far more information and far more reliable results at a far smaller expense than all sorts of half-hearted efforts at getting experts and equipment into the highlands. . . . [Patients] have so many relatives about that autopsy is refused (though in Okapa [Zigas] can talk them into it), and a case really shifted out of the region would neither bother them nor would autopsy in the event of death disturb them unduly.[23]

This effort to ensure a "good clinical workup" was later abandoned as Gajdusek grew more confident in the field and the problems of arranging the transfer proved insuperable. The suggestion nevertheless is revealing. It recalls a long tradition of European explorers abducting the ideal types of native and displaying them to a metropolitan audience.[24] There could be few interventions in social life more dramatic. It is just one of the ways in which medical field-workers seem to take on the colonial power's ability to control native bodies and to move them about.

The medical investigator did not seek immersion in the culture: he extracted from it the raw material that only a foreign group of researchers could fully understand. The most important material proved to be the brains, which showed the crucial evidence of a degeneration of unknown cause in patients with clinical kuru. By August 1957, Gajdusek had complete charts on over 150 patients. The significance of these documents depended most of all on the attached laboratory results. "From them we can study all that has been done, all that our laboratory tests have shown, and make all the analyses we wish from kuru." Not just the social life, but the bodies, too, of the Fore could thus be reduced and translated into assemblies of signs and numbers. Specimens could be interrogated on the hospital bench and in foreign laboratories and forced to speak in a reproducible manner. Photographs also helped to capture this material; and, later on, Gajdusek would make several revealing videos of the neurological deterioration he had observed. The vividness of these representations, and their insertion into a series of medical articles, helped to confer a sense of reality on the *disease* of kuru.

Yet, a reality for whom? Perhaps not for the Fore, whose bodies were never quite as submissive as the final medical papers implied: "We have not yet had them shoot at us, although they are not always very intrigued by our studies and tell us so in no uncertain terms. . . . they know full well that it's all sorcery, [but] that it is best to humor our skepticism."[25] Although the investigations had established kuru as a disease, its cause and treatment were still unknown. Gajdusek and Zigas had tested a vast range of medicines, all to no avail. From Gajdusek's letters and journals one gets an idea of the despair and the resistance gathering in the bodies of the Fore. It was always the attempt to obtain the corpse of the loved one, and particularly to secure the brain, that seemed to arouse most resentment. "Naturally, everyone would like to get their hands on kuru brains," Gajdusek wrote, in reference to competition among pathologists. "We were lucky to get two and may get further ones, but our ex-cannibals (and not 'ex-') do not like the idea of opening the head."[26]

Still, the pathologists had looked at enough brains to convince medical authorities that kuru was indeed a disease. The result of Gajdusek's fieldwork in 1957 was the "discovery" of a definite clinical entity that correlated with obvious neuropathology.[27] Yet it was a "disease" before the pathology showed up, and it remained a disease without a cause long after all the specimens had been examined. Negative virus studies and the chronic character of the illness caused Gajdusek to drop the idea of an infectious etiology; and no toxic elements could be identified in any of the samples he collected. But the distribution of the cases was consistent, he thought at the time, with an hereditary or a familial explanation. It was, in fact, this lack of a definite cause that made the problem more important to Western medical scientists. Much of Gajdusek's early correspondence was taken up with the allocation of research rights over the specimens. "If we can't 'crack' kuru — with hundreds of cases available for full study during any 3–6 month period, I see little hope for Parkinsonism, Huntington's chorea, multiple sclerosis."[28] Kuru was not just a new curiosity of medicine, it had become essential to an understanding of neurology.

Kuru was not sorted out medically until the next decade, and it did not happen in the New Guinea Highlands. Not until 1963, with laboratory evidence of transmission of the disease to chimpanzees from cell-free suspensions of human brain, did Gajdusek conclude that kuru was the first human slow virus infection.[29] With the disappearance of kuru from the youngest age group in 1964, and then progressively from older cohorts, cannibalism of those who had died of kuru seemed the most likely means of virus transmission.[30] Women and children appeared to have taken a more active part than men in mourning rites and cannibalism, and so were more likely to become contaminated. Thus, the evolving medical understanding of kuru required a renewed attention to ethnography. We go from ethnography to disease to ethnography again. But it is no longer the cultural anthropology of the Berndts: it has become the medical anthropology of Shirley Lindenbaum and others.[31]

Since cannibalism had long been suppressed in the controlled areas, more for reasons of civility than science, the disease had almost disappeared before it could be identified. The isolation of a slow virus did little or nothing to prevent or curtail the phenomenon. All the same, the detection of a new, exotic disease and the tracking down of a peculiar etiological agent had been a notable medical achievement. In 1976, Gajdusek would receive the Nobel prize in Physiology or Medicine for his contribution.

The Meaning of Kuru

What does the construction of kuru as a problem tell us about the knowledge communities of anthropology and biomedical science, and how they come to create and reshape the frames in which they insert people and phenomena? For one thing, the technique (and sacrament) of medical fieldwork was, it seems, a matter of participant extraction, rather than the participant observation of anthropology. I say "sacrament" because the taking of blood and the "consuming" of the body seem to have far more ritualized power than the taking of notes. The specimen collection is also a form of coerced speech, diminishing the influence of informants on the investigation. One can interrogate specimens as harshly as one likes in the controlled environment of the laboratory. In anthropology, the outsider must seek rapport with the people among whom he moves; but the scientist has, in effect, the ability to appropriate bodies and to read them in a laboratory, so rapport is less crucial. Unlike the anthropologist, the medical field-worker makes little attempt to acquiesce in the reality of his informants. The medical approach relies much more on intervention and control in order to construct meaning, and thus is more akin to missionary work where the aim is to achieve conversions and to extract confessions.

The anthropologist can study only that which is susceptible to explanation by a trained onlooker. The ethnographic text thus concentrates on the experience of the observing scholar though it still speaks to a community of colleagues. In contrast to this, the medical field-worker goes on patrol in search of the unseen and unheard, captures it, reduces it, and then allows it to be revealed, often in a distant laboratory. By then the exercise of assigning meaning has become more of a collective endeavor, with the emphasis, in the end, on the laboratory correlates of the field. The social role of the medical investigator is eclipsed in the final text, and the interactional aspects of fieldwork, so obvious in most ethnographic interpretations, are disguised.

In 1952 the shaking of the bodies of the Fore had a spiritual or social meaning; by 1957 medical scientists had invested the phenomenon with quite another significance. Reading the various documents generated in the Highlands during that decade, we move from a world wracked with hysteria and emotional insecurity to a world in the grip of disease. They are worlds constructed from diverse though ultimately intersecting field practices. None of the explanations has any impact on prevention or treatment of the phenomenon. Whether we prefer possession by invisible spirits, or by invisible psychological processes, or by invisible slow viruses — and just how we cobble together segments of these explanations — depends on our own cultural allegiances.

Conclusion

I have emphasized the distinctness of two grammars of practice, each on the boundary of its usual context, each nuanced and inflected to gain explanatory power. But it should also be obvious that the investigators appropriated knowledge derived from other practices and incorporated it into their own accounts: the virology of kuru makes no sense without reference to Fore mourning rites; and the medical anthropology of kuru could not, in the end, ignore the biological characteristics of a slow virus. Yet the reinflections of these competing knowledge claims are significant: thus cannibalism, for instance, is no longer simply a primitive mourning rite, it has become also a means of contamination. One has the sense of an interaction occurring in which medical scientists try to make the competence of the anthropologists their own. Thus the language of kuru is not the pure, authoritative laboratory language one might expect, but rather it lives on the boundary of its own context and the context of anthropology.[32] But then again the boundaries between behavior and biology, body and mind, have always been rather tenuous conventions.

In the beginning, Gajdusek had been confident that all he needed to assign a stable meaning to kuru was access to a laboratory. But in the late 1960s, it was impossible for anyone to understand the phenomenon of kuru without taking account at least in some measure of each of the two grammars of practice that the medical scientists and the anthropologists were using, sometimes aggressively, in the trading zone. Despite its promoters' appeals to pure technical reason, biomedical science in this setting was, in practice, neither disinterested nor culturally isolated.

Acknowledgments

For their comments on earlier drafts of this chapter, I am grateful to Robert Kohler, Henrika Kuklick, Charles Rosenberg, and Bruno Latour.

Endnotes

1. On the scientific field as the locus of a competitive struggle with the goal of protecting competences, see Pierre Bourdieu, "The Specificity of the Scientific Field and the Social Conditions for the Progress of Reason," *Social Science Information* 14 (1985):19–47.

2. In these trading zones, or points of contact, rather than simple exchange, there occurs a symbolic appropriation of other disciplines' problems, practices, languages. Peter Galison has recently developed the idea of trading zones, though he is interested as much in language as practice, and tries to identify a functional pidgin or creole language connecting disciplines; see Peter Galison, *Inside the Trading Zone: The Post-War Realignment among the Three Subcultures of Physics* (Unpublished paper, presented at the History and Sociology of Science Department Seminar, University of Pennsylvania, December 10, 1990). In the case of kuru, there was no identifiable pidgin connecting anthropology and medical science—rather, there were efforts to appropriate problems and practices into a distinct grammar of practice, and so reproduce and extend the cultural standards of a particular knowledge community.

3. Neuropathology was, after all, the first medical neuroscientific discipline, and the neurosciences have continued to expand through neuropathological means; see Franz Seitelberger, "The Role of Neuropathology in the neurosciences," in F. Clifford Rose and W. F. Bynum, eds, *Historical Aspects of the Neurosciences,* (New York: Raven Press, 1982), 265–272.

4. R. M. Berndt to R. F. R. Scragg (director, Public Health Dept), 9 July 1957, in Judith Farquhar and D. Carleton Gajdusek, eds., *Kuru: Early Letters and Fieldnotes from the Collection of D. Carleton Gajdusek* (New York: Raven Press 1981), 89–90.

202   Warwick Anderson

5. D. C. G. to J. E. Smadel (associate director, NIH), 15 March 1957, in D. Carleton Gajdusek, ed., *Correspondence on the Discovery and Original Investigations of Kuru: Smadel-Gajdusek Correspondence, 1955–58* (Bethesda, Md.: National Institute of Neurological and Communicative Disorders and Stroke 1976), 50.

6. R. M. Berndt, "A Cargo Movement in the Eastern Central Highlands of New Guinea," *Oceania* 23, no. 1 (1952):48.

7. B. Malinowski, The *Sexual Life of Savages* in North-Western Melanesia (London: E. Routledge and Sons 1929), 137. The classic study is still E. E. Evans-Pritchard, *Witchcraft, Oracles, and Magic among the Azande* (Oxford: Clarendon Press 1937).

8. Berndt, "Cargo Movement," 49, 50.

9. Ibid., 44.

10. Ibid.,

11. R. M. Berndt, "Reaction to Contact in the eastern Highlands of New Guinea," *Oceania* 24, no. 3 (1954):191.

12. Ibid.,

13. A sense of what Steve Shapin has called "virtual witnessing" ("Pump and Circumstance: The Literary Technology of Robert Boyle," *Social Studies of Science* 14 (1984):481–520). On ethnographic authority, see James Clifford, *The Predicament of Culture: Twentieth Century Ethnography, Art, and Literature* (Cambridge, Mass.: Harvard University Press 1988).

14. V. Zigas to J. Gunther, director, PNG Public Health dept. (until early 1957), then assistant administrator of PNG, 26 December 1956, in Farquhar and Gajdusek, *Kuru: Early Letters,* 1.

15. D. C. G., "Introduction," in Farquhar and Gajdusek, *Kuru: Early Letters,* xxiii.

16. Ibid., xxii.

17. Ibid., xxiii.

18. See William Coleman, *Yellow Fever and the North: The Methods of Early Epidemiology* (Madison: University of Wisconsin Press 1987).

19. D. C. G. to R. F. R. Scragg, 20 March 1957, in Farquhar and Gajdusek, *Kuru: Early Letters,* 22.

20. D. C. G. to J. E. Smadel, 3 April 1957, *Kuru: Early Letters,* 29.

21. D. C. G. to J. E. Smadel, 8 July 1957, *Kuru: Early Letters,* 87.

22. R. M. Berndt, "A 'Devastating Disease Syndrome': Kuru Sorcery in the Eastern Central Highlands of New Guinea," *Sociologus* 8, no. 1 (1959):11n.

23. D. C. G. to F. M. Burnet, 13 March 1957, *Kuru: Early Letters,* 6. Ian Wood was the director of the Clinical Research Unit, a division of the Hall Institute, located in the Royal Melbourne Hospital.

24. See, for instance, Charles Darwin's story about Jemmy Button in *Voyage of the Beagle* (London: 1844). Also see Bernard Smith, *European Vision and the South Pacific 1768–1850* (Oxford: Clarendon Press 1960).

25. D. C. G. to J. E. Smadel, 25 August 1957, *Kuru: Early Letters,* 121.

26. Ibid., 67.

27. V. Zigas and D. C. Gajdusek, "Kuru: Clinical Study of a New Syndrome Resembling Paralysis Agitans in Natives of the Eastern Highlands of New Guinea," *Medical Journal of Australia* 2 (1957):745–54; D. C. Gajdusek and V. Zigas, "Degenerative Disease of the Central Nervous Syndrome in New Guinea: The Endemic Occurrence of 'Kuru' in the Native Population," *New England Journal of Medicine* 257, no. 30 (1957):974–78. and D. C. Gajdusek and V.

Zigas, "Kuru: Clinical and Epidemiological Study of an Acute Progressive Degenerative Disease of the Central Nervous System among Natives of the Eastern Highlands of New Guinea," *American Journal of Medicine* 26, no. 3 (1959):442–469.

28. D. C. G. to J. E. Smadel, 6 August 1957, *Kuru: Early Letters,* 103.
29. D. C. Gajdusek and C. J. Gibbs, "Attempts to Demonstrate a Transmissible Agent in Kuru, Amyotrophic Lateral Sclerosis and Other Subacute and Chronic Nervous System Degenerations of Man," *Nature* 204 (October 1964):257–259; D. C. Gajdusek, C. J. Gibbs, and M. P. Alpers, "Transmission of Experimental 'Kuru' to Chimpanzees," *Science* 155 (1967):212–214.
30. J. D. Mathews, "A Transmission Model for kuru," *Lancet* 1 (1967):821–25; J. D. Mathews, R. Glasse, and S. Lindenbaum, "Kuru and Cannibalism," *Lancet* 2 (1968):449–452.
31. Anthropologists who studied the Fore after Gajdusek looked at the effect of the *disease* of kuru on society; see, for example, S. Glasse, "The Social Effects of Kuru," *Papua and New Guinea Medical Journal* 7, no. 1 (1964):36–37; Shirley Lindenbaum, "Sorcery and Structure in Fore Society," *Oceania* 41, no. 4 (1971):277–87; and E Richard Sorenson, *Edge of the Forest: Land, Childhood, and Change in a New Guinea Proto-Agricultural Society* (Washington, D.C.: Smithsonian Institution Press 1976). But see Berndt, A "Devastating Disease Syndrome" for initial resistance to the medical model.
32. M. Bakhtin, *The Dialogic Imagination,* ed. Michael Holquist, trans. Caryl Emerson and Michael Holquist (Austin: University of Texas Press, 1981).

14

The Skin, the Skull, and the Self: Toward a Sociology of the Brain

SUSAN LEIGH STAR

Introduction

The brain is a troubling object for sociologists. This chapter was written at a time in which sociologists are reexamining topics such as the body, material culture, and the nature of intelligence. We speak easily of the body in a Foucauldian sense as laced through with discourses, constituted intersubjectively, a true body politic with a sociology of its own. We are beginning to venture, trailing anthropology by some distance, to examine the materials that shape human work and interaction: tools, scientific materials, various sorts of architectures and physical arrangements.

But the brain as such has been forbidden territory. To us as sociologists it appears to stand firmly at the juncture of two alien lands: individual psychology and human biology. As sociologists, we first learn the only thing that most sociologists will ever agree upon — in the words of father Durkheim: *"The determining cause of a social fact should be sought among the social facts preceding it and not among the states of the individual consciousness"* (1966 [1938], p. 110). Social facts are *sui generis*. Whatever sociology is, it is about more than the sum of individual psychologies. And a turn in the other direction helps even less. Most of us, especially those of us in the sociology of science, have spent many years fighting the various sorts of biological reductionism that abound in every academic discipline. Nurture, not nature . . . or if a bit of both, then the nature is egalitarian and somewhat meek.

In the context of modern sociology, then, for a sociologist to venture into the domain of the brain might seem to mean a double defection: toward individualism *and* toward biological reduction. So best we avoid either talking about brains, or only do it informally, outside the pages of our sociology.

Even I, who have spent several years studying neurophysiologists and brain researchers of various sorts, have been afraid to talk directly, as a sociologist, about the brain. Listen to how I sidestepped the question of taking sides in a historical study of the localization of function controversy:

"Were they right? Are you a localizationist? The answers to these questions are the same, in essence. I am not a neurophysiologist, and at best I am a rank amateur in the world of neurons and reflexes. The *organization* of neurophysiology is my concern here" (1989, p. 29).

While this is true, I have realized recently that it leaves out much of what I think is important about the sociological aspects of the brain research situation. I hope in this chapter to feel my way through to a different kind of answer — not a credo about functional localization, but a statement about the sociological relationship between the material and the abstract in both science and social science. If we as sociologists of science are moving toward an ecological approach — one that admits of the importance of stuff and matter — and as well toward a deeper understanding of things like abstractions and formal representations, then it is also time we admitted the brain to sociological standing, without making quislings of ourselves.

A codicil: none of us really come as strangers to the brain, since the foundational metaphors of brain science pervade popular culture, and have for some time. I have seen books, in both popular and university bookstores, that describe everything from right-brain sexuality to right-brain approaches to cooking (these are not hypothetical examples). These are something of a burlesque of what I mean, but nearly any shelf of any bookstore will yield similar examples. Compelling images of the brain as director of behavior, behavior located *within* (and often in special places within) the brain, are absolutely everywhere in Western scientific and pop scientific writings.

Metaphors, as Donald Schon (1963) elegantly sketched in his book, *The Displacement of Concepts,* are a way of making a bridge between two unlike things. These are things that would be unlike in common parlance, or perhaps things that inhabit two different worlds: "Your eyes are deep pools of pitch" (where "eyes" and "pitch" are not normally thought to inhabit the same world), or "Society is a maze with a minotaur in the center" (which transposes a startling image from one domain onto another). Lakoff's work on metaphors also shows that, like long-standing arguments, they can define the limits and boundaries of a way of looking at the world — in a sense they beg certain epistemological questions through the act of conjoining that which is unlike, and *without questioning the act* (Lakoff, 1987; Lakoff and Johnson, 1980). So to ask the question "In what way are eyes like pitch" is to unframe the metaphor at its root, and, in the process, to undermine much of its power.

Some metaphors are so deeply entwined with basic assumptions about nature that they assume a foundational sort of character. The idea of "the Great Chain of Being" in the Middle Ages, for example, was a vivid metaphor for a highly ramified hierarchical social and theological order. In our times the foundational metaphor of evolution as a kind of battleground has something of that same power, as does the metaphor of human development as a ladder up which one ascends.

Our image of the brain is a complex, multifaceted foundational metaphor involving psychophysical parallelism, or the idea that the brain and the mind operate in separate spheres but still somehow in tandem. I think that the modern form of this metaphor, and its institutional base, was established in the late 19th century, and draws its power from developments in brain surgery and medical technology, and on a whole structure of invisible work and attribution.

In simple form the metaphor reads something like this: there is a lump of tissue and fluid inside the skull that is responsible for making us do things, *in the same way that:*

A typesetting compositor picks out keys for typesetting (Dodds, 1877–1878, p. 357)

The terms of one series of numbers may be used to map onto and predict the sequence in a second series (Benham, 1879)

Electricity runs through an electric wire, and may be formed in certain patterns

Competing political territories may influence each other in trade: "There are thought territories as well as cell territories, and these merge into each other, and exert reciprocal influences" (Crichton-Browne, 1872)

A musical instrument produces music:

With an equalized circulation in the brain the mind has the best opportunity of expressing itself vigorously; with an interruption of that circulation we shall find diminished manifestations of mental utterance, and lastly in shock or exhaustion, we shall see the mind unable to express itself, though all the while conscious of the impediments which block its pathway to the outer world . . . this most perplexing break in that cerebral harpsichord, through which the mind produces the simple language of infancy, and daily converse, or the immortal utterances of a Cicero and a Burke. (n.a., *Journal of Insanity,* 1870)

This central metaphor set is joined by a series of supporting metaphors that make analogies between experimental injuries inflicted on animals and those naturally occurring in humans, both of which point to physical causes of behavior. It is worth noting that even in the 19th century the computational/ informational metaphor is beginning to be used. Coulter (1991) has an incisive analysis of its contemporary form.[1] Here is a description of the canonical event at which the patient/experiment equation was developed:

I was present at the meeting of the Physiological Section of the International Congress of Medicine in 1881. Ferrier's work was brought prominently into notice at this meeting. There was a Homeric contest of transcendent moment to the advancement of knowledge, and vital to the interests of mankind, over the dog shown by Goltz and the two monkeys exhibited by Ferrier. Experimental injuries had been inflicted upon the cerebral cortex of each of these animals. The condition of the dog was supposed to prove that localisation of function in the cortex cerebri

---

[1] I am grateful to Mike Lynch for bringing this to my attention.

did not exist. One of the monkeys had characteristic cerebral hemiplegia; as it came into the room, Charcot remarked, "It is a patient!" The other monkey showed no signs of hearing when a percussion cap was snapped in its immediate vicinity; indeed, it was the only mammal in the room that did not jump as the detonation occurred. Three years before this demonstration I listened to a lecture in which we were told that the brain functioned as a whole, whatever that may mean; but now all doubt as to the truth of the great doctrine of cerebral localisation was laid to rest, and the ground was prepared for the immense progress of the coming years. (Ballance, 1922, pp. 169–70)

Joining the Abstract and the Concrete

One of the things that is going on in the conduct of brain research, and in all sorts of theorizing about the brain, is that two unlike realms, the abstract and the concrete, are being joined. People are talking about something that is concrete – the fleshly brain – and joining it with something abstract – the mind. But it is much more complicated than that.

The concrete is actually what you have direct experience of in a given situation; it is what gives situated action its specificity, contingency, local knowledge, historicity, call it what you will. When you go away from that situation, and attempt to represent what is no longer immediately there, or when you have a vague general idea about what is coming next, you are abstracting. That is, you are dropping away details, or sketching out a situation that will be modified with concrete particulars later. Generalizations are claims to be able to find commonalities between several such remembered or imagined situations.

In a certain sense the abstract has priority over the concrete, in the sense that we form a guess or have a general image about a situation before we know the details, and we can never remember or represent *all* the concrete details of a situation. The concrete grounds, instantiates, and changes our abstractions, and so on, endlessly and in great complexity.

However, when the concrete object is difficult to see (feel, hear, etc.), then we perforce are dealing with either abstract or concrete *substitutes:* signs, delegates, actants, symbols, pointers, or other kinds of relationship (Griesemer and Wimsatt, 1989, have an elegant discussion of related issues). In the case of brains, we take those substitutes to be a wide range of things: behaviors, movements, changes in the body, reports of emotions or thoughts – all are thought to reflect changes in the brain.

This is the place where sociology can legitimately make a contribution to the study of the brain. For it is always the case that choosing substitutes and, in general, the whole tricky business of lining up the abstract and the concrete, is a collective process. So a statement like "This petit mal seizure can be localized from its origin in a lesion in the left temporal lobe" is really a report about the collective work of a community of scientists, patients, journal publishers, monkeys, electrode manufacturers, and so on, over a period of some 100 years.

Let us return again to the relationship between the abstract and the concrete, but this time think of it collectively. The concrete is what you have at hand, but when you consider a collective, it has numbers of hands, no one of which has it all at hand. So information is constantly being passed back and forth between members of a collective[2] about what it is they have concretely at hand—and they do this via a series of substitutes meant to stand for the concretion. Two of the kinds of substitutes are *formalisms* and *signals*. And as collectives develop routine ways of sharing their experiences, the routines become the basis for delegating the communication—let us call these *technologies*.

A collective abstraction is a formalism, and a collective concretion is a technology. They are meant in the one case to stand in for invisible abstract things, like an aggregation of statistics, or in the other case to collect data from invisible or remote things, like brains. *Technologies are a form of collective forgetting*—the process is *so* routine that I can mimic it, and the other people in the collective do not have to keep repeating themselves.

When a collective tries to join the concrete and the abstract, a special sort of *zone of ambiguity* is opened up. Let us suppose that a group of scientists creates a mathematical model of human behavior. The abstract mathematical model is meaningful in this sense when it is applied to concrete instances of behavior, whether they be instantiations, predictions, or patterns. Immediately a zone is opened for discussion about what will be retained and what will be dropped in the attempt to "validate the model." There are struggles about defining terms, the limits of the model, to what it does and does not refer, and so on. If the model is said to represent more obvious political issues, such as the distribution of intelligence by race, other fights are propagated which also come to populate this zone.

It is often in such zones that the subtler, invisible forms of social stratification take place. This is especially true when the zone is heavily patrolled, when the stakes are very high, when entrance to the zone is via an arduous or elite training process. The zone is always vulnerable and delicate, for many reasons: we may lie to each other about our concrete experience; we may silence the voices of others' concrete experiences; we may be unable to conceive of the zone except in the terms dictated by an experience imprisoned, circumscribed, or itself violated, and so on. A technology may be used collectively to forget something that harms another—lobotomy, and its roots in the foundational brain/mind metaphor, for example.

I will make a generalization at this point and say that, in general, people find the zone created by this joining of abstract and concrete very confusing, and they do not often think of it as delicate or vulnerable, and

---

[2]And if not, then it is not a true collective.

especially not as *populated*. I think this is because the zones are easily made occult. Abstractions are slippery things. Even when they are collective, even when we are not talking of silencing or deliberate violations, links must be made between the concrete particulars of a situation and the abstraction in use. These are frequently semiprivate, if for no other reason than that the process of making, checking, revising them is distributed over space and time, and recursively, information about *that* process is asynchronous in distribution.[3]

We can begin to think sociologically about the brain if we recognize the properties of a large zone of negotiation over the nature of the zone in which people attempt to link their experience of a concrete brain with the abstract representations of mind. What are the properties of this zone? When did it develop? How has it changed in the last, say, 100 years? Who controls it? Whose experiences are included and whose are excluded? Are there gatekeepers, strategies, technologies, routines, and silences? What shape does this zone have, and can we understand the historical specificity of that shape? These are all familiar questions for sociologists.

Let me ask you to make a gestalt switch, and think for a moment of brains quite literally as pieces of stuff, as concrete matter. What are their material properties? The brain is hidden, delicate, well protected, incredibly variable in shape and size across individuals, and until fairly recently, and even now most of the time, only accessible in a visible, concrete way after death, and then only to a tiny number of people. In other words, the brain is very hard to get at.

How then to choose the substitutes for this concrete thing, in trying to make some useful generalizations about it? How to manage the work of lining up the concrete and abstract realms?

Brain History: Useful Phenomenology

It is much easier to see the unfolding of these questions historically. In the late 19th century many lines of work joined together to investigate the localization of brain function: surgery, neurology, pathology, hospital administration, physiology.

In creating a common object, researchers often make the assumption that the boundaries of the phenomenon as established by the several lines of work coincide. Surgeons, neurologists, pathologists, and physiologists were all addressing the problem of localization of function in the nervous system. Because their results were used to legitimate one another's findings, a common boundary for the functions they addressed was (often tacitly)

---

[3]What Carl Hewitt would call the "open systems" properties of information/ knowledge-processing communities.

established. The emergence of coincident boundaries here is important in understanding another aspect of the theory's success and entrenchment.

Localizationists approached the mind-body relationship from within the constraints imposed by the materials with which they worked. Neurologists investigated and treated paralysis, epilepsy, and various other disorders. Surgeons attempted to locate tumors within the brain and nervous system, and needed to understand nervous anatomy in order to avoid serious damage inside the skull. Physiologists observed muscle movements. They made correlations between the application of electrical current or deletion of an area and such movements. Pathologists, like surgeons, observed the physical characteristics of the brain, the differential distribution of types of cells, and the composition of tumor cells.

As localizationists coordinated findings from these different fields to make their arguments for localization of function, they tacitly agreed about many things. Among these were tacit common boundaries for the phenomena addressed by their several lines of work.

These boundaries were the skull and the skin. Quite simply, these were the physical limits faced by surgeons and physiologists, who had to cut through them. They were the informational barriers for neurologists, who had to guess what was inside and exactly where it was located. The skull and skin were demarcators of data for pathologists. They were supplying information about the interior of the brain to neurologists and physiologists.

Pragmatist Arthur Bentley (1975), in his essay "The Human Skin: Philosophy's Last Line of Defense," provides a brilliant analysis of the philosophical implications and limits of adopting the skin (and similarly, by extension, the skull) as a boundary for philosophical analysis. He argues that there is no logical or analytic a priori reason for adopting the skin as the border for analyzing many kinds of behavior. Rather, we take it for granted that the skin bounds many kinds of activity. When we view human beings as ending at the skin, we create an analytic chasm that can never be bridged — not by interaction, society, history, or knowledge. This is the chasm between *organism* and *environment,* in which both are reified. Many questions about the nature of knowledge, in fact, get begged because we take the skin for granted as the outer limit, for example, the nature of interpersonal communication. Thus, the skin as boundary has provided an enduring basis for both dualism and physiological reductionism.

Localizationists found themselves facing a similar philosophical chasm. For them, the skull became the edge of the mind. It was the physical boundary for mental functions. These researchers did say that there was a distinct difference between mental and physical realms, as we shall see below. However, the triangulation of their work results forced them to act as if these realms had a common boundary.

There was support from other sources for the idea that the individual mind was contained in the individual nervous system. Developments in medicine helped support this view. After the 1860s, the germ theory of disease, coupled with an emphasis on the individual patient, prevailed over group or epidemiological models. Increasingly, investigators focused upon diseases as entities located exclusively inside the individual. I contrast this with the view of disease held by doctors earlier in the century. They often mixed patients with different diseases within hospital wards so as not to "concentrate" one type of disease in one place. It was thought that so doing might make the disease more powerful. This was not a theory of contagion per se as we now think of it. Rather, it was an approach to disease that did not take the boundary of the skin for granted; disease was conceptualized more like the weather than like an event occurring inside separate skins. The rise of surgery also supported this individualist emphasis as antisepsis became more successful. The germ theory of disease strengthened the organism-environment distinction; germs *invaded*. When surgeons opened the skin for surgery, they created an opening for germs. When the skin was a barrier against germs, and a barrier to be crossed in the course of the surgical process, the boundary was reinforced in another sphere. Again, the skin was the outer limit for analysis. Meanwhile, the mind also became more firmly lodged in the skull. As the individual patient became an ever stronger unit of analysis in medicine, other events influenced the adoption of the foundational metaphor of parallelism by localizationists.

But returning to the contingencies of everyday work, there was no doubt that for the localizationists, physical realms were preferable to mental realms for getting work accomplished. Mental phenomena were messy, imprecise, and not easily discovered or replicated. As Hughlings Jackson said "Our concern as medical men is with the body. If there be such a thing as disease of the mind, we can do nothing for it" (1932b, p. 41).

Wherever possible, localizationists opted for the explanatory primacy of the physical realm. The brain caused the mind, and not the other way around. If only they could understand the brain, understanding of the mind would eventually follow. Yet they all understood the import of trying to decipher psychological cues and mental events. Such phenomena could not be dismissed altogether, for they often formed the immediate clues to nervous disease found by the neurologists (e.g., see Crichton-Browne, 1872; Ross, 1882).

Particularly interesting here is the study of aphasia, a complex impairment of speech or language ability. Aphasia may range from total muteness, to a restricted ability to speak some words, to a globally distorted speech capacity. Some form of aphasia is often a symptom of brain disease. Not all inability to speak is aphasia per se; neurologists distinguish it from a simple physical impairment, as with impaired vocal cords that prevent enunciation. But when the vocal cords were not damaged, yet patients were unable to either speak, write or understand language normally, the symptom or diagnosis was aphasia. Localizationists in England saw aphasia as a type case of a mental phenomenon caused by physical brain damage (e.g., Seguin, 1881).

Localizationists sought to correlate these impairments with physical damage or brain tumors. As evidence from different realms accumulated, researchers developed an increasingly strict doctrine of psychophysical parallelism. Again, this is the idea that the brain and the mind operate in tandem but as completely separate and sovereign realms. As Jackson scholar Engelhardt (1972) states:

> The choice between these options was important, for it would define what factors and indeed what "facts" would be of immediate significance to neurology. For example, a materialism would dismiss the importance of psychological events, while an interactionism would require such events to be acknowledged as causal factors. But a theory of the concomitance of mind and body would allow one to attend to purely physiological explanation without denying the significance of psychological reality. In developing his notion of neurology, Jackson considered that he was making a pragmatic choice between these options. (p. 23; Levin, 1960)

Jackson's claims for the separation of the two spheres were quite strong. They were based on the reliability of the physiological parameters: "The trustworthy symptoms in the diagnosis of actual and primary disease of the organs of the mind are physical, and the untrustworthy symptoms for that diagnosis are mental" (1875, p. 492). Most explicitly, he (1932c) says that:

> The doctrine I hold is: first, that states of consciousness (or, synonymously, states of mind) are utterly different from nervous states; second, that the two things occur together — that for every mental state there is a correlative nervous state; third, that, although the two things occur in parallelism, there is no interference of one with the other. This may be called the doctrine of concomitance. (p. 72)

Another way of looking at this statement is that Jackson and his associates cleared the zone from anything messy: work, failure, patients who did not fit the proper categories. With what, then, was it filled?

The Contradictions

Localizationists recognized that material and immaterial realms could not, without serious philosophical difficulties, simply be posited as causing action in one another. They also recognized that in principle "correlation is not causation," although they sometimes used correlation as proof.

The major conceptual difficulties thus caused by parallelism were how the two realms (mind and brain, or mind and body) were brought together, and by what mechanisms they were made to operate in tandem. Again, it is not surprising to find that the responses of localizationists to these problems were neither unified nor consistent. They were facing multiple, incommensurate audiences: philosophy, medicine, physiology, antivivisection and evolutionary biology. In addition, their everyday work posed serious technical difficulties and uncertainties.

In order to resolve the conflicting demands of the several audiences, localizationists adopted several general strategies. The first strategy was to refer philosophical problems to an expert within their ranks. This was someone who understood their daily work concerns, but who would speak as a philosopher for them. The person elected to do this was John Hughlings Jackson. As he addressed many of the contradictions posed by parallelism and the mind-brain relationship, Jackson became a kind of symbolic leader for localizationists.

Second, localizationists developed theories and concepts that could act as plausible bridges between the realms of the mind and the brain. These explanations were not, strictly speaking, philosophically accurate. However, they were, in fact, good enough theoretical explanations to allow work to continue respectably.

But as a final resort, when problems could not be resolved, localizationists would simply jettison intractable problems into other lines of work. That is, those difficulties that could not easily be addressed by some physical or medical model were relegated to "mind"-related lines of work, such as psychiatry and psychology. As problems were jettisoned into other lines of work, psychophysical parallelism was reinforced on an organizational level. Such a division of labor effectively obscured many of the epistemological problems arising from the mind-brain gap. The contradictions were thus eradicated from immediate concern.

I consider each of these strategies in turn in the following sections.

John Hughlings Jackson as Symbolic Leader

Symbolic leaders are individuals who embody or represent one kind of activity or position in a conflict (Klapp, 1964). The participants in a conflict vest such a figure with power to speak for a viewpoint or debating stance. They often invoke her or his image to embody one side or another of the conflict. (For example, "Gloria Steinem" is often invoked as a broad reference to feminism, only partly in reference to her actual work.)

Jackson became "the great unifier" for the philosophical difficulties faced by localizationists in trying to define the mind-body relationship. So pervasive and powerful was this designation that it has persisted to the present day. There are dozens of testimonial articles, references, and honorifics accorded to Jackson—"the father of neurology"; "the most brilliant neurologist the world has ever seen"; "a goldmine of theoretical astuteness" (Angel, 1961; Bishop, 1960; Brain, 1957; Critchley, 1960a, 1960b; Levin, 1965; Lassek, 1970; Smith, 1982).

Jackson was the oldest of the important localizationist theorists. In many ways, he formed the key bridge between the older physicians' more diffusionist approaches and the newer localizationism (Greenblatt, 1970). He had been a student of Brown-Séquard's, although he disagreed vehemently with Brown-Séquard about localization. Still, Jackson's language was more familiar to those in the old school: he spoke of forces moving through the body, and balances, at the same time as he spoke of localization of function. He said, wryly, of himself, "I am neither a universaliser nor a localiser. . . . In consequence I have been attacked as a universaliser and also as a localiser. But I do not remember that the view I really hold as to localisation has ever been referred to" (1932b, p. 35).

Jackson's medical methods and many other aspects of his theory derive from assumptions and training practices prevalent throughout British medicine. He adhered to many of the older methods of diagnosis and treatment, even while bringing these methods into the new arena of localization of function. Here is another reason he became a symbolic leader for the localizationists.

Like Brown-Séquard, Jackson is perhaps best known in medicine, then and now, for his work on epilepsy. He described epileptic seizures in great detail, and was a pioneer in collecting clinical details about the disease. He created an evolutionary theory about epilepsy, as well as for the whole nervous system (Jackson, 1878, 1879). He argued that epilepsy represented, in perfect ordinal correspondence, a process of "dissolution" that was the opposite of evolution. This was a concept directly borrowed from Herbert Spencer (e.g., see Jackson, 1932b, 1932c). Spencer saw the entire natural and social world in terms of evolution and dissolution. Jackson was a staunch Spencerian throughout his career.

Historians of neurology have often referred to Jackson as exclusively influenced in theory-making by Spencer. For example, Head (1926) states that "Jackson derived all his psychological knowledge from Herbert Spencer" (p. 31). Yet as we have seen above, Jackson was also an exemplar of contemporary diagnostic forms, and a crucial bridge between old and new systems of medicine. Use of Spencer's language proved important in cementing alliances with, and making plausible arguments to, evolutionary biologists. Localizationists and evolutionary biologists and philosophers had become political allies during the antivivisection debates and trials of the 1870s (French, 1975). The two groups had formalized many of these professional alliances through the formation of the Physiological Society (Sharpey-Schafer, 1927). In addition, Charlton Bastian, physician at Queen Square, was Herbert Spencer's literary trustee, and a friend and coworker of Darwin, Russell Wallace, and Thomas Henry Huxley (Holmes, 1954, p. 39).

But Jackson's claims to strict parallelism, and his attempts to demarcate a strictly physiological sphere of investigation, were not entirely successful (Engelhardt, 1975). Jackson was forced to address the philosophical problem of linkage between the realms he distinguished. His writing on the subject is dense, rich with contradictions and subtleties, and embodies all the complexly interwoven processes described throughout this book. He weaves together clinical and basic evidence. He absorbs anomalous information by incorporating objections and changing their framework into a localizationist-oriented one. He couples abstract concerns about the nature of cognition with very practical clinical advice based on years of working with epileptics and victims of nervous disorders (Star, 1986, 1989).

The result is writing with many of the attributes of scripture: infinitely interpretable, inspirational, filled with both hardheaded advice and ambiguity. Anne Harrington (1987) has called him a "Rorschach" for subsequent generations, and I agree. His writing is thick, convoluted, full of metaphors and images (Mitchell, 1960). Jackson's work became a final authority on the mind-body problem in the same way that any such writing becomes authoritative—quoted for a wide variety of purposes, interpreted by many, and ultimately, a repository for unanswerable questions.

Plausible Bridges

In addition to using Jackson's work for holding unanswerable questions, localizationists also formed several kinds of temporary, workable theoretical bridges over the mind-body chasm.

One such type of bridge was *concomitance theories.* These were theories that were based on the idea of correlation of activity between the mind and the body. They lodged the connection in temporal simultaneity. If a tumor was discovered in the brain of a former patient who had aphasia, then the tumor was affecting the speech area. Both speech and the brain were affected at the same time; therefore, there must be a connection between the brain (tumor) and the mind (speech). These theories relied heavily on the concept of substrata, either anatomical or physiological, which underlie functions or behaviors. The major philosophical difficulty here was with discovering the nature of connections between the presumed substrate and its products. The nature of learning also presented difficulties: how did learning occur, involving as it does both a substrate and a behavior?

Building-block theories attempted to find "basic units" whose assembly would constitute either brain or mind. We find primarily here those reflex theories that saw actions as built up of particulate sets of reflexes. The primary difficulties implied by these theories were problems of storage of nervous impulses and coordination between basic units (Dewey, 1981 [1896]).

Representation theories tried to link the realms of mind and brain by positing that one realm "stood for" and indicated or implied the other. These theories were the most abstruse, and much of Jackson's work relied on these sorts of concepts. According to representation theories, the gap between mind and brain was bridged by a kind of calling back and forth, or imaging, between realms. Representation theories included theories of ordinal representation, automaton theories, and some theories about muscle sense. In discussing representations across realms, they also often addressed behavioral problems in terms of consciousness and unconsciousness. A major source of difficulty for these theories was how to account for mechanisms of perception between or across realms: what symbolic "language" could bridge the gap between mind and brain, for example?

A practical concern about the usefulness of concomitance often appeared in the early volumes of the localizationist journal *Brain,* which was founded in 1878. Here, writers stated that even though the connection between the realms of brain and mind was impossible to understand philosophically, concomitance was the only practical solution to the difficulties. In short, it was a theory that would allow them to get on with their work. This example from Cappie (1879) is typical:

In studying the causation of mental phenomena we need not be deterred by the fact that our knowledge, at the best, is likely to be always remote. It may not be the less positive so far as it may pretend to go. We must ever be content to accept the fact of consciousness as ultimate. The physiologist will never be able to perceive why the brain's activity should be associated with its manifestations. Yet, if a correlation be assumed, he may be able to determine why one form of consciousness rather than another is present; why it is present with a felt amount of intensity; or why in certain circumstances it is obscured or suspended. The nexus between molecular activity of brain and activity of thought and volition may forever remain "unthinkable," but many of the modifying conditions of the molecular activity may certainly be determined, and the remote but necessary influence of these on the mode and intensity of mental action may thus to some extent be recognized. (p. 373)

A review of G. H. Lewes's *The Study of Psychology* in the same issue makes the same point:

It is necessary to adopt a new method of research—to translate the facts of consciousness into terms of another class. . . . The new method follows from the assumption, for long tacitly, though fitfully, made by all, that organic and mental processes are strictly correlated; so that a complete analysis of the one would give an equally complete acquaintance with the other: and then, while some terms of the one series may be beyond our reach, the corresponding members of the other series may be accessible. (Benham, 1879, p. 392)

Another corollary to the doctrine of concomitance was the establishment of the idea of an anatomical substratum that somehow (and the mechanism or bridge proposed varied considerably) "gave rise to" behavior or mind. This was perhaps the most important outcome for the subsequent history of neurophysiology. As daily work practices subsumed the subtler philosophical questions about the mind-brain connection, they became institutionalized in neurophysiology at this time. Philosophical questions gave way to working assumptions about substrata (e.g., see Benham, 1879; Ferrier,

1879). For example, in Ferrier's *Functions of the Brain* (1876), he refers to "the anatomical substrata of consciousness" (p. 225), and says that "In order that impressions made on the individual organs of sense shall excite the subjective modification called a sensation, it is necessary that they reach and induce certain molecular changes in the cells of their respective cortical centers" (p. 257).

Mind, or "mental operations in the last analysis must be merely the subjective side of sensory and motor substrata" (p. 256) and the mechanism of connection between the mind and the brain Ferrier calls a "molecular thrill" (p. 258). In a later work (*The Localisation of Cerebral Disease* [1878]), he similarly states that "the physiology and psychology are but different aspects of the same anatomical substrata, is the conclusion to which all modern research tends" (p. 5). But in this work, it is not a "molecular thrill," but rather gross pathological processes that are said to be the source of concomitance. This is also true of his lesion experiments, yet the switch in levels of analysis from molecules to lesions is not explicitly discussed.

Similarly, there was disagreement and ambivalence in localizationist writings about the composition of the hypothesized anatomical substratum. Ferrier (and others at times) said that: "The various regions of the cortex are linked together by systems 'of associating fibres', which furnish an anatomical substratum for the associated action of different regions with each other (1876, p. 11). (See also Dodds, 1877–1878, for a review of other similar opinions.) In some contrast, Jackson (1932a) saw the substratum as electrical in nature:

We have, as anatomists and physiologists, to study not ideas, but the material substrata of ideas (anatomy) and the modes and conditions of energising of these substrata (physiology). Where most would say that the speechless patient has lost the memory of words, I would say that he has lost the anatomical substrata of words. The anatomical substratum of a word is a nervous process for a highly special movement of the articulatory series. That we may have an "idea" of the word, it suffices that the nervous process for it energises; it is not necessary that it energises so strongly that currents reach the articulatory muscles. (p. 132)

And he goes on to say, in a passage vividly illustrating his commitment to attend to physical realms:

How it is that from any degree of energising any kind of arrangement of any sort of matter we have "ideas" of any kind is not a point we are here concerned with. Ours is not a psychological inquiry. It is a physiological investigation, and our methods must be physiological. We have no direct concern with "ideas," but with more or less complex processes for impressions and movements. (p. 133)

Here again we have a disclaimer of involvement with "mental" processes, coupled with an untestable mechanism ("energizing") that joins two realms.

By contrast with concomitance theories, building-block theories saw mind or consciousness as more directly emergent from combinations of physical units. Maudsley (1890), for example, articulates this view from a philosopher's point of view. He sees mind as what is left over, or even excreted, after the building blocks of reflexes and learned reflexes combine to form nervous action.

Building-block theories, because they are particulate and modular, had several engineering-like advantages. Units could be distributed over the nervous system in different amounts or densities, thus providing a model for localization of function. These theories also answered in a direct way many of the anomalies of noncorrelation faced by all localizationist experimenters and pathologists.

Those who held to building-block theory saw consciousness, or mind, as emerging from the building blocks as alternately "sense datum plus memory" (Ferrier, 1876, p. 44), or, as Gowers put it (1885), as a residue or discharge of memory:

Memory, like other mental actions, has its psychical [sic] side. Every functional state of the nerve elements leaves behind it a change in their nutrition, a residual state, in consequence of which the same functional action occurs more readily than before; and this residual disposition is increased by repetition. (p. 117)

Dodds, in a detailed summary of localizationist doctrine (1877–1878, p. 357) exemplifies many of the problems faced by localizationists in using a building-block model to solve the problems of parallelism. He also shows how they came to resolve anomalies. Dodds wonders about what explanation could be given of Ferrier's experimental evidence for a functionally differentiated motor region. It is not, he thinks, that there is actually a center for the sensory phenomena; rather:

The only probable explanation appears to us to be that the cells of this motor region of Ferrier form, as it were, the motor alphabet of the will. In a way as yet not understood, the will is able to decide upon certain movements, and then, compositor-like, to pick out the area, by whose stimulation the desired result is accomplished. (p. 357)

Dodds's compositor image reflects a concern with the idea of coordination. What coordinated the parts of action? A key passage from Ferrier (1876) illustrates the major change in emphasis on this question brought about by the localizationists:

Hence we are not entitled to say that mind, as a unity, has a local inhabitation in any one part of the encephalon, but rather that mental manifestations in their entirety depend on the conjoint action of several parts, the functions of which are capable, within certain limits, of being individually differentiated from each other. (p. 42)

The problems of the origin and consequence of conjoint action were never fully resolved by localizationist researchers. Rather, concomitance and building-block models became themselves increasingly robust as they proved useful in addressing clinical issues and in doing experimental work.

The problems of storage, coordination, and the mechanism of connection between the two realms were handled by representation theories.

Representation theories are the most philosophically sophisticated of the mind - body bridges constructed by localizationists. They are also the most complete in terms of the number and variety of phenomena addressed. They bridge the gap between the mind and body by postulating that activity from one realm is translated across via some sort of algorithm. Indications are made across realms which represent either movements, ideas, words, or physiological processes. As Jackson said, "We understand a speaker because he arouses our words" (1932d, p. 206).

Representation theories arose in part from the observations made in treating the different forms of aphasia. One of the core philosophical problems addressed by representation theories is that of "consciousness versus unconsciousness" or automatic versus reflexive, conscious activity. Early on, clinicians were confronted with patients who apparently could think but who could not speak. These patients could intelligently (by gesture) answer questions, showing that they heard and understood words. Yet they were mute. Jackson and other writers analyzed this phenomenon to try to distinguish word understanding and word articulation capacities. This is turn raised the question of whether there was, in Jackson's terms, "inner" and "outer" speech. Were the patients talking to themselves inside their heads? Was this a different "speech" than that which is articulated aloud? These questions would continue to fascinate neurologists for some years (Head, 1926; Riese, 1959).

Jackson (1932d) used the concept of representation to resolve these questions in two ways. First, he postulated the existence of inner and outer domains of experience, most notably inner speech and outer (articulated) speech, but also inner muscle sensations versus outward movements:

A word is a psychical [sic] thing, but of course there is a physical process correlative with it. I submit that this physical process is a discharge of cerebral nervous arrangements representing articulatory muscles in a particular movement, or, if there be several syllables, in a series of particular movements. (p. 205)

In speaking aloud there is a strong discharge of the same nervous arrangements with a correlative vivid psychical state — in this case the discharge is so strong that lower motor centres are engaged, and finally there are movements of the articulatory muscles in a particular sequence. (p. 207)

Second, he mapped evolutionary changes onto the individual, saying that processes of dissolution were ordinally represented, beginning in the larger natural spheres and descending to the individual. These representations could be represented faintly or vividly in the individual nervous system, mere echoes or actual imprints of external events (Walshe, 1961).

The concept of representation solved some of the contradictions and gaps that remained with the doctrine of concomitance. Specifically, it helped serve to transform static functions into terms more compatible with physiology, and to avoid some of the previous philosophical objections to the idea of a substratum. Jackson's (1932f) definition of anatomy here was unique: "To give an account of the anatomy of any centre is to draw what parts of the body it represents, and the ways in which it represents them" (p. 96).

Engelhardt (1972) says that Jackson transformed psychological notions of cerebral localization (which he said were outside the province of medicine) into physiological and sensory-motor concepts via the use of the concept of substratum. Jackson then redefined anatomy in terms of representation, and transformed the physiological concepts into anatomical ones. This was in turn tied back to parallelist, localized models of the brain. In Jackson's (1932b) words:

It is an anatomical division. By anatomical I mean that it is after the different degree of complexity, cf., or representation of parts of the body by different centres. The nervous system is a sensorimotor mechanism, a coordinating system from top to bottom. The evolutionist can take a brutally materialistic view of disease of any part of the nervous system, for the reason that he does not take a materialistic view of mind—does not confound nervous states with psychical states. The highest centres are only exceedingly complex, and special sensorimotor nervous arrangements representing, or coordinating . . . the whole organism. The doctrine of evolution has nothing whatever to do with the nature of the relation of psychical to the physical states of these centres; it simply affirms concomitance of psychical states with states of these centres. (pp. 40–41)

Again, Jackson here addressed the important audience of evolutionary theorists. Evolutionary biology had long concerned itself with its own species of mind-body problem: how did "consciousness," or that which makes humans uniquely human, evolve? Huxley (1904 [1874]) and others had postulated a progression from automatic to reflexive or voluntary activities based on the evolution of the brain. Human beings retained automaton-like movements in the lower part of the nervous system, the animal part. Higher functions were contained in, and localized within, the higher cortical centers. Jackson (1932c), drawing on Spencer's ideas of increasing heterogeneity with evolution, saw an increasingly complex, localized human brain as the inevitable outcome of evolutionary processes. His 1884 Croonian Lectures to the Royal College of Physicians (1932e, pp. 45–75) concluded that the highest centers, which make up the "physical basis of consciousness," are the most complex and the most voluntary (although the least organized in the sense that they are naturally ordered). They draw, via representation, on lower centers. An 1873 article of Jackson's had stated that:

Lesions . . . discharge through the corpus striatum. I suppose that these [cerebral] convolutions represent over again, but in new and more complex combinations, the very same movements which are represented in the corpus striatum. They are, I believe, the corpus striatum "raised to a higher power." (p. 200; see also 1932e, pp. 218–219)

In sum, then, representation theories are a kind of Cartesian sign language, where the body and mind mutely gesture across the epistemological gap of parallelism.

Jettisoning Unsolvable "Mind" Problems

It was in part the emphasis on the practical primacy of the physical sphere that made possible the third strategy used by localizationists to explain the contradictions of parallelism: jettisoning unsolvable "mind"-related problems into other lines of work. This division of labor ensured that neurologists and surgeons could concentrate on the physical basis of disease that had been their basis for success.

The creation of "garbage categories" is a process familiar to medical sociologists. When faced with phenomena that do not fit diagnostic or taxonomic classification schemes, doctors often make residual diagnoses. One function of such diagnoses is to shunt unmanageable, incurable, or undiagnosable patients into other spheres of care where they will not interfere with the ongoing work. Hysteria, senility, and depression, for example, have been criticized as such categories (e.g., see Henig, 1981).

Localizationists created such categories for problems that did not have an identifiable localized referent or the possibility of a physical treatment. These patients were diagnosed as hysteric or neuraesthenic. Early in the period studied here, patients so diagnosed were treated by the hospital at Queen Square. They received much the same treatments as patients with brain tumors, stroke, or other diseases with known physical referents.

However, by the turn of the century, patients diagnosed with "mental" as opposed to physical problems. began to be referred to psychiatrists and to be jettisoned from the jurisdiction of localizationist neurologists and surgeons. This practice both reinforced the physical orientation of these workers, and mirrored the philosophy of parallelism on an organizational level. In an active sense, it helped solve the contradictions posed by parallelism by lodging "mind" and "body" in the ordinary division of labor and referral systems in medicine.

Invisible Work

To return to our earlier questions about the relationship between the abstract and the concrete, or the material and the formal, what can we say about the experience of localizationist researchers, and how did that experience get translated into contemporary metaphors still powerful in brain research? How was the zone organized? Who or what held it together?

Joins between the concrete and the abstract are unstable. This is because they are metaphoric, and also because of the ambiguity and nondeterminism involved in the "winnowing" and "implementation" processes. People manage the instability in a variety of ways, including work, technology, routines, and the making of more metaphors. In the use of metaphors, particularly those as pervasive as psychophysical parallelism, it is easy to forget that someone must provide the transport from domain to domain, someone must do the winnowing, and these and other processes must be continually renewed in order to remain meaningful.

What gets left out of the story of brain research is precisely this collective process of transporting, of making and use of technology, of reviewing the metaphors, and of the specific circumstances of discovery and knowing involved. Here I would ask of the reader a willing suspension of disbelief — step outside the dominant metaphor of the brain as puppeteer, and think instead of the work processes involved in brain research. This is a gestalt switch. Some of the things that rarely get spoken of in discussing psychophysical parallelism are the conditions of work of early neuroscientists, the circumstances of patients, the work of patients, nurses, and kin, career pressures and other political pressures on neurophysiologists, and their emotional involvement with particular philosophical points of view. I can give only a brief glimpse of some of these aspects here, but they include many sorts of uncertainty, a myriad of ingenious technologies that made surgery safer and more widespread, the defeat of antivivisection, and certain widespread cultural changes that made patients more involved in scientific medicine.

I would like to conclude with some images of those circumstances, and begin to shape how socially, politically, and ethically the inclusion of these circumstances might inform neurosciences and metaphors about the brain.

First, I wish to invoke the voice of David Ferrier, writing in his notebooks about the animals from which he was coaxing information. The difficulty of working with reacting subjects, particularly monkeys, is vividly conveyed in his laboratory notebooks. His notes were often written in an obviously hasty, shaky hand, as he tried to record minute-by-minute events in the laboratory. The pages of several notebooks are spattered with blood stains. Ferrier noted that the monkeys were often "mischievous," hostile, or affectionate, constantly trying to run away from him, climb up his pants' legs, bite or scratch. In Ferrier's words:

Apparently monkey disinclined to move. Could see somewhat as he when making a push away from being pursued did not knock except occasionally. . . . Difficult to say whether right extended or not as being disinclined to move it — at any rate we had few methods of testing . . . After this we tried hard to get the bandage off the left eye. Was very unwilling to move at all. When kicked would run against anything. Taken into the other room. Sat still with head down. Would not respond to when called. Gave him a piece of cracker and he put it in his mouth. Took him back into laboratory. Got quite still and grunted or made a rush anywhere when distracted. (Ferrier: MS246/5, January 5, 1875)

In addition to being reactive in a behavioral sense, the animals were physiologically fragile. Damage to animal subjects, including operative complications, was common. The notebooks record frequent accidental deaths in the laboratory from hemorrhage or chloroform overdose. Ferrier's notes depict these difficulties:

No application could be made nor could the electrolisation be made to be localised.

This movement was very difficult to analyse as the brain very speedily lost its excitability.

Hard to distinguish where the electrodes were due to the bleeding.

He speaks of animals exhausted, in states of stupor. He pokes and prods the animals, gives them smelling salts and electrical shocks, bangs on the water pipes to see if they react. An experiment in 1879 records an operation whose sequelae Ferrier is not "sure if existed before" (Ferrier D: MS 246/2, August 8; MS246/4, December 1874; MS246/7, October 1879).

Clinical experience was equally uncertain. Diseases of the nervous system are very labile, and rarely limited either in their location or in their effects on the organism. The tumors found by investigators did not grow in neat patterns nor conform to topographical diagrams of the brain. Rather, they expanded across the analytic boundaries created by researchers. They spread not only to different areas of the cortex, but descended (or originated) in the lower parts of the brain, thus violating the often fragile boundaries of functional neurological theories. Individuals also produced widely varying "pictures" of different nervous system diseases, adding another potent source of uncertainty. Buzzard (1881), for example, describes one patient with a huge tumor who appeared entirely unaffected by it. Upon his death, Buzzard notes in a puzzled tone: "It is remarkable that a person suffering from so extensive and grave an intracranial lesion should have been able to enjoy a long day's hunting within a week of his death" (p. 132).

Both surgeons and neurologists also faced uncertainty because many diseases shared symptoms in common with brain tumors: lead poisoning, thyroid diseases, muscular dystrophy, and so on. Among these diseases was advanced syphilis. As with brain tumors, tertiary syphilis could cause paralysis, epileptic-like seizures, and blindness. Conservative physicians wanted all patients with these symptoms to undergo a prophylactic course of treatment for syphilis before even considering a diagnosis of brain tumor. This meant a wait of at least 6 weeks, often decisive in the clinical trajectory of a tumor patient.

This letter from neurologist Gowers to surgeon Horsley attests to some of the difficulties: "possible syphilis and urgency to act as soon as ever the absence of result from treatment is just definite enough. I hear you think it is very likely a large tumour. I suppose you will do the second part, here, as soon as is proper. Can you tell me when?" (quoted in Lyons, 1965, pp. 263-264).

Neurosurgeon Cushing (1905) lucidly describes some of the difficulties:

The neurologist spends days or weeks in working out the presumable location and nature of, let us say, a cerebral tumor. An operator is called in; he has little knowledge of maladies of this nature and less interest in them, but is willing to undertake the exploration. The supposed site of the growth is marked out for him on the scalp by the neurologist; and he proceeds to trephine. The dura is opened hesitatingly; the cortex is exposed, and too often no tumor is found. The operator's interest ceases with the exploration, and for the patient the common sequel is a hernia, a fungus cerebri, meningitis and death.

In conclusion, I would like to venture some remarks about how this group of researchers, and most neuroscientists following them, structured the zone of ambiguity that kept together the brain and the mind. On the one hand, near-certain death, labile materials, lack of standardization, many sorts of political pressures: the concrete. On the other, a mind divisible into components, each guarding an aspect or aspects of behavior. On the one hand, success, and work proclaimed in the literature: solving the problem of shock, antisepsis, inventing new kinds of bone wax for surgery, and saving people with brain tumors who had previously been doomed to a certain death. On the other hand, failure, and silence about certain kinds of work: failure to cure epilepsy, the silent work of mothers, wives, attendants for the desperately chronically ill, failure to achieve consistency of results in localizing function. Fear of antivivisectionists and of the unruliness of animals.

Over a long historical moment, the work that was visible, with its embedded successes, produced representations that came to stand for the invisible material, the brain itself, and all the forms of invisible work and failure hidden in the zone of ambiguity. (See Star, 1991, for a further discussion of these issues.) Maps and atlases, stained prints of brain cells, and orderly divisions of the mind became the representations of the foundational metaphor of psychophysical parallelism—the brain as puppeteer.

If we are now asking foundational questions about material culture, we should include questions about these metaphors, as well as the activities of those maintaining it. One contribution of sociology of science to science itself is to frame those metaphors as human constructions with human consequences. It makes a difference whether we think of ourselves as isolated selves with functionally localized brains, or as a collective of actors engaged in understanding how our concrete experiences aligns with our abstract experience. On the one hand, Cartesian individualism; on the other, shared fate and respect for all the work involved in a situation.

So what would an alternative metaphor look like, and what would it include? I speak here not only of 19th-century localizationists, but of us today, because I think we are still in the grip of a very important and very dangerous fundamentalism, resting on the brain-mind and organism-environment "great divides."

What I want is for us to "unbeg" the question of what is inside and what is outside. What we need, I think, are metaphors that are ecological and dialectical, and we need to go back and question those that are not. To do this requires some prior work. The persistent image of the brain as inside the head and controlling behavior and abstract mind outside is too powerful to overthrow without the links to work practices and political organization that evolved with it in the first place (Star, 1990). Let me end, then, with a list of beginnings.

First, there is the problem of aloneness. Solitude is the condition that stops us from "seeing" mind as communal and jointly constructed. As one sociologist said to me a couple of months ago, "If I wake up in the morning speaking Chinese, and go into a locked room alone and stay there all day incommunicado, then come out at night, I still speak Chinese. Doesn't that imply that the brain somehow has in it my ability to speak Chinese?" At that time I answered him that our relationship to others was a bit like our relationship to oxygen — just because we could hold our breaths did not imply that we could survive without air. I would also now say that you cannot understand a table by trying to pull the atoms away from the empty spaces that are said to surround them. We need all kinds of spaces between ourselves and other people, which does not violate the primacy of us to each other. Pulling the brain out of that composition would be like trying to pull an atom out of a table in order to explain why it is not really part of the table.

Second is the problem of seeing the social as trivial and infinitely malleable. We are imbued with images of ourselves as sturdy little physical beings, somehow clothed in "society." "Socialization" comes on top of the biological givens, and as such, can be changed in much the way you would change a set of clothes. We need new words to speak of embeddedness, of cocreation and codeterminacy, of the ways we are deeply embedded.

Metaphors to speak to these issues are difficult. I like to think of concepts that are like a knife with a perfectly spherical blade. Cutting everywhere, or nowhere? But not cutting apart into great divides.

Acknowledgments

A fellowship from the Fondation Fyssen in 1987–1988 allowed me to think and talk through many of these ideas with colleagues at the Centre de Sociologie, École des Mines, Paris. Adele Clarke and Joan Fujimura have contributed to many of the ideas about material here, and Geof Bowker, Alberto Cambrosio, and Gail Hornstein made many helpful suggestions. Mike Lynch made extensive and very useful comments on the original manuscript. I gratefully acknowledge their support.

References

Angel R (1961): Jackson, Freud and Sherrington on the relation of brain and mind. *Am J Psychiatry* 118: 193–197

Ballance C (1922): A glimpse into the history of the surgery of the brain. *Lancet* 202: 165–172

Benham FL (1879): Review of *The Study of Psychology* by GH Lewes, *Brain* 2: 390–400

Bentley A (1975): The human skin: Philosophy's last line of defense. In: *Inquiry into Inquiries: Essays in Social Theory,* Ratner S, ed. Westport, Conn.: Greenwood Press [Original work published 1954]

Bishop WJ (1960): Hughlings Jackson (1835–1911). *Cerebral Palsy Bull* 2: 3–4

Boring EG (1950): *A History of Experimental Psychology,* 2nd ed. New York: Appleton-Century-Crofts

Brain R (1957): Hughlings Jackson's ideas of consciousness in the light of today. In: *The Brain and Its Functions.* Oxford, England: Blackwell, Wellcome Historical Medical Library

Buzzard T (1881): Pain in the occiput and back of neck. *Brain* 4: 130–132

Cappie J (1879): On the balance of pressure within the skull. *Brain* 2: 373–384

Coulter J (1991): The informed neuron: Issues in the use of information theory in the behavioral sciences [Cognitive Science Technical Reports]. Department of Computer Science, SUNY, Buffalo.

Crichton-Browne J (1872): Cranial injuries and mental diseases. *West Riding Lunatic Asylum Med Rep* 1: 97–136

Critchley M (1960a): Hughlings Jackson, the man and the early days of the National Hospital. *Proc R Soc Med* 53: 613–618

Critchley M (1960b): The contribution of Hughlings Jackson to neurology. *Cerebral Palsy Bull* 2: 7–9

Dewey J (1981): The reflex arc concept in psychology. In: *The Philosophy of John Dewey,* McDermott JJ, ed. Chicago: University of Chicago Press [Original work published 1896]

Dodds WJ (1877–1878): On the localisation of the functions of the brain: Being an historical and critical analysis of the question. *J Anat Physiol* 12: 340–363

Durkheim É (1966): *The Rules of Sociological Method,* New York: Free Press [Original work published 1938]

Eccles J, ed. (1982): *Mind and Brain: The Many-Faceted Problem,* Washington, D.C.: Paragon House

Engelhardt HT (1972): *John Hughlings Jackson and the Concept of Cerebral Localization.* Thesis for M.D. degree, Tulane University, New Orleans, Louisiana

Engelhardt HT (1975): John Hughlings Jackson and the mind-body relation. *Bull Hist Med* 49: 137–151

Ferrier D (1876): *The Functions of the Brain,* London: Smith, Elder

Ferrier D (1878): *The Localisation of Cerebral Disease.* London: Smith Elder

Ferrier D (1879): Pain in the head in connection with cerebral disease. *Brain* 1: 467–483

French RD (1975): *Antivivisection and Medical Science in Victorian Society,* Princeton, N.J.: Princeton University Press

Gowers WR (1885): *Lectures on the Diagnosis of Diseases of the Brain,* London: Churchill

Greenblatt SH (1970): Hughlings Jackson's first encounter with the work of Paul Broca: The physiological and philosophical background. *Bull Hist Med* 46: 555–570

Griesemer J, Wimsatt WC (1989): Picturing Weismannism: A case study of conceptual evolution. In: *What the Philosophy of Biology Is,* Ruse M, ed. Dordrecht, The Netherlands: Kluwer

Harrington A (1987): *Medicine, Mind, and the Double Brain,* Princeton, N.J.: Princeton University Press

Head H (1926): *Aphasia and Kindred Disorders of Speech,* 2 vols. New York: Macmillan

Henig RM (1981): *The Myth of Senility: Misconceptions about the Brain and Aging.* New York: Anchor/Doubleday

Holmes G (1954): *The National Hospital, Queen Square.* Edinburgh, Scotland: Livingstone

Huxley T (1904): On the hypothesis that animals are automata, and its history. In: *Method and Results: Essays.* New York: Appleton [Original work published 1874]

Ireland W (1879): Review of *The Relations of Mind and Brain,* by H. Calderwood *Brain* 1: 535–540

Jackson JH (1873): Observations on the localisation of movements in the cerebral hemispheres, as revealed by cases of convulsion, chorea and aphasia. *West Riding Lunatic Asylum Med Rep* 3: 175–339

Jackson JH (1875): A lecture on softening of the brain. *The (London) Lancet* nvn: 11 (November 1875): 489–494

Jackson JH (1878): On affectations of speech from disease of the brain (Part 1). *Brain* 1: 304–330

Jackson JH (1879): On affectations of speech from disease of the brain (Part 2). *Brain* 2: 323–356

Jackson JH (1932a): On the nature of the duality of the brain. In: *Selected Writings,* London: Hodder and Stoughton. 2: 129–145.

Jackson JH (1932b) Some implications of dissolution. In: *Selected Writings,* London: Hodder and Stoughton, 2: 29–44

Jackson JH (1932c) Evolution and dissolution of the nervous system. In: *Selected Writings,* London: Hodder and Stoughton, 2: 45–75

Jackson JH (1932d): Words and other symbols in mentation. In: *Selected Writings,* London: Hodder and Stoughton, 2: 205–212

Jackson JH (1932e): Notes on the physiology and pathology of the nervous system. In: *Selected Writings,* London: Hodder and Stoughton, 2: 215–237

Jackson JH (1932f): Remarks on evolution and dissolution of the nervous system. In: *Selected Writings,* London: Hodder and Stoughton, 2: 92–118

Journal of Insanity (1870): Review of *An Aphasia, or Loss of Speech, and the Localization of the Faculty of Articulate Language,* by Frederick Bateman

Klapp O (1964): *Symbolic Leaders.* Chicago: Aldine Lakoff G (1987) *Women, Fire, and Dangerous Things: What Categories Reveal about the Mind,* Chicago: University of Chicago Press

Lakoff G, Johnson M (1980) *Metaphors We Live By* Chicago: University of Chicago Press

Lassek AM (1970): *The Unique Legacy of Doctor Hughlings Jackson,* Springfield, Ill.: Charles C Thomas

Laurence S (1977): *Localization and Recovery of Function in the Central Nervous System.* Ph.D. dissertation, Clark University, Worcester, Mass

Laycock T (1857): *Lectures on the Principles and Methods of Medical Observation and Research,* Philadelphia: Blanchard and Lea

Levin M (1960): The mind-brain problem and Hughlings Jackson's doctrine of concomitance. *Am J Psychiatry* 116: 718–722

Levin M (1965): Our debt to Hughlings Jackson. *J Am Med Assoc* 191: 991–196

Lyons JB (1965): Correspondence between Sir William Gowers and Sir Victor Horsley. *Med Hist* 9: 260–267

Maudsley H (1890): *Body and Mind: An Enquiry into Their Connection and Mutual Influence,* 2nd ed. New York: Appleton

Mitchell EG (1960): Writings of John Hughlings Jackson, *Cerebral Palsy Bull* 2: 34–35

Riese W (1959): *A History of Neurology.* New York: MD Publications

Ross J (1882): Review of *Studien über das Bewusstsein,* by S Stricker, *Brain* 5: 99–105

Schon D (1963): *The Displacement of Concepts.* Cambridge, Mass.: MIT Press

Schmitt FO, Worden F, eds. (1979): *The Neurosciences: Fourth Study Program.* Cambridge, Mass.: MIT Press

Seguin E (1881): A second contribution to the study of localized cerebral lesions, *J Nerv Ment Dis* 8: 510–552

Sharpey-Schafer E (1927): *History of the Physiological Society during its First Fifty Years, 1876–1926,* London: Cambridge University Press

Smith CUM (1982): Evolution and the problem of mind: Part 2. John Hughlings Jackson. *J. Hist Biol* 15: 241–262

Sperry R (1980): Mind-brain interaction: Mentalism, yes; Dualism, no. *Neuroscience* 5: 195–206

Star SL (1986): Triangulating clinical and basic research: British localizationists, 1870–1906. *Hist Sci* 24: 29–48

Star SL (1989): *Regions of the Mind: Brain Research and the Quest for Scientific Certainty,* Stanford, CA: Stanford University Press

Star SL (1990): What difference does it make where the mind is? Some questions for the history of neuropsychiatry. *J Neurol Clin Neuropsychol* 2: 436–443

Star SL (1991): The sociology of the invisible: The primacy of work in the writings of Anselm Strauss. In: *Social Organization and Social Processes: Essays in Honor of Anselm Strauss,* Maines D ed., Hawthorne, N.Y.: Aldine de Gruyter, pp. 265–283

Von Bonin G, ed. (1957): *Some Papers on the Cerebral Cortex,* Springfield, IL: Charles C Thomas

Walshe FMR (1961): Contributions of John Hughlings Jackson to Neurology. *Arch Neurol* 5: 119–131

Wimsatt W (1976): Reductionism, levels of organization, and the mind-body problem. In: *Consciousness and the Brain* Globus G, Maxwell G, Savodnik I, eds. New York: Plenum, pp. 199–267

Young RM (1970): *Mind, Brain, and Adaptation in the Nineteenth Century: Cerebral Localization and its Biological Context from Gall to Ferrier.* Oxford, England: Clarendon Press

15

Other "Ways of Knowing": The Politics of Knowledge in Interwar German Brain Science

Anne Harrington

> Since the ultimate process of life can never be fully explained through causal-mechanical analysis, the question arises whether the physician . . . may not also have at his disposal some other means of knowledge. At once such concepts as empirical knowledge, consideration of the whole, and intuition spring to mind.
>
> — Hans Löhr (1935)

A Nostalgic Introduction

The year is 1968, and the American psychologist Gardner Murphy is looking back 4 decades and wistfully recalling a Camelot-like time in the brain and mind sciences. It was a time filled with excitement about (in Murphy's words) a "new and vital way of viewing organisms, a view 'from above.'" In the United States, Karl Lashley's work on laboratory mice had contributed to a "renaissance of belief in the wholeness of central nervous system function." In England, Henry Head's post–World War I studies of brain-damaged soldiers had raised interest in the dynamic reorganizing capacities of the human brain following localized damage. In the lab, Sir Charles Scott Sherrington's work on the "integrative" nature of reflex nervous action was overturning the 19th-century view of the organism as a bundle of sensory-motor reflex arcs and pointing the way toward a hierarchical, biologically purposeful alternative.

In spite of these international reorientations, however, it was clear, looking back, that something particularly rich and urgent was going on in German-speaking Central Europe during this period. There was Hans Driesch, who resurrected teleological, vitalistic thinking in biology with his supramaterial concept of the *entelechy;* there was Jakob von Uexküll, who called attention to the irreducible unity of an organism in an environment characterized by biological meaning (an orientation that would influence both ethology and psychosomatic medicine in Germany); there was the holistic *(Ganzheit)* school of biopsychology under Felix Krueger that emphasized the primacy of affect and irrationality in the unfolding of perception and cognition; there was the school of Gestalt psychology in Berlin under Wertheimer, Köhler, and Koffka that stressed the way in which perceived reality was ordered by active contributions of the perceiving mind/brain; there was the dynamic, teleological neurology of Constantin von Monakow that emphasized the regenerative capacities of the brain; and there was the psychosomatic "organismic" neuropsychiatry of Kurt Goldstein that celebrated health as biological "value." The list could be extended, but enough has been said to make the point.

The nostalgia pervading the Murphy assessment of all this work 40 years after the fact is striking (and his essay is just one example of a type that was widespread). He invites us to gather together to honor men who, bucking the mechanistic tide, had sensed instinctively the correctness of a view "in which life, instead of being interpreted from beneath in terms of supposed physical and chemical processes of accretion and blind interaction, somehow sets the stage upon which physical and chemical realities can occur" (1968). We look back at a world in which Big Truths were being affirmed in the face of the pettiness of the past. The old inadequacies of the "machine model" of brain and mind were being rejected; the new watchwords "dynamism" and "wholeness" were ringing in everyone's ears; and the philosopher Christian von Ehrenfels's famous formulation, "the whole is greater than the sum of the parts," served as the unofficial clarion call of the movement (Ehrenfels, 1891). In both postwar America and postwar Germany, the convictions of this Camelot were fading or had faded, but for one brief shining moment, we had rarely all felt so good about ourselves.

Beyond Nostalgia: Locating Holistic Brain Science in Cultural Politics

Now this inspiring story of reformism has its own truths, but it is not the whole story. One of the jobs of the historian is to set different *kinds* of stories, different levels of truth against each other, and to see what patterns emerge. Thus this chapter aims to ask some preliminary questions about the alliances and larger cultural aims mostly obscured by the Murphy et al. "feel-good"-type accounts of holism's history. Here I am interested in exploring what holism as a research program actually "meant" for the historical actors who were involved in its creation, and particularly in trying to understand the peculiar passion driving the *German* holistic reformist effort in early 20th-century brain science and psychology.

The story this chapter will try to tell — and it is only one of many that could and should be told about German holistic mind and brain science[1] —

---

[1] I am in the process of completing a book-length study on German holistic life and mind science, provisionally entitled *Reenchanting Life and Mind: The Wholistic Perspective in German Psychobiology from Wilhelm II to Hitler*.

is concerned with what one may call the *politics of epistemology*.[2] In telling this story, I am concerned in the first place with the problem of what sorts of insights scientists (specifically neurologists and psychologists) in interwar Germany were prepared to accept as legitimate knowledge within science. I am concerned in the second place with how these scientists understood the collective ideological or value-laden implications of their epistemological choices.

The 19th-Century Legacy of Associationism

I begin, then, where the holists generally started themselves: with a hard look at the enemy to be vanquished. Fortunately, the images of this creature are clearly drawn for us by the holists — generally in bold caricature for strategic purposes — and can be easily reconstructed. In psychology, the enemy of holistic ideals is unanimously recognized as *associationism,* a philosophy of mind that had its origins in 18th-century British empirical philosophy. It was the Englishman Thomas Carlyle who, in his influential essay "Signs of the Times" (1829) had first identified the spirit of the "Age of Machinery" with the associationist philosophy of John Locke. As he saw it, Locke's doctrine of the mind was mechanistic because it emphasized the accumulation and sorting of external data according to fixed laws and played down, if it did not outright deny, an active, synthetic role for the mind. The doctrine, Carlyle said, "is mechanical in its aim and origin, its method and its results. It is not a philosophy of the mind; it is a mere . . . genetic history of what we see *in* the mind" (cited in Marx 1978 [1988], p. 168). As such, it was an epistemology that encouraged individual fatalism and collective passive adherence to the instrumentalist, physicalist goals of the emerging machine age.[3]

The enemy denounced with a vengeance! But we need all the same a somewhat more differentiated perspective by which we can understand the way in which associationism would become the nemesis of holistic enemies

---

[2] Sandra Harding, in her feminist analysis of scientific methodology, defines an epistemology as "a theory of knowledge. It answers questions about who can be a knower . . .; what tests beliefs must pass in order to be legitimated as knowledge . . .; what kinds of things can be known . . ., and so forth. Sociologists of knowledge characterize epistemologies as strategies for justifying beliefs: appeals to the authority of God, of custom and tradition, of "common sense," of observation, of reason, and of masculine authority are examples of familiar justificatory strategies" (1987, p. 3).

[3] As Carlyle stingingly noted elsewhere, not only are we meant to believe that the universe was "one huge, dead, immeasurable Steam-engine, rolling on, in . . . dead indifference"; we must turn away from all contrary evidence of the "Stupendous" in "stupid indifference," each of us having learned from associationism to see ourselves also as "a mere Work-Machine, for whom the divine gift of Thought were no other than the terrestrial gift of Steam is to the Steam-engine; a power whereby cotton might be spun, and money and money's worth realised" (Carlyle, 1833 [1974], pp. 926, 949).

of the machine mind and brain, particularly in the German-speaking world. To do this, we must begin by noting one crucial feature of the associationist model of mind: its rejection of the "metaphysical" notion of innate "faculties" of mind (that would have allowed for "intuitive" knowledge of such things as the existence of God, good and evil, etc.), and its assertion that *all* human knowledge and experience has its origin in sensation (the associationist companion doctrine of *sensationism*). Following this reception of sense data by the special senses, various physiological processes were then supposed to occur that combined and stored the information in such a way that it could later be mechanically revived by the brain in the form of "representative images" or primitive ideas. These ideas—these atomistic units of thought—once revived, were held to "associate" with one another in accordance with certain fixed, rational "laws" (of resemblance, contiguity, etc.). All interactions between elements happened in a necessary, predictable manner comparable to the way in which material atoms were believed to interact and combine in three-dimensional Newtonian space—even while it was clear that the "laws" of mind and the "laws" of matter ruled separate realms. The units of thought themselves—the mental atoms—were pretty grim creatures that would have gratified any Coketown-style ("fact, fact, fact!") Thomas Gradgrind. While clearly one step removed from reality itself, they were nevertheless reliable missives from the external world, variously defined by the associationists as "copies," "pictures," or "representative images" of real-world experience. In sum, we have here to do with a model of mind that was a theory of *knowledge* first (empiricism), and a description of the "workings" of thought only secondarily.

How did it come about that an empirical philosophy developed in Britain would, by the late 19th century, have begun to trouble the sleep of *German*-speaking wholists? At first glance, it would seem that a country that had been dominated since the late 17th century by such rationalist, idealist philosophers as Leibniz, Kant, and Hegel would not have taken up company with such alien associates as Locke and Hume. Certainly some 20th-century holistic advocates would look at the late-19th-century swing toward empiricism as an aberration, if not a willful betrayal, of German cultural identity. Yet, if we look at the matter from their point of view—that is, using "authentic" German categories of historical explanation—we must wonder why Hegel himself did not warn his followers that his rationalistic thesis would necessarily carry in it the seeds of an empiricist antithesis (cf. Peters and Mace, 1967, p. 18).

Whatever the most plausible explanation, there was little dispute among later holistic advocates in the sciences of mind that the 19th century had witnessed a steady process of accommodation to the mechanistic mental philosophy of the 18th-century British empiricists. Classic signposts identified along the way included the work of Johann Friedrich Herbart (1776–1841), who grafted associationist dynamics onto a more traditional unifying soul; the work of Ernst Heinrich Weber (1795–1878) and Gustav

Theodor Fechner (1801–1887), who advanced the use of quantitative methods in psychology, focusing on incremental, elemental sensations; and the work of Wilhelm Wundt (1832–1920), who leaned heavily on associationist philosophy in developing an experimental psychology devoted to precise identification of the shifting furniture of consciousness (using observers trained in the exacting methods of introspection). Associationism, for better or worse, turns out to have been the model of mind most central to the practical development of experimental research in Germany.

In the 1870s, this eminently practical model of mind was given a further dramatic lease on life through its incorporation into the language and practical activities of neurology. The neurologist Carl Wernicke in Breslau "neurologized" the mental atoms of the associationists by turning them into primitive units of sensory-motor "memories." In place of the associationist mind "container," he offered a brain that stored all sensory-motor impressions in special localized regions. The back of this brain was specialized for processing and storing sensory data, taking in information from the outside world; the front of this brain was motor, reacting to the world in response. The basic architecture of Wernicke's brain, in other words, was built on the model of the sensory-motor reflex loop. Within the brain, sensory-motor units communicated with each other along so-called association fibers. They ran up and down them like so many electrical pulses along a telegraph line, interacting in accordance with the established psychological "laws of association," and generating in this way the full complexity of mind and consciousness (Wernicke, 1874).

This model of human brain functioning was an immediate success, providing a framework for the creation of multiple new clinical entities (the aphasias, the agnosias, the apraxias) and triggering what has come to be regarded as a "golden age" in the history of the study of the human brain. The most immediate advantage of this approach to brain/mind functioning was the way in which it offered neurologists a practical visualizing technology for making sense of the *breakdown of the human machine*. Generations of medical students — armed with their paper-and-pencil diagrams of centers and connections — learned to listen to the fragmented or chaotic language utterances of their patients, not for the human stories they could have been representing, but as indirect indicators of this or that type of "disconnection" or primary damage. In this non dialogue, the final doctor-patient encounter generally occurred at the patient's autopsy, when the neurologist would check the accuracy of his predictions against the revealed physical state of pathology or injury in the patient's brain.

Then, sometime in the first decade or so of the 20th century, people began increasingly to voice their discontent with this spectacularly successful and useful model of mind and brain functioning. Why? Gardner Murphy and other guardians of the nostalgic history of holism stress that the essential causes for the early 20th-century reaction against associationism in psychology and neurology were experimental and empirical.

Mention is typically made of the experimental attacks on associationism marshalled by the Würzburg school of psychology (e.g., Külpe, 1893) and, later, by the Gestalt school (e.g., Wertheimer, 1912). Also important to this story are the clinical objections to the associationist model of brain function and breakdown raised by such men as Arnold Pick (1913), Kurt Goldstein and Adhémar Gelb (Gelb and Goldstein, 1925), Constantin von Monakow (1911), and Henry Head (1926). Again, there is significance to all these developments, but they do not add up to the story I want to tell here.

Associationism Revealed as the Epistemology of Positivism

I invite us instead to consider the following: sometime before the holistic psychologists and neurologists had begun their attack on the associationist ideas, the idealistically oriented post-Kantian philosophical community had already attacked it on *epistemological grounds* (Ringer, 1969, pp. 90–93). The specific objections of the philosophers centered on the fallacies of the so-called commonsense view of reality. The naive man on the street supposes that his knowledge of the external world is derived from trust-worthy perceptions. By adding up all his various perceptual experiences, he believes himself in a position to arrive at an ever more complete knowledge of external reality.

This same "commonsense" epistemology, stressed the idealists, was fundamental to associationist psychology—with its stress on the sensory basis of all knowledge and its additive, passive view of cognition. In fact, associationism was often even more naive than the man on the street, since it supposed that even our subjective impressions about the outside world could be explained in terms of that same outside world; namely, as end products of various physiological processes in our brains.

All this (said the neo-Kantians) began to explain why the associationist model of mind had been such a success in the mid–19th century. This was (as all holists equally knew) the era of the hard-nosed scientist—the era of those uncompromising "facts-and-dust" men like Darwin and biophysicists Hermann von Helmholtz and Emil Du Bois-Reymond. Of course, it would just suit the materialist, secularist ideology of these men if it should turn out that only units of earthly sense experience, atomistically combined and analyzed, should "count" as legitimate data in the game of knowledge.

To give some sense of the emotional stakes, let me remind you what this ideology of "facts and dust" felt like by quoting from one of its more uncompromising expressions in modern European literature, a passage in the 1913 novel *Jean Barois,* by Roger Martin du Gard:[4]

---

[4]I am indebted to Ruth Harris for first calling my attention to the significance of this novel and urging me to read it.

I do not believe that mind and matter are mutually exclusive entities. . . . I know that my personality is but an agglomeration of particles of matter, whose disintegration will end it absolutely.

I believe in universal determinism; that we are conditioned by circumstances in all respects.

Good and Evil are mere arbitrary distinctions.

I believe that though all the phenomena of life have not yet been analysed, they will be analysed one day.

(Martin du Gard, 1913 [1949, pp. 255–256])

Now, it is possible that some neuroscientists among us today would find such statements relatively uncontroversial, but for the generation growing up at the turn of the century, they represented the blackest face of nihilism. Fortunately, however, this was not to be last word on the matter. As early as the late 18th century, Kant had suggested an apparent answer to the spiritually starved. He had made it clear that perception, far from being a straightforward passive experience, in fact involved an active structuring of phenomena according to various hard-wired mental categories and intuitive judgments. This meant that what we call outer "reality" is, in fact, a complex construction of our own subjectivity. Lo and behold, it turns out that empirical knowledge—the knowledge of "dust and facts," of science— is a limited, human thing, and in no sense absolute truth. And this discovery seemed to let a considerable amount of air into the cramped scientific world picture of the late 19th century; it seemed to raise the possibility that one could legitimately turn to "other forms of knowing" for insights into what it means to be a human being with a human mind and brain. Empiricism could no longer justify its claim to exclusive authoritative status.

Holistic Neurology and Psychology as a Revolt against Positivism

This said, let us shift scenes, and look now again at our holistic neurologists and psychologists in early 20th-century Germany, busy with their experimental and clinical work. Let us now ask: given what we know at this point about the central place of associationist psychology in the naive-realist epistemology and positivist ideology of the late 19th century, can we find a further layer of significance in the holistic rejection of the associationist model of mind?

Yes. The argument of this chapter is that German holistic neurology and psychology, especially after 1918, was not only about discovering or constructing organicist or nonatomistic models of mind and brain; it was also about challenging epistemological rigidity in the natural sciences—and challenging it under the conviction that epistemological reform alone could offer a way out of the dust-choking world of *Jean Barois*. The political and cultural imperatives seemed to many more urgent than ever before. Let us consider the background: a bitterly lost war that had betrayed middle-class German youth far more profoundly than our own Vietnam war, political and spiritual lack of direction, international humiliation, a weak government widely believed to have sold out to the enemy, a hyperinflated, unstable economy, fear of Bolshevik revolution, and all manner of beliefs in Jewish and Freemason conspiracies.

In the midst of this sense of cultural crisis, we find a society simultaneously seeking scapegoats *and* groping for new truths that would point the way to renewal. One of the most important of these new truths focused on epistemological opportunities beyond the strictly positivistic. Even before the war, some people had found it possible to argue that emancipation from mechanism and atomism meant no longer being prepared to *think* mechanistically and atomistically, meant a willingness to explore mental possibilities beyond those of dry empiricism and passive association of ideas. On the popular political level, hostility toward what were perceived as the soulless piecemeal truths of natural science began to broaden in the 1890s, and finally crystallized in the postwar years with the rise of the ubiquitous *Lebensphilosophien* (philosophies of life), which celebrated the moral and epistemological superiority of natural, unanalyzed experience over dry, dissected abstractions. "Life is more than rational *Wissenschaft*," said the philosopher Müller-Freienfels in 1921, "and to me philosophy is not merely *Wissenschaft*, but knowing, knowing even also of that which does not enter into *Wissenschaft*. Indeed, philosophy is more even than knowing; philosophy is itself life" (cited in Forman, 1984, p. 337).

Now inasmuch as the mind and brain sciences were clearly concerned both with knowing and life — and hence with the world beyond the "merely" rational — scientists in these fields found themselves potentially in an odd position of particular authority in the collective cultural struggle against epistemological and experiential limits. The message from holism was that mind and brain properly studied — *not* mechanistically, but with an intuitive *(anschaulich)* "feeling for the whole" — could reveal truths suited not just for the head, but for the heart.[5] Thus local stories about minds, brains, and life processes were expanded by the holistic scientists themselves into existential, moral, and political stories for use and consumption on other levels of the culture.

How did all this work? I begin with a name that at least will be familiar

---

[5]Certainly there was also resistance to this idea of incorporating intuition *(Anschauung)* into the methodologies and epistemologies of a new, more "holistic" mind and brain science. Although no friend of mechanism and atomism himself, Max Weber did not hesitate to dismiss the advocates of *Schauung* with an untranslatable, sarcastic pun: *"Wer schauen will, gehe ins Lichtspiel"* ["Whoever wants to *Schauen* (with its double meaning of *intuit* and *look*), let him go visit the cinema!"] (cited in Wyatt and Teuber, 1944, p. 231n.).

to many neuroscientists today: Kurt Goldstein. Goldstein was a direct product both of the Wernicke school of associationist-connectionist neurology and the Kraepelin school of biological psychiatry, known for its so-called nihilistic approach to the therapeutics of mental illness. In 1913, one finds Goldstein writing a monograph on racial hygiene, in which he blames increases in nervous and mental disorders, mounting suicide attempts, and social dissatisfaction on the strains of modernism, industrialization, and technological living. The overvaluation of the natural sciences had had especially pernicious consequences, leading to depression and confusion, "an overvaluing of the external, the material and . . . an undervaluing of the intellectual and the spiritual" (Goldstein, 1913, p. 52). The urgency of this sort of rhetoric should not be underestimated: the behavioral biologist Jakob von Uexküll tells the story in his memoirs about a fellow university student who was so distressed on reading Darwin that he ultimately committed suicide (Uexküll, 1936).

Looking at Goldstein's subsequent career as a clinical neurologist, we can see a gradual process of grappling with limits in his chosen field — theoretical limits, certainly, but limits that were also epistemological and value-oriented. The process took a while: it was not until the early 1930s — on the dawn of the rise of Hitler that would drive him out of Germany — that Goldstein would find the intellectual strength to come to the decision that a "holistic" approach to the phenomena of mind and brain meant that the neurobiologist must ultimately become something more than a natural scientist if he hoped to come to terms with what was most important in his subject matter. In the final analysis, he must choose the penetrating intuitive, holistic *"Schau"* of a Goethe over the mechanistic analysis of a Newton (this comparison between Newton and Goethe is his — see Goldstein, 1933). In his words, "The knowledge we need can be comprehended only by a special mental procedure which I have characterized as a creative activity, based on empirical data, by which the 'nature' comes, as a *Gestalt* [italics added], increasingly within the reach of our experience" (Goldstein, 1959a; see also Goldstein, 1939).

What did this mean for the practical business of clinical neurology? Goldstein's point is self-consciously radical: looking with this nonatomizing, Goethean gaze at his patients, the neurologist becomes able to "see" (directly intuit) that humans are not simply made up of physics and physiology; they possess as well a nonreducible "essence" or essential personality. Because human organisms have "essence" as well as organs, sickness and health are not simply issues of malfunction and function. They become issues of free choice and value. The so-called catastrophic reaction of Goldstein signals an existential encounter of the essence of the individual with his diminished capacity, and the biological adaptation of the organism to such malfunctioning becomes an issue of life-affirming value (Goldstein, 1959b).

The neurological existentialism of Goldstein represents one solution to the limits of the 19th-century scientific worldview. Let us turn now briefly to a second—very different—solution by a neurologist whose name is also familiar to many neurologists today: Constantin von Monakow. In the 1950s, a collection of Monakow's interwar writings were collected and published in a book entitled *Gehirn und Gewissen* (which can be translated somewhat awkwardly as "brain and moral sense"). This tells you right away where we are. The writings represent Monakow's attempt to show how biology (specifically neurobiology) might offer mankind a source of enduring values in a world torn apart by nationalist rivalries and alienated from its true roots. Deeply shaken by the events of the First World War, Monakow came in the interwar years to share the deep pessimism and sense of crisis of so many of his contemporaries. His search for answers took him back to his science, and in both the limits and the power of the nervous system he thought he had discovered a beacon with which to cut through the moral darkness of the postwar years and reveal the road to spiritual renewal.

To begin, neurobiology taught Monakow that the sort of reasoning employed in natural science represented only one of several possible human (biologically moderated) approaches to truth, and was in no sense a road to Truth itself (Monakow, 1930a). This was important, because it then became possible to conclude that membership in the scientific community need not be inconsistent with belief in mystical-religious realities.

But that was not all. Monakow's understanding of the brain was rooted in a dynamic hierarchical model of nervous system functioning (influenced by Jacksonian evolutionism tempered by Bergsonian vitalism and teleology). In this model, different levels of instinctive functioning were arranged in a hierarchical relationship to one another from the most primitive (basic survival drives) to the most sophisticated (parental love, social instincts, and beyond). Each level of the instinct system was not only functionally significant for the whole, but possessed a differentiated *natural* (biologically given) value within it as well, corresponding to its place on the hierarchy (see Monakow, 1916). It turns out that the biological instinct to believe and hunger after spiritual truths stood in the Monakow system at a higher level than the impulse to understand and master the world through science and reason. The nervous system, in short, not only contained within it a source of insight into authentic ("natural") moral values; it also contained within it an epistemological hierarchy, a hierarchy of relative Truth. How did Monakow know this? The answer introduces us to his most radical epistemological break from the scientific traditions in which he had been trained. For Monakow, insight into the biological hierarchy of values and truth was not possible through the rationalistic, analytical methods of the natural sciences. It was instead a product of "biological," tacit knowing. Monakow called the source of this knowledge the *syneidesis* or biological conscience (Monakow, 1927). The moral decisions of daily life, he concluded, were equally not things to be decided rationally or according to convention, but were properly things of biological "inner knowledge." This faith in the mysterious ways of the *syneidesis* would lead Monakow to suggest in 1930 (two years before his death) that the irrationality of the war had perhaps been the birth pains of a spiritual revival – "a stormy preliminary stage in a powerful spiritual world movement" that was only now making its presence felt, especially among the young (1930b, p. 80).

My last example of what I have been calling the "politics of epistemology" focuses not on holistic neurology, but rather on its close cousin, Gestalt psychology. The nostalgic story of holistic mind and brain science stresses how the so-called Gestalt revolution was a conceptual response to the inadequacies of 19th-century atomistic ideas, particularly in the field of perception. Yet psychologists in the 1920s and 1930s also often attached important epistemological imperatives to the Gestalt concept. To understand why this was so, one must know that the idea of "Gestalt" was part of a much larger cultural tradition in Germany. In Goethe's teachings, Gestalts had been the primal forms or ideas out of which nature built all her variations – they were the timeless Platonic forms of pure thought behind the flux of external being. By observing and comparing the various metamorphoses of these forms, the primal form of the type could be deduced (as Goethe would attempt to demonstrate with his description of the primal plant or *Urplanze*). To begin to appreciate the appeal of such a system for an antimechanist philosophy in search of a method, it is important to understand that these *Ur-Gestalten* postulated by Goethe were at once very real *and* very much nonmaterial; they were self-actualizing ideas or Platonic potentialities that were realized *in,* but not created *by* matter and mechanical law. In other words, Goethe's morphology postulated a generative aesthetic blueprint *behind* the variety and apparent disorder of material life (cf. Schneider, 1979).

For *völkisch* writers in the early 20th century who made extensive use of the Gestalt concept – men like Oswald Spengler (1918) and Houston Stewart Chamberlain (1928) – it was clear that these Gestalts could be grasped only though holistic intuition. The atomizing methods of reason and analysis were useless here. This is because, try as they might, such methods were inherently only suitable for seeing mechanism and "dead" law. One is reminded of the old saw that points out, that if the only tool you have is a hammer, you tend to treat everything like a nail. Chamberlain made the case for intuitionism more sharply in the course of identifying racial purity with purity of Gestalt: "Descartes pointed out that all the wise men in the world could not define the color 'white'; but I need only to open my eyes to see it, and it is the same with 'race' " (*Grundlagen,* cited in Field, 1981, p. 216). Thus we begin to see the noninnocence of these only apparently esoteric debates.

As Gestalt psychology developed in Germany in the 1920s, we see the rise of similar epistemological concerns and claims. It turns out that the internationally recognized Berlin school of Gestalt psychology (the school that would largely emigrate to the United States after 1933) was widely criticized within Germany by other holistic psychologists for its failure to grasp the irrationalist, intuitionist spirit of the Gestalt tradition. The Berliners, it was said, claimed to be holists but, epistemologically, they were still fundamentally "positivistic" — too experimental, empirical, and relying on a "dogmatic phenomenology" that ultimately (in the words of one holistic psychologist, Ferdinand Weinhandl) "was just as much a product of the true Jewish spirit as Husserlian phenomenology" (Weinhandl, 1940, p. 5).

In contrast, the Leipzig school of holistic psychology under Felix Krueger (that saw Gestalts as the end point of a psychological genetic process) was committed to the necessity of a certain epistemological irrationalism — the primacy of immediate, nonverbalized experience — for grasping the ultimate meaning of psychological and social events (see Geuter, 1985). During the Hitler years (after all the founders of the Berlin school of Gestalt psychology had emigrated), such intuitionist forms of Gestalt and holistic psychology would be channeled in Germany in increasingly practical directions. In one development, it merged with the German "character" psychology where it was used in the military — to "holistically" identify men of officer caliber (cf. Weinhandl, 1931; Geuter, 1984).

Conclusion: Politics and Epistemology in Holism

This final example of Gestalt psychology will begin, I think, to make it clear that the "politics of epistemology" in holistic mind and brain science could in some cases be quite a public form of politics indeed. In concluding this chapter, I would therefore like to take a moment first to linger over this question of the relationship between *personal* politics in our story (that is, a scientist's linking of epistemological questions to his personal cultural or nonscientific concerns) and the formal spectrum of Big Politics, from liberal republicanism to socialism to fascism. Do the personal politics and the public politics of holistic epistemology map onto each other in any reliable fashion?

There is a respected tradition in the literature that argues for a link between antiscientific, irrationalist philosophies of "life" and fascist organicist thinking (Viereck, 1965; Lukács, 1957; Stern, 1961; Mosse, 1964). Was holistic mind and brain science epistemologically and politically part of this network of alliances? Certainly, there is no question that certain irrationalist strains of German holistic psychology and neurology did find a home with National Socialism (especially in the early years). The Leipzig holistic psychologist Friedrich Sander would go so far as to declare that politically and epistemologically:

Present-day German psychology and the National Socialistic world view are both oriented towards the same goal: *the vanquishing of atomistic and mechanistic forms of thought.* The one [focuses on] the structure of volkish life, the other [on the] researching of spiritual reality—but both aim to conquer through *organic thinking.* . . . If, with believing heart and thoughtful sense, one intuites the driving idea of National Socialism back to its power-source, one rediscovers two basic motives behind the German movement's colossal struggles: the longing after *wholeness* and the will towards *Gestalt.* . . . [Each German] must eliminate ["Ausschalten"] all those alienated from Gestalt; above all, he must nullify the power of all destructive foreign-racial influences. (Sander, 1937; emphases added)

Yet to admit a (complex) historical relationship between holistic science and fascism is *not* to admit any sort of logical inevitability in the directional arrow between personal and public politics. Let us consider the case of Kurt Goldstein. A Jew and a Socialist imprisoned by the Nazis in 1933, he used his epistemological holism as a doorway to a philosophy of existentialism and individual choice in an inevitably imperfect world. Or consider Constantin von Monakow: mystical and elitist though he was, deploring the superficiality and emptiness of modern life and celebrating the higher epistemological status of intuited truth, he was nevertheless a cosmopolite who rejected nationalism and worked for international cooperation in science. To the end of his days, he spoke proudly of his role in heading the 1904 International Brain Commission for cooperation in the neurosciences (Monakow, 1970).

On other fronts, our attention to the epistemological dimension of the holistic reaction in psychology and brain science points us toward a number of broader contextual themes we might also fairly call "political." For example, it is well known that, partly in reaction to a postwar international boycott on German science, some German critics began increasingly to cast doubt on the very possibility of a science that transcended national borders, and sought to expose the Enlightenment lie of value-free scholarship (Forman, 1973). Given that the holistic reaction in German psychology and neurology was reaching its peak of influence during these same years, an analysis of influence and relationship here would seem both logical and potentially quite provocative. While space does not allow me to pursue the point here, I believe that careful analysis of the interwar German holistic movement in psychology and the brain sciences would reveal clear cases of science formally contributing to the "biologization" and "sociologization" of its own knowledge-claims. In such cases, nationalistic rivalries in science would show themselves to have interacted with epistemological skepticism in a variety of critical ways.

Clearly, no analysis of the cultural values shaping the 20th-century holistic reaction against psychological and neurological atomism would want to *reduce* it to epistemological conflicts. At the same time, we may be surprised, the longer we look, at just how pervasive and insistent the epistemological element turns out to be. And then we should perhaps wonder why we were ever surprised. As Sandra Harding says, pointing out the impact of feminist critiques of knowledge on familiar beliefs about

242 Anne Harrington

women, men, and social life, "How can it be otherwise when our ways of knowing are such an important part of our ways of participating in the social world?" (Harding, 1987, p. 189).

References

Carlyle T (1833): *Sartor Resartus.* In: *The Norton Anthology of English Literature,* 3rd ed., Abrams MH, ed. New York: Norton, 1974, 2:922–953

Chamberlain HS (1928): *Natur und Leben,* Uexküll J von, ed. Munich, Germany: F Bruchmann AG

Du Gard RM (1913): *Jean Barois,* Stuart Gilbert (trans). New York: Viking Press, 1949

Ehrenfels C von (1890): Über Gestaltqualitäten. *Vierteljahrsschrift für wissenschaftliche Philosophie* 14:249–292

Field GG (1981): *Evangelist of Race: The Germanic Vision of Houston Stewart Chamberlain.* New York: Columbia University Press

Forman P (1973): Scientific internationalism and the Weimar physicists: The ideology and its manipulation in Germany after World War I. *Isis* 64:151–180

Forman P (1984): Kausalität, Anschaulichkeit, und Individualität, or how cultural values prescribed the character and lessons ascribed to quantum mechanics. In: *Society and Knowledge: Contemporary Perspectives in the Sociology of Knowledge,* Stehr N, Volker M, eds. New Brunswick, N.J.: Transaction Books, pp. 333–347

Gelb A, N Goldstein K (1925): Psychologische Analyse hirnpathologischer Fälle. XI. Über Farbennamenamnesie, nebst Bemerkungen ueber das Wesen der amnestischen Aphasie überhaupt, etc. *Psychologische Forschung* 6:127–186

Geuter U (1984): *Die Professionalisierung der deutschen Psychologie im Nationalsozialismus.* Frankfurt: Suhrkamp Verlag

Geuter U (1985): Das Ganze und die Gemeinschaft — Wissenschaftliches und politisches Denken in der Ganzheitspsychologie Felix Kruegers. In: *Psychologie im Nationalsozialismus,* Graumann, ed. New York/Berlin: Springer-Verlag, pp. 55–88

Goldstein K (1913): *Über Rassenhygiene.* Berlin: Julius Springer

Goldstein K (1933): Die ganzheitliche Betrachtung in der Medizin. In: *Einheitsbestrebungen in der Medizin,* Brugsch T, ed. Dresden/Leipzig: Th. Steinkopf, pp. 143–158

Goldstein K (1939): *The Organism: A Holistic Approach to Biology Derived from Pathological Data in Man,* foreword by Lashley KS. New York: Americam Book Co.

Goldstein K (1959a): Notes on the development of my concepts. *J Ind Psychol* 15:5–14

Goldstein K (1959b): Health as value. In: *New Knowledge in Human Values,* Maslow A, ed. New York: Harper, pp. 178–188

Harding S (1987): Introduction: Is there a feminist method? In: *Feminism and Methodology,* Harding S, ed. Bloomington: Indiana University Press, pp. 1–14

Head H (1926): *Aphasia and Kindred Disorders of Speech.* New York: Hafner Publications

Külpe O (1893): *Grundriss der Psychologie auf experimenteller Grundlage dargestellt.* Leipzig

Löhr H (1935): The physician must come to terms with the irrational. In: *Nazi Culture: Intellectual, Cultural, and Social Life in the Third Reich,* Mosse GL, ed. New York: Grossett and Dunlap, 1966, pp. 227–234

Lukács G (1957): *Die Zerstörung der Vernunft.* Darmstadt/Neuwied: Luchterhand, 1973

Marx L (1978): The neo-romantic critique of science. Reprinted in: *The Pilot and the Passenger: Essays on Literature, Technology, and Culture in the United States.* New York: Oxford University Press, 1988. [First published as "Reflections on the Neo-Romantic Critique of Science" in *Daedelus*]

Monakow C von (1911): Lokalisation der Hirnfunktionen. In: *Some Papers on the Cerebral Cortex,* Bonin Guon (ed and trans). Springfield, Ill.: Charles C Thomas, 1960, pp. 231–250

Monakow C von (1916): Gefühl, Gesittung und Gehirn [Drei Vorträge, gehalten im Psychiatr.-neurologischen Verein in Zürich am 23. Oct., 13 & 27 Nov., 1915]. Reprinted in: *Gehirn und Gewissen. Psychobiologische Aufsätze,* Minkowski M, ed. Zurich: Morgarten Verlag Conzett and Huber, 1950, pp. 97–230

Monakow C von (1927): Die Syneidesis, das biologische Gewissen. Reprinted in: *Gehirn und Gewissen. Psychobiologische Aufsätze,* Minkowski M, ed. Zurich: Morgarten Verlag Conzett and Huber, 1950, pp. 231–282

Monakow C von (1930a): Wahrheit, Irrtum, Lüge. Reprinted in: *Gehirn und Gewissen. Psychobiologische Aufsaetze,* Minkowski M, ed. Zurich: Morgarten Verlag Conzett and Huber, 1950, pp. 283–340

Monakow C von (1930b): Religion und Nervensystem. Reprinted in: *Gehirn und Gewissen. Psychobiologische Aufsaetze,* Minkowski M, ed. Zurich: Morgarten Verlag Conzett and Huber, 1950, pp. 341–373

Monakow C von (1970): *Vita Mea/Mein Leben,* Gubser AW, Ackerknecht EH, eds. Bern/Stuttgart/Wien: Hans Huber

Mosse G (1964): *The Crisis of German Ideology.* New York: Grosset and Dunlap

Murphy G (1968): Personal impressions of Kurt Goldstein. In: *The Reach of Mind: Essays in Memory of Kurt Goldstein,* Simmel M, ed. New York/Berlin: Springer-Verlag, pp. 31–34

Peters RS, Mace CA (1967): Psychology. In: *The Encyclopedia of Philosophy.* New York: Macmillan/The Free Press, 7:1–27

Pick A (1913): *Die agrammatischen Sprachstörungen: Studien zur psychologischen Grundlegung der Aphasielehre.* Reprint. Berlin: Springer-Verlag

Ringer FK (1969): *The Decline of the German Mandarins: The German Academic Community, 1890–1933.* Cambridge, Mass.: Harvard University Press

Sander F (1937): Deutsche Psychologie und nationalsozialistische Weltanschauung. *Nationalsozialistisches Bildungswesen* 2:641–661

Schneider MA (1979): Goethe and the structuralist tradition. *SiR* (Fall 1979):453–478

Spengler O (1918): *The Decline of the West,* 2 vols., Atkinson CF (trans). New York: Alfred A. Knopf, 1926, 1928

Stern F (1961): *The Politics of Cultural Despair: A Study in the Rise of the Germanic Ideology.* Berkeley/Los Angeles: University of California Press

Uexküll, J von (1936): *Niegeschaute Welten. Die Umwelten Meiner Freunde. Ein Erinnerungsbuch.* Berlin: S. Fischer Verlag

Viereck P (1965): *Metapolitics: The Roots of the Nazi Mind.* New York: Capricorn Books

Weinhandl F (1931): *Characterdeutung auf gestaltanalytischer Grundlage.* Langensalza: Hermann Beyer and Söhne (*Friederich Mann's Paedogogisches Magazin,* Heft 1324).

Weinhandl F (1940): *Philosophie— Werkzeug und Waffe.* Neumünster: Wachholtz

Wernicke C (1874): *Der Aphasische Symptomencomplex. Eine Psychologische Studie auf Anatomischer Basis.* Breslau: M. Cohn und Weigart

Wertheimer M (1912): Experimentelle Studien ueber das Sehen von Bewegung (Habilitationsschrift, Akademie für Sozial- und Handelswissenschaften zu Frankfurt a. Main). Leipzig: Johnann Ambrosius Barth. Reprinted in: *Drei Abhandlungen zur Gestalttheorie.* Erlangen: Verlag der Philosophischen Akademie, 1925, pp. 1–105

Wyatt F, Teuber HL (1944): German psychology under the Nazi system: 1933–1940. *Psychol Rev* 51:229–247

PART 5

Knowledge and Values across
Disciplines: Reconstruction
and Analysis of an
Interdisciplinary Dialogue

16

At the Intersection of Knowledge and Values: Fragments of a Dialogue in Woods Hole, Massachusetts, August 1990

RECONSTRUCTED BY ANNE HARRINGTON*

Morning Roundtable Discussion: Friday, August 3, 1990

Afternoon Roundtable Discussion: Friday, August 3, 1990

Morning Roundtable Discussion: Saturday, August 4, 1990

Afternoon Roundtable Discussion: Saturday, August 4, 1990

*The following conversations are based on transcript material that was extensively cut and stylistically reworked for clarity and thematic coherence. The goal was not an eidetic rendering of every remark made at the Woods Hole workshop, but a tour through the range of debates covered. The substance of individual remarks was not altered in the course of editing; however, participants named in these discussions were all given the opportunity to clarify their points and correct misunderstandings before publication. I am particularly grateful to Paul MacLean for his input to this editing process. All remaining errors are my own responsibility.

Evening Roundtable Discussion: Saturday, August 4, 1990

Concluding Roundtable Discussion: Sunday, August 5, 1990

Participants

Judith Anderson Woods Hole Marine Biological Laboratories, Woods Hole, MA

Warwick Anderson History and Sociology of Science Department, University of Pennsylvania, Philadelphia, PA

Jason Brown Department of Neurology, NYU Medical Center, New York, NY

Jean-Pierre Changeux Laboratory of Molecular Neurobiology, Pasteur Institute, Paris, France

Terrence Deacon Department of Biological Anthropology, Harvard University, Cambridge, MA

John Dowling Department of Biology, Harvard University, Cambridge, MA

John Durant The Science Museum Library and Imperial College, London, England

Alan Fine Department of Physiology and Biophysics, Dalhousie University, Faculty of Medicine, Halifax, Nova Scotia

Anne Harrington Department of the History of Science, Harvard University, Cambridge, MA

Richard Held Department of Brain and Cognitive Sciences, Massachusetts Institute of Technology, Cambridge, MA

H. Rodney Holmes University of Chicago, Chicago, IL

Ruth Hubbard Department of Biology (emeritus), Harvard University, Cambridge, MA

Stephen M. Kosslyn Department of Psychology, Harvard University, Cambridge, MA

Rodolfo Llinas Department of Physiology and Biophysics, NYU Medical Center, New York, NY

Paul D. MacLean National Institutes of Mental Health, Poolesville, MD

Edward Manier Department of Philosophy/Reilly Center for Science, Technology and Values, University of Notre Dame, Notre Dame, IN

Godehard Oepen McLean Hospital and Harvard Medical School, Cambridge, MA

Diane Paul Department of Political Science, University of Massachusetts, Boston, MA

Massimo Piattelli-Palmarini Center for Cognitive Science, Massachusetts Institute of Technology, Cambridge, MA

Detlev W. Ploog Max Planck Institute for Psychiatry (emeritus), Munich, Germany

Robert Richards Department of History, Philosophy, and Social Studies of Science and Medicine, University of Chicago, Chicago, IL

Judy Rosenblith Wheaton College (emeritus), Norton, MA

Walter Rosenblith Massachusetts Institute of Technology (emeritus), Cambridge, MA

Londa L. Schiebinger Department of History, Pennsylvania State University, University Park, PA

James H. Schwartz Columbia University Medical School, New York, NY

Susan Leigh Star Department of Sociology and Social Anthropology, University of Keele, Keele, England

Frank Sulloway Science, Technology Society Program, Massachusetts Institute of Technology, Cambridge, MA

Elliot S. Valenstein Neuroscience Laboratory, University of Michigan, Ann Arbor, MI

Morning Roundtable Discussion: Friday, August 3, 1990

Target Speakers: JEAN-PIERRE CHANGEUX,
MASSIMO PIATTELLI-PALMARINI, DETLEV PLOOG

1. What Are the Biological Constraints of Human Knowledge?

The Argument: Epistemology is that branch of thought that examines the presuppositions and basis of human knowledge. Here neuroscientists are challenged to reflect on the contributions their discipline might be able to make to epistemological problems. In particular, they are asked whether our ability to achieve accurate knowledge of the external world is constrained by the unique neurobiology of the human brain.

STEPHEN KOSSLYN: I would like to start off with some observations and a question that may lead us to confront a key topic of this conference: What is the relation between the brain and human knowledge? Listening to the papers this morning, I was reminded of a couple of points made by Wittgenstein in his *Philosophical Investigations.* First, in one place in this text he talks about looking at a photograph of a man walking uphill with a dog, and notes the fact that we don't see this as a photograph of a man sliding downhill backward. But a Martian might interpret the photograph differently than we do. Wittgenstein says something like "I don't have to go on to tell you why we see things the way we do." What he's talking about is the fact that pictures are under description, that our brains are built in such a way that we interpret the world in certain ways and not others. At another place in the *Investigations,* he says "If a lion could speak, we could not understand it." That is the second point. What he's driving at is the fact that the lion's form of life, the fact that he runs around on four legs in high grass and all that, has caused him to evolve a different neurocognitive structure than that possessed by humans, who are bipedal and so on.

Now it has begun to become clear in the course of this morning's session that the difference between the cognitive world of the human and the cognitive world of the lion is not just a function of differences in isolated anatomical structures—more of this, less of that. It's a function of a whole system. The frontal lobe has evolved onto this structure presumably. You've got to have higher programmable commands in order to make use of the pathways, and so forth.

This takes me to my question: What do neuroscientists really know about the brain that actually reflects on the nature of knowledge? That is, from a Wittgensteinian perspective, what is it *out there* that shaped things *in here* and constrains what we can know and how we construe or experience the world? Is there anything we know about that? What facts are known that put constraints on what a human being can know versus what a lion can know?

JAMES SCHWARTZ: I don't want to answer the question, but just to add to it, or maybe sharpen it a little further. Although I'm a neuroscientist, I did take a lot of philosophy in college, and as I remember it, John Locke said that the mind is a tabula rasa. The problem with this notion is that people began to ask, "Well, what about reality? How do you know when you're being fooled? What is real?" If your mind is a tabula rasa, how do you know what's happening? As I understand it, Kant tried to solve that problem by saying that the mind comes equipped with devices that organize external reality into trustworthy patterns, even while ultimate reality remains finally unknowable. In the 1830s, Comte came along and said, "The hell with all this stuff. We're just going to look at things and see what happens." And that worked its way through science, until we got to William James, who said "Anything that works is okay." And that's practically the whole story — a very simple history of the philosophy of the last 3 centuries. All right. Now we're faced with a problem, because nobody had solved the initial dilemma: What inside the head corresponds to external reality?

STEPHEN KOSSLYN: No, that wasn't the question. Let me try again. The question was: What inside the head constrains the way that you construe, when you know what you're experiencing is real?

JAMES SCHWARTZ: Well, how can we know what constitutes an explanation to a question like this if we don't know what's out there?

TERRENCE DEACON: I'll take a stab at what I think is an answer to what neuroscience findings can contribute to the attempt to understand what we as human beings can know about ourselves and external reality. I think I contribute to the process. I start by taking a paraphrase of something that Bertolt Brecht in his play *[Galileo]* has Galileo saying to his Inquisitors. They ask him what he's trying to show about the truth of the universe. They think he's trying to get at truth, and, of course, for them, only God knows truth. He basically says, "No, no, my purpose here is to decrease the infinitude of error in the world, not to come up with truth." And I see neuroscientists as doing this same thing: providing us with an expanding picture of our constraints and limitations. The more we learn about the circuits, about what's connected to what, about our ability to stimulate one circuit and activate one system, or damage that circuit and disrupt a system, the more we have learned about the way the mind is limited by the substance that it is and by the kind of system that it is.

Now that means that all our assumptions about what self is, or even what such lofty terms as "emotions" or "ethics" are, have also to be fitted within the constraints we have discovered. As we elaborate those constraints, it seems to me, what we are doing is decreasing the infinitude of variety of theories about what minds can do or be. So my answer to your question is not that we're providing the truth or the answer, or that knowing a connection tells us definitely we've found a function. I would rather say that the work we do eliminates a lot of alternatives.

JASON BROWN: It seems to me that there are inherent limitations in thinking of anatomy as a constraining factor on function. We are groping for a way of thinking about the brain as a mature system that reads off function. And I think that we have a real opportunity at this meeting, where we have a lot of evolutionary theorists, to try to reintroduce into our anatomical thinking a dynamic of growth and change and novelty. All too often, anatomical structures are viewed as static, frozen things. However, if we start to think of structure as a continuously dynamic process in its own right, then we are no longer asking the question how far anatomy must constrain speculations in psychology. Instead we are looking directly into anatomical or physiological process for insights about the way in which new patterns of cognitive process may be formed. I assume there has to be a link between the cognitive realm and the neurophysiological process mediating cognition.

2. Is Human Knowledge a Product of Human Biological History?

The Argument: The problem of the constraints that the brain imposes on the human ability to achieve knowledge of reality continues to be considered. It is suggested that the particular ways in which our brains come to know the world have been dictated in our evolutionary history by the practical value of certain representations of reality over others for our survival as a species (so-called evolutionary epistemology). Some participants are uncomfortable with an approach that appears to want to define "truth" for humans as anything that aids their biological survival. Some suggest that the brain as an organ has evolved into something that has sprung the bare survival calculus of its evolutionary history.

JEAN-PIERRE CHANGEUX: I think there is another way in which we must address this issue of biological constraining factors on human knowledge. The standard Darwinian concept is that the final value of anything is the survival of the species, and of the human species in particular. If this is the case, then in principle we know things in the way we do because they have value for the survival of our species. But, as we all know, knowledge can also be used to the detriment of life, and the life of our species. That means that there is bad knowledge, false knowledge, knowledge we should not trust or accept as corresponding to our biological needs.

ROBERT RICHARDS: Can I add to that? In the Darwinian model, of course, the environment for a given animal provides the criteria for what will work for it and what will not work. And the criterion of "success" is naturally reproductive success. So it would strike me that, in the case of knowledge, the criteria for what works and what doesn't work are, at least initially, biologically constrained. We have certain representations of the world. Some of them will work to solve the problems that biology sets for us, namely, getting food, escaping predators, reproduction, and so forth. One can imagine a building up of a hierarchy of meanings, in which the effectivity of higher level meanings is established by the whole structure, the organism's whole biological way of being-in-the-world.

MASSIMO PIATTELLI-PALMARINI: It seems to me that the problem with all that is the following: You can go up to a certain point with this way of thinking—biological constraints, evolutionary constraints, what works and what doesn't work. Eventually, though, you get to a point where you have something that is *"a tiger,"* or *"the animal that my grandmother prefers,"* and so on, and so forth. At this point, you have to understand under which description the thing you are talking about is going to eat you up, or not eat you. It is not just the matter of having decent descriptions. It is a matter of under which description what I do coincides with what is valuable to think or describe. The classic selective Darwinian story can take you up to a certain point but cannot lead you all the way up to an understanding of the nature of full-blown description, truthful description, and meaning.

To cut a long story short, it is impossible to derive truth from mere success. You have to describe success as that which works *because* it is true, or at least approximately true. And this chronology, from the point of view of evolution and epistemology, leads inevitably into a vicious circle.

ROBERT RICHARDS: That remark is the kind of objection that was made to Darwin: that, for example, you can explain the characteristics of plants and even mobile animals, but you really can't give a biological or an evolutionary account of man. Well, it strikes me that, if you do it with persistence and care, you can come pretty close.

JOHN DURANT: I want to suggest another way of looking at this dilemma from an evolutionary perspective. Assuming that we've got some extra stuff up here in our heads that other animals don't—as I suspect many of us think—the question then becomes: What is that extra stuff for? For what adaptive traits was that stuff selected in our evolutionary history? That's a question that I think is enormously important for what neuroscience is trying to do. Once you could come up with some robust ideas about that problem, then you might want to ask our neuroscientists what range of functions does the extra stuff natural selection has put there actually *commit* us to do? And here a crucial issue is buried. I'll try to get at it.

It seems to me that one of the things that might have been implied in Stephen Kosslyn's question is this: Have we got a brain mechanism that is *only* good for certain things? Or have we got a brain mechanism put there by selection for the reasons that our robust theories suggest, but which actually permits us to do a very much larger—perhaps an indefinitely large—range of things? It's the choice between saying what we have in our head is a lot of thermostats (I'm using "thermostats" as a metaphor for things that are good only for a very restricted range of purposes), or what we have is something more like a variety of universal Turing machines. Turing machines, although they are simple, are capable of an enormous — in fact, an indefinitely large — range of problem-solving activities of a certain well-defined sort. I do think those are *in principle* answerable questions. I also think that at present science is nowhere near answering them.

EDWARD MANIER: Stephen Kosslyn's questions were, "What do neuroscientists know about the brain that reflects on the nature of knowledge?" and "What facts are known to put constraints on what a human being can know versus what a lion can know?" These are huge, even audacious questions.

I wonder if we know enough even about the bat's brain to say which of its circuits constrain the echo locations and discriminations the bat can make. I am a philosopher, not a neuroscientist, but it seems to me we work with rough-and-ready analogs, assuming that we will someday know which circuits in the bat brain map and distinguish the bat's discriminations of types of prey, obstacles, and so on. Do we know enough about the circuitry of any organism's brain to say how that brain maps the world, the environment, the specific ecological niche inhabited by that organism?

Things get very much more complicated if we try to distinguish what a human being can know and what a lion can know. Jimmy Schwartz rightly turns our attention to the specifically epistemological dimension of the human world, the questions raised by Locke and Kant and William James. But how could a knowledge of brain circuitry help with these questions? How does, how could, knowledge of circuitry, of molecules, of membranes, help us distinguish the visual images of real past experience from our most fantastic visual fiction? How does, how could, our knowledge of the brain help us understand why we think we're behaving rationally when we actually are not? Why we think we're telling a true story about our most important beliefs and desires, when we're actually confabulating a fiction to put the best face on our behavior? There are other basic issues: What does it mean to describe an event in my recent past? To have self-knowledge? To be a human self? These questions have to be clarified while we are asking if it is possible to construct neural models or even animal models of what a human being can know.

I think neurobiologists and psychologists and philosophers have to work together on these questions. In that spirit, I seek clarification of the question under discussion. Are we asking if our knowledge of the brain enables us to sort the various things we think we know into the class of things we really *do* know, that is, that correspond to some external reality, and the class of things we only *imagine* we know? If that is the question, I would be amazed if there's somebody in the room who can answer it affirmatively. Who can tell us if our current understanding of the human brain even points in the direction of an explanation of the distinction between justified true belief, mere opinion, and outrageous fantasy?

3. What Makes the Human Brain "Human"?

The Argument: Returning to neuroanatomy and neurophysiology, a special role is proposed for the frontal lobes in the ability to access or "know" a reality that is peculiarly "human." A subtext here is the question of what distinguishes the human brain in its anatomy and/or functioning from the brains of other animals.

STEPHEN KOSSLYN: I'm sorry if Ed Manier found my original question unclear. May I try to sharpen the question a little bit? One of the striking things we see when we look at the monkey brain versus the human brain, is that there is a huge hunk of stuff attached to the front of the human brain. When you look and compare the human versus the monkey brain, it is striking how much frontal lobe is involved in the former case. So the question is: Given this huge development in frontal cortex and accompanying structures elsewhere, what kind of things can we do as humans, what kinds of things can we *think,* that many would not be prepared to consider possible in animals? I'm really trying to hear what things the neuroscientists have to say that may make contact with some of the concerns of people interested in the nature of knowledge.

JASON BROWN: The frontal lobe issue raises interesting questions. One of the things that is frequently commented on in discussions about frontal lobe patients is that they know what to do. They understand the strategy for performing a test: they just don't do it. So here we are confronted with a very pretty break between knowledge and behavior. These are people who "have" the right knowledge, but somehow cannot access it or translate it into action. Maybe how we understand that difference can help us in some way to understand the way knowledge compels behavior and what the frontal lobe regions might have to do with this story.

JEAN-PIERRE CHANGEUX: I would also like to respond to the part of Stephen Kosslyn's question that focused on the correlation between the expansion of the frontal cortex with behavior. It is clear that there is a very close relationship; you can even demonstrate that the appearance of certain functions are related to the timing of development for that part of the cortex. One test we have to show this is the delayed-response task, which is a learning task of some sort or another, consisting of memorization of features of an object that is perceived as rewarding. People have studied this in human infants, using, for example, the A/not-B test of Piaget. Before between 9 and 12 months of age, we find that babies cannot perform this task. Then something seems to click and certain kinds of skills seem to become more possible. It is known that this same period of time in the infant's development is accompanied by extensive maturation of prefrontal cortex areas and an enormous expansion of their connectivity with other regions. After about 9 months, the connectivity of the whole cortex picks up enormously. The exact correlation between cortical connectivity and the appearance of certain functions is not precisely known, of course. Still, it is possible to say that there are very well defined functions that can be related to the development of these areas. If you now look at the monkey and do experiments on it, you can show that if you remove the frontal cortex or if you make lesions, then these tasks don't get performed.

RODOLFO LLINAS: Stephen, we each understand your question in different ways. As it also gives us a chance to reply in accordance with our particular concerns, we're grateful to you for posing it. There are two categories of answers that I would consider. One says: Human beings have a bigger brain (relative to body size) than other animals, so they can implement a *richer form* of "computation," one having many more parameters than is present in other forms. The other view would say: No, no, human beings have a bigger brain because they have to specialize in some areas (i.e., language — abstract thinking) and these areas require special hardware; that is, human brains have special modules that allow a different *quality* of "computation."

I think this quandary is one of the central issues to be addressed in neuroscience. The modular view assumes that something occurred in the brains of humans (a miracle) that allows a special-purpose neuronal entity to appear (a deus ex machina not related to the brain's previous history). The second view considers that such "modules" are nothing other than a specialization of preexisting properties; that is, nothing truly new, just more of the same allowing better or more detailed analysis.

Some of us believe that the CNS may be regarded as a system capable of transforming function vectors from one coordinate framework into another (a vector–vector transformation); that is, it is a geometric functional space. The larger your brain is, the larger the functional space. Indeed, such vectors are the product of a large set of vector components, each representing a parameter in brain function. My short answer, then, would be that human brains are larger because evolution has implemented a more general problem solver, rather than a more specialized one. Maybe another way of saying it is that no region of the brain is so specialized that it can do only one thing except for motoneurons and primary sensory pathways.

4. On Cultural Constraints in Neuroscience

The Argument: The historians of science now begin to probe their way into the debate, but they approach the issue of human knowledge and its constraints from a different perspective. It is first suggested that discussions of human brain functioning need to take into account the fact that all human brains function as agents imbedded in a cultural setting. A further point is then made. There is no place "outside" the cultural system from which the human brain can contemplate its own functioning. For this reason, there would seem to be a need to ask questions about sociocultural constraints on the naturalistic knowledge-claims produced by the brains of neuroscientists (including knowledge-claims about the neurobiological impossibility of absolute knowledge).

FRANK SULLOWAY: I also wanted to respond to Steve Kosslyn's question, but I've gotten the impression that we're all answering it in somewhat different ways or understanding it in different ways. And so I want to be clear about what I thought the question was before I try to answer it. I thought you were concerned, Steve, about what sort of tacit conventions in culture constrain the brain to respond and to know the way it does? In what ways would an understanding of these processes of constraint illuminate both the subhuman side of understanding and the brain sciences themselves?

I was just reading a book, *Changing Order,* written by the British sociologist of science Harry Collins.[1] He starts off with the following sort of question: "What is the appropriate response to the sequence '2, 4, 6, 8' "? He volunteers a series of possible responses. One response would be "10, 12, 14, 16." In another culture, however, the response might be "82, 822, 8222," and so forth. And in yet another context, the appropriate response to "2, 4, 6, 8" would be the answering phrase "Who do we appreciate?" And so on. Collins then goes on to describe what it was like for an experienced laser builder to build another laser, and all the things that went wrong even though he knew all the theory behind laser construction. It took an enormous amount of tacit knowledge to actually construct the laser. The point of all this, of course, is to get people to understand that culture, even scientific culture, is composed of an enormous amount of enculturated tacit knowledge that cannot be anticipated and that confounds even the best instrument makers and experimentalists. Indeed, all of science is dominated and constrained by the enormous cultural tacitness of knowledge.

It strikes me that, if you start to think about why the human brain is so large compared to lower organisms, one answer may have to do with our need to cope with the tremendous amount of tacit information we have to deal with as uniquely cultural animals.

Anyway, this all leads to what sociologists of science consider a radically different view of science, namely, the social construction of science. The world is not an empirical world where the brain just works out problems; it is a world where an enormous amount of tacit and subtle understanding is taking place in a cultural context.

ANNE HARRINGTON: I'm glad that Frank has raised this issue of the cultural imbeddedness of scientific knowledge, with all the inevitable questions it raises about the epistemological status of that knowledge. I would like to try to build on his remarks a little, and put on the table a possible different way of conceptualizing these issues about constraints on knowledge. I wanted to direct the question to Jean-Pierre Changeux, but feel it might have relevance to a number of you. I wanted to ask you, Jean-Pierre, what it is you believe you do when you engage in model building? In your presentation this morning,[2] you presented us with a lucid description of your

---

[1]Collins HM (1985): *Changing Order: Replication and Induction in Scientific Practice.* Newbury Park, Calif.: Sage.

[2]Unfortunately, this presentation was the only one that was not revised for publication in the present volume. But see Changeux's (1985) *Neuronal Man: The Biology of Mind* (1st American ed.) New York: Pantheon Books.

Darwinian selective model of the brain, and stressed that this model is a way for us to increase our knowledge. But, as you stress yourself, model building is not about discovering truth in any absolute sense. It may not only be about decreasing error, either. You talked about usefulness: models are valuable if they are useful. I wanted to push you on this, or see if maybe you wanted to go beyond this, and ask you about the things that motivate you—motivate brain scientists in general—to choose the models that they do. I want, in a sense, to exploit your model to my own ends for the moment, and say something a little provocative in order to hear your reaction.

You claim in your model, if I understand you correctly, that classic Darwinian evolution gave rise to a system, the human brain, that works like its creator. There is almost a macrocosm/microcosm vision implied in what you are saying. What occurred to me, listening to you, is that one could read a certain amount of ideology into this vision of yours. Maybe I am mistaken, but historians of science are trained to be alert to these kind of possibilities. For example, if the brain works like a Darwinian system, there's no question of our absolute "naturalness" as human beings. We're solidly grounded in a biological system based on chance and necessity, with no extras added on. And if this is so, then the question of human values, for example, begins to look rather different than if this were not the case. Do you, in fact, intend to imply anything like this in your model? Could you share some of the biography or philosophy behind your model selection? It seems to me that knowing about *this* human part of the scientific process is also a critically important part of our attempt to understand the nature of the knowledge we promote and are prepared to accept as scientists. It is at least as important as the question about to how far the human brain constrains our knowledge, and perhaps even a little easier to get a handle on.

JEAN-PIERRE CHANGEUX: Let me say something about the usefulness of models. As scientists, all we do, or can do, is make representations of the world; in my case specifically, living objects or processes. The study of human brain functions is part of the life sciences—at least I am prepared to say it is, although that may be an issue we will have to discuss. Human beings are living organisms, and our brain is part of our organismic nature; it produces objects that are perhaps abstract, but are nevertheless produced by our biology.

Now why use evolutionary models and Darwinian vocabulary to explore the brain and its processes? The reason is that these things offer a very simple material and distinct way of conceiving "creativity," the development of "novelty" in a living system. They do not have any kind of metaphysical limitations that require you to bring external forces into the system. So they are, let us say, the simplest kind of model one can make of a self-evolving or a self-organizing device. As I said, we are working with models, and I don't think there is any absolute truth. We try to compare our models with an external reality whose existence we have to accept, and we try as scientists to approach external reality as closely as we can. And the value of our models lies in the extent to which they map onto reality. If our models are good, they are preserved, they are perpetuated, they are utilized by other scientists. In this way, there is an accumulation of knowledge. Because there is only one reality we are moving towards, there is also no contradiction between the different elements of knowledge that we are articulating. I am a strong believer in the unity of science. There should not be any contradiction between the different units of knowledge that we are accumulating. Taken altogether, this leaves us with a notion of the value and the truth of our knowledge as scientists that is based on the fitness of our scientific representations with the real constraints of the outside world.

Afternoon Roundtable Discussion: Friday, August 3, 1990

Target Speakers: JASON BROWN, STEPHEN KOSSLYN, PAUL MACLEAN

1. The Neuropsychology of Value Choice

The Argument: The question of values is examined from the perspective of neuropsychology. Participants focus particularly on how far our understanding of individual values as effective forces in the world is compromised by the neurological imperative to deny the possibility of individual "free choice" in the common sense and/or metaphysical sense of the word.

RODNEY HOLMES: This is a question for Jason Brown. Apparently, if one asked you, as a neuroscientist, what is the source of freedom, of volition, you would say it was a delusion.

JASON BROWN: Yes, a sort of . . . psychic illusion, if you will.

MASSIMO PIATTELLI-PALMARINI: If this is so, are you concerned with the extent to which neuroscience can address the problem of everyday choice? If there is no free will in any metaphysical sense, can you say anything about the range of psychological possibilities that the brain faces in situations of choice?

JASON BROWN: I imagine that you have a whole series of nested possibilities, with one fractionating out of another. Ultimately, the kinds of choices that one makes reveal something about the nature of one's self, because choices all flow ultimately out of subsurface constructs. The alternative to my model would be one in which well-demarcated, isolated modular units of representation are scanned, searched, and selected among by a conscious self. This model confronts you with the different problem of trying to unify across different contents, and explaining how a conscious self motivates choice among competing outcomes. Actually, I think a good deal of current work argues for decisions being made prior to conscious awareness.

JEAN-PIERRE CHANGEUX: I have a different sort of question about your attempt to understand value choice from the neurobiological perspective. We know that the *types* of values we will tend to honor or defend are linked with the given culture in which we were raised. Culture dramatically limits the freedom of our value choices by telling us what we should consider right or wrong. What is the role of culture in your neurobiological model?

JASON BROWN: From the intrapsychic standpoint, the problem of value is how feelings attach to ideas. I believe that idea and feeling are part of the same original matrix, so that every cognition has an affective element. The model I have presented is solipsistic in that values are seen from inside-out as stages in the outward migration of concepts to facts. On this view, will fractionates into the drives that build up or articulate the primitive self-concept. This deep construct is gradually subdivided into part-concepts and the feelings derived with them. These concepts result from environmental constraints on the free expression of will. Culture acts on the individual to inhibit maladaptive behaviors, thus parcellating the self-concept into values that will be socially acceptable. This is similar to the role of the environment in evolutionary theory in the elimination of unfit organisms.

Now, in order for choice to occur, there has to be the possibility of action. If you are living in a police state, you lack the ability to express your own nature and self-promptings. External control is complete. Alternatively, if you are not confronted by choice, internal pressure (self-interest) is complete. Ultimately, human societies that are successful strike a balance between the outward movement of the individual and the claims and pressures of society. There is not just one solution. What works in one society may not work in another. Microgenesis, like evolution, can tell us how behaviors (or values) arise and the nature of the fit between organism and environment, but only in terms of adaptation. Such theories do not, it seems to me, provide a basis on which a system of values can be founded.

ROBERT RICHARDS: Leigh Star was saying at lunch that she wondered whether there was anything in the conversation that took place earlier this morning that couldn't have been said in 1870. Frankly, your theory bears a strong resemblance to the ideomotor hypothesis of William James, Carpenter, and others active in the 1870s and 1880s. Let me just make sure that I understand what you're saying. Your notion is that there is somehow a content or a structure that wells up from the unconscious, and that gives rise to parallel outcomes. One of these outcomes is an act, and one of these outcomes is an idea. It would be a mistake, I think you say, to suppose that ideas initiate acts. They don't really initiate acts: they're *representations* of acts and conducts. This is why volition becomes a "psychic illusion." The act takes place, and then you have the idea of it.

It seems like there are some obvious difficulties with this theory. It seems as though I can have an idea about what I had for breakfast this morning, which is not associated with any action. "What I had for breakfast" is not a representation of any action that I am producing.

JASON BROWN: I'm not saying every idea leads to an action. An idea is an outcome. Cognition does not proceed from one outcome to another but through a process of bottom-up replacement. A conscious idea or proposition is a product of subsurface cognition no less than an act that has unfolded. My point was that underlying and generating these conscious outcomes are deeper concepts. Actually, it's much more complex than this because I believe, in fact, that the action that we're aware of in a motor act—say, moving a finger—is really a perception. We do not actually *experience* the motor patterns of an act. You mentioned William James. James felt that motor action left no trace in consciousness except maybe a feeling of agency, but otherwise nothing. My position is similar: I believe that we're aware of perceptions of movements, but we have no direct access to the fact of our actions in the world.

ALAN FINE: If I understand you correctly, you want to sweep the prospect of ideas causing acts under the rug of the unconscious or the subconscious. That may be possible. But it's not obvious to me how you would then account for the way in which *external* ideas influence our actions. In other words, we get a message from someone, which leads us to take some active answering steps. The information in that message, transmitted to our consciousness, does seem to play a causal role in our action.

JASON BROWN: If you are asking how the external world feeds into an endogenous cognitive process, my view is that it occurs through successive levels of constraint; that is, the unfolding of acts and ideas is an autonomous process in which contents are selected by the inhibition of alternative routes. It's either some variant of this theory or a sensory psychology in which contents are pieced together from elements. Moreover, one of the messages of clinical work is that meaning is mediated by deeper or more archaic structures in the brain, not by the so-called higher levels, the neocortical systems. We get disturbances of word or object meaning from limbic–temporal lobe damage, not from cortical damage, where we see form-based errors. Limbic pathology gives you meaning deviation, neocortical pathology gives you abnormalities of form detection or of fine discrimination. This suggests that meaning is encoded early on in the system. I would say that recognition of a message is an early phase of the perception of a message. What comes into consciousness is the final modeling.

2. Consciousness and the Human Brain

The Argument: From the topic of free choice, the debate moves perhaps inevitably to human consciousness: whether neuroscience needs to pay attention to it and whether the principles of the brain research enterprise could ever in theory allow subjectivity any effective or functional (nonreducible) role in human behavior.

JOHN DURANT: I wanted to stay with this issue of consciousness, but turn the fire onto one of the other speakers. Stephen Kosslyn has given us a fine analysis of the neural-network approach to brain functioning. I was sorry, however, that he was so ready to say that what he was describing were *mental* events. It seems to me, in fact, that the whole mind-body question is begged by his analysis. It is perfectly possible to imagine that the systems

Stephen Kosslyn is modeling could do what he describes without any vestige of conscious awareness at all. I wonder if you recognize or acknowledge that the mind-body problem is untouched by your approach.

STEPHEN KOSSLYN: Well, I would first answer by saying that there are at least two mind-body problems. There's the problem of how mental events relate to the brain, the part of the body that appears to be relevant to mental events. This is the problem that I address. Then there is the problem of *qualia,* of the nature of introspectively evident conscious experience and how it relates to the brain.

JOHN DURANT: That's the problem I meant.

STEPHEN KOSSLYN: Right. I can only come at this from the perspective of the other issue, that is, how particular states of the brain may have nonaccidental properties related to the way they feel. Something sharp poking you feels sharp. And there is a sense in which this "sharpness" maps neatly onto the functional representation we find in the brain.

As for consciousness, which is what you're really interested in, at this point, I have nothing to say about it at all. I'm also not sure that anybody else does either. Let me actually ask: Jason, is consciousness epiphenomenal in your view? Does it have any functional role at all?[3]

---

[3]PAUL MACLEAN: Since the discussion did not return to this topic, referred to in the 19th century as the doctrine of psychophysical parallelism, I would like to add a retrospective footnote. Many people have argued that subjective experience is a useless epiphenomenon, claiming that the brain can do everything it does without our being there as subjective individuals. But my own answer to that (MacLean, 1960, 1990) has long been that the mere existence of subjectivity

> means that the brain has an additional source of information for influencing behavior. It is as though the brain could hold up before itself the mirror of subjectivity as a means of reflecting information. The most persuasive illustrations are provided by human beings in conveying information about their subjective states through verbal communication, as witness how poor our vocabulary and literature would be without words expressive of subjective states. Were it not for the communication of subjective feelings, immobilizers, rather than anesthetics, might have been developed for the relief of pain. (1990, p. 575)

As Stephen Kosslyn indicates in his comments, some philosophers today have come to refer to the problem at hand in terms of *qualia,* things for which there are no physical measurables. An example of such an Immeasurable would be the sensation of something green: "Nothing in the domain of physical science is green; therefore, nothing in the materialist universe is green. . . . By rejecting any other domain, materialism is saddled with the problem of groundless appearances" (Cooney, 1991 pp. 208–209).

References

Cooney B (1991 [in press]): *A Hylomorphic Theory of Mind.* New York: Peter Lang
MacLean PD (1960): Psychosomatics. In: *Handbook of Physiology, Neurophysiology III,* Field J, ed. Washington, DC: American Physiological Society, pp. 1723–1744
MacLean PD (1990): *The Triune Brain in Evolution: Role in Paleocerebral Functions.* New York: Plenum Press

JASON BROWN: Well, I think it demands certain things of our brain models, and in that sense is not entirely irrelevant. There is, for example, the fact that a conscious mental state exists over time, over a passage of physical time. Consciousness confronts us with the problem of mental duration over pure physical succession. Mental states span succession. We have a "now," a present in which we live. This present is active. And to me this is really one of the deepest issues that needs to be confronted by any model that would relate the mind and the brain. To just talk about "memory" is begging the issue.

PAUL MACLEAN: I just wanted to ask people here how familiar they are with Jason's brain model of *microgenesis*. Jason, you've got to do something about your terminology. Take Friedrich Sanders's definition of the phenomenon you want to describe. *Actualgenesis,* he called it. That means "realization," "actualization." There's something about this term "micro" that obscures the richness of the concept. The root "micro" is very important from the standpoint of "microtime." But the full concept of "microgenesis" is just bubbling with possibilities to address the issues we're discussing here, and it gets lost on most audiences because they don't understand all you're implying.

STEPHEN KOSSLYN: I wanted to add that I don't have a theory of consciousness, because I don't understand what we're talking about. I understand the notion "conscious" as an adverb, that something can be more or less conscious. But I don't understand the concept of conscious*ness.* I did not mean to say that nobody has anything worth saying about the topic, just that I don't.[4]

JEAN-PIERRE CHANGEUX: My feeling is that the words *mind* and *consciousness* are used like the word *life* was used in the past, in the 19th and early 20th century. It was thought then that there were some properties of life that were irreducible, but that were necessary to defining the essential nature of the organism. Today we reject that idea, and deny that there is any substance to discussions about "life," "life forces," and so on. Similarly, I don't see the necessity of talking about "mind" as some entity beyond its physical properties.

3. What Constitutes Explanation in Brain/Behavior Studies?

The Argument: Since it had been established that the capacity of the human brain to know itself is probably constrained by both biological and cultural factors, debate now turned to the pragmatics of obtaining knowledge under these less than-Olympian conditions. What should be our justificatory strategy for granting epistemological status or truth value to one kind of claim over

---

[4]Since engaging in this discussion, Stephen Kosslyn felt himself sufficiently challenged to conclude that it would be worthwhile to develop some views on the problem of consciousness. The fruits of his post-workshop reflections can be read in his final contribution to this volume, "Cognitive Neuroscience and the Human Self."

another kind of claim? How can a neuroscientist recognize when he or she has succeeded in achieving something called "knowledge"?

JAMES SCHWARTZ: I'm a neurochemist, and I study individual synapses, and synaptic plasticity. And I can demonstrate that a number of biochemical reactions take place in a certain cell when I do something to that cell. But my main problem is to convince people, peers and colleagues, that what I'm observing has anything to do with what I'm trying to explain. If I were to look for other biochemical reactions, they might happen also, and it's not clear to me how I know that what I am choosing to display has anything to do with what I think I'm trying to explain.

EDWARD MANIER: I'd like to pursue this interesting question about how one knows when one's explained something in science. Clearly *explanation* and simple *correlation* are two separate things. I have an idea about what would have to be accomplished to be able to claim, for example, that modulation of channel properties is an "explanation" for learning, or that PET scans of activities of certain areas of the brain is an "explanation" for image formation. There are some standard things that one always looks for in the history of science, when one is trying to decide why a certain theoretical or explanatory position has won out in contest with other comparable positions. One of these involves the superior scope and range of the victorious position. That is, are there phenomena within the scope of the science of which the new position offers a superior account? At the same time, the new position must be able to account for the limitations of the defeated position: it must be able to take all the generalizations from the old position, and show their limitations, why they break down under certain circumstances. It must be able to resolve controversies. In short, an explanation based on data produced, say, by the technology of PET scan must be judged successful if it can generate a hypothesis about imagery that would otherwise be unavailable to folks who were just dealing with behavioral studies or with computational mental networks. Only if an approach possesses some of this rich hypothesis-generating power are we justified in saying that we're doing more than simply lining different kinds of phenomena up against each other. Only then can we say, "We're developing a neurobiological explanation of visual imagery."

STEPHEN KOSSLYN: Well, if those are the criteria, let me give an example of a new prediction that came out of our PET-scan work, a prediction that has begun to clarify a phenomenon people had been ignoring because they couldn't make sense out of it. When Shepard and Metzler did the original three-dimensional rotational imagery experiment,[5] I remember Jackie Metzler telling me that people reported that they were twisting their arms or fingers as they "mentally rotated" imaged objects. Nobody knew what to think about this finding at that point, so we all just ignored it. I've seen

---

[5]Shepard RN, Metzler J (1971): Mental rotation of three-dimensional objects. *Science* 171:701–703.

patients with brain damage who were given a mental rotation task, and they would twist and move all over the place. This made no sense at all in terms of the standard theories of imagery, including my own.

However, looking at the PET scans, I noticed that parts of cortex involved in motor control were lighting up (along with the basal ganglia). So I thought, what if it's not the case that the parietal lobes are used to register *location* per se, the way that Mishkin has argued?[6] What if they're used for guiding *action,* so that we're in fact dealing with the storage of a kind of motor program that helps you navigate or shift your eyes systematically? Imagine walking into the entrance of your living rooms. Now tell me where the furniture is. When you ask people that question, their eyes start moving like crazy.

So then I thought, "What would that predict about mental rotation of images?" Well, if it's true that mental rotation involves updating location representations of eye movement patterns (which specify locations of parts of imaged objects) that are essentially motoric, then if you were moving your foot or hand or arm in a direction opposed to the direction of your mental rotation, you should find interference. The interference should increase the closer the motor movements approximate the kind of motoric representations used in mental imagery, namely, eye movement patterns. However, if you rotate a limb in the same direction, it should not interfere as much with the mental rotation. Preliminary results suggest that this prediction is correct. And I doubt that anybody would have predicted this before the PET work. In addition, other PET work suggests that there are two ways to form mental images, one based on activating stored visual memories and one based on allocating attention to regions in space. Preliminary results are also supporting this distinction.

4. Knowledge and Values in Neurobiology: The Case of the Triune Brain

The Argument: Using the example of the triune-brain model of Paul MacLean, new questions about the relationship between knowledge and values in the brain sciences are now raised. How far is the enormous success of the triune-brain model in our time less a function of its explanatory power — what it helps us "know" — and more a function of its resonance with certain values we share as a culture — its ability metaphorically to articulate our collective fears of an irrational "beast within"?

JOHN DURANT: I wanted to get back to the agenda of the day, which has the words "knowledge and value" in the title. I have a little difficulty saying

---

[6]Ungerleider LG, Mishkin M (1982): Two cortical visual systems. In: *Analysis of Visual Behavior,* Ingle DJ, Goodale MA, Mansfield RJW, eds. Cambridge, Mass.: MIT Press.

what I want to say now, because Paul MacLean is a major contributor to postwar studies of brain behavior. However, I want to criticize what he's been saying, and I want to suggest that the theory he offered is more a *metaphor* than a theory. The triune-brain concept is enormously attractive. We should address it because I would judge it to be probably the single, most-influential idea in brain science in the postwar period, at least in terms of public or popular perceptions of what brain science has to say about the human condition. Paul MacLean himself mentioned that Carl Sagan used the idea in a popular book back in 1977. Arthur Koestler also used the idea in a number of popular books. Indeed, Koestler even used it as the rationale for suggesting that the solution to the problem of the human condition would be to find a drug that could reconcile two of the three triune brains which he believed to be out of kilter with each other. This was his reaction to the idea of so-called *schizo-physiology,* central to the triune brain concept.

Now I want to hazard a proposition to which I would invite the neurobiologists present to respond. My proposition is that the doctrine of the triune brain makes no sense in evolutionary terms (and it is fundamentally an evolutionary doctrine) because it amounts to a neurological version of the theory of *recapitulation.*

The theory of recapitulation, which was popular in the 19th century, suggests that successive adult stages in the evolution of life are preserved intact in sequence in the bodies of more advanced organisms. This implies that by looking at more advanced organisms you can see within them the intact adult relics of more primitive, earlier stages. I hope that you can see that this theory sounds very similar to the doctrine of the triune brain.

The trouble is that the theory of recapitulation is defunct. It is dead. And it is dead for a very good reason: evolution works on organisms, whole organisms. We have no reason to expect — in fact, we have very good reason *not* to expect — ancient brain structures to be preserved intact over long periods of time with new structures simply being added on top of them. Instead, we have every reason to expect natural selection to produce adaptive, functional behavior by molding the whole brain. If I'm right, then there is no evolutionary reason to look for archaic reptilian and primitive mammalian brains in our own brains. The evolutionary rationale is just not there.

Let me quote a comment that Paul MacLean has made in print: "Nature, despite all her progressiveness, is . . . more tenacious than the curator of a museum in holding onto her antiques." That does not describe the evolutionary process as I understand it. In nature, the antiques, when they're no good, are discarded.

Now I say all this against the backdrop of having claimed that the idea is still very, very influential in public, and I would suggest again that the reason for its success lies in the fact that it is a very powerful metaphor. In a sea of technicalities, the idea has a powerful and attractive simplicity.

Moreover, it appeals to ideas that are deeply rooted in our culture, ideas which I have elsewhere summarized under the notion of "the beast within." This notion refers to the belief that we all carry primitive beasts around inside ourselves, which can be released under certain circumstances and which are responsible for all kinds of bizarre, violent behavior. Ultimately, this idea goes back in Western culture to the doctrine of original sin.

In sum, we have here, in a nutshell, something in brain science that is a potent value-laden description of the human condition, but which, as I would now suggest, actually has a very tenuous hold technically within biology as a whole. I would therefore invite us to address what I see as a problem right at the center of the agenda of this meeting.

PAUL MACLEAN: One of the worst things in this world is to be "discovered" by the popular media. I remember that when Carl Sagan raised the question of my appearing on his program, I said, "When people see me on television, then they'll know I've gone over the hill."

But I would respond to the criticism with a question of my own: How do you explain directional evolution? I mean, this is a marvelous thing. During Permian times, there were therapsids (mammal-like reptiles) and there were several lines of them all developing in similar directions. One of the inventions they were all independently developing at this time was thermo-regulation. So, as the paleontologist E. C. Olson has asked: "What do we have here? Two systems? One for now, and one for the future?" The late Harley Shands used to say that the wonderful thing about language is that it allows us to invent the future; nature also seems creative in this way. She doesn't just pick out one individual [species] and say, "You develop thermoregulation." You find instead a range of unrelated lines all doing it at the same time and over millions of years. Now, for me, that's very hard to explain.

With mammals, the problem gets worse. How do you explain the fact that you have all these species of mammals all developing cerebral changes along the same lines? And why did humans end up (so far) getting additional changes and the greatest amount of change? As Olson says, "How does genetic-selective neo-Darwinism explain this?" It doesn't; it can't. So with that, I've said what I want to say.

5. The Conservatism of Function and Structure in the Human Brain

The Argument: The discussion focuses on the scientific criticisms made of the triune-brain model. Various arguments in defense of the model are offered; indeed, it is suggested that the concept of the "beast within" is no "mere" cultural metaphor, as was suggested, but an evolutionary and neurobiological reality.

DETLEV PLOOG: I would like to say that I think some claims have been made about the triune brain that are misleading. First of all, you spoke of Ernst

Haeckel's "phylogenetic laws," the idea that stages in earlier evolution are recapitulated in ontogeny. In its naive form, this idea has, of course, been long miscredited. Nevertheless, I think we must be cautious. Take the homologies between the lower midbrain in animals and the midbrain and pons in humans. You find that an experienced comparative anatomist can extrapolate from one to the other. If he knows the human dimensions of a particular structure, he can identify its anatomy in the lizard brain.

JOHN DURANT: That, of course, is structure; it's not function necessarily. There's a very big difference.

DETLEV PLOOG: You can take this issue up in functions, too. Of course, we're not talking about the same function and it's clear that, in the higher organisms, more primitive functions are embedded in other functions. I would never say that the snapping and tongue flicking of the frog is the same as the tongue flicking of the newborn human baby. But if you look at anencephalic creatures, for example, you see that they carry out all kinds of pontine and mesencephalic functions, much as animals functioning normally at that level do. The lizard yawns, the baby yawns, you yawn. I could give you a lot of behaviors, scratching and so on, that have been preserved over millions of years.

I remember sitting in an anatomy course where the instructor projected onto the screen a picture of a decorticated chimp brain and a decorticated squirrel monkey brain. Even experts could not distinguish the decorticated chimp brain from the decorticated squirrel monkey brain. And these creatures are 50 million years apart. So I think the idea of the triune brain is a *metaphor,* as you say, but it is a metaphor that continues to retain explanatory power for us.

STEPHEN KOSSLYN: If you're building a car while it's in motion, you *can* remove the gas pedal. It may be a little risky, but you can do it. You can certainly redecorate the inside and so on. However, you're not going to take out the engine and put a new engine in, especially if you're going uphill. My point is that some of the basic functions that are performed by lower level areas must be conserved, or the animal is dead. Whether you'd rather have a new, improved model or not, you're sort of stuck with at least your basic machinery, if you plan to keep driving.

JOHN DURANT: But you can tinker with all your parts, you can keep tinkering with them.

STEPHEN KOSSLYN: It's not so clear. Some of these things are highly integrated systems.

JEAN-PIERRE CHANGEUX: I am also not convinced of this criticism. First of all, it's important to remember that DNA represents the basic building block of our body, and DNA has, in fact, not evolved as such since its introduction. Similar genes have been found in *Drosophilia,* the fruit fly, and in mammals, in vertebrates in some cases. We are working at the moment, for example, with a molecule taken from a very primitive fish, the torpedo fish, and we have found that the genetic correspondence between this primitive fish and man is an 80-percent sequence. So there is no question that you have fantastic conservation of structure in evolution, which I think is what Paul MacLean has been trying to argue.

JOHN DURANT: Yes, but you're talking about genes, and the doctrine of recapitulation is specifically about organic structure. It is about the product of development, not the blueprint.

TERRENCE DEACON: I think we have to distinguish between two things. I think we all agree that there are conservative features in evolution. The question is, does the triune brain assume something more? Recapitulation includes other assumptions. One of those is the notion of *terminal addition*.[7] I think that is what John Durant is really objecting to here. Can we at least agree that terminal addition is dead?

JEAN-PIERRE CHANGEUX: No, no. The statement was made — and challenged — that we have a primitive beast in ourselves. I say, yes, of course we have a primitive beast. We have plenty of beasts living in ourselves. Whether all this is a product of recapitulation by means of terminal addition is another question. But across the evolutionary scale, you have the same category of genes being pooled.

6. Ideological Factors in Evolutionary Thinking

The Argument: The debate swings back now to the extent to which the triune brain, whatever its explanatory merits, nevertheless has been and continues to be more than dispassionate model building. Again it is affirmed that this model of the brain resonates with, if it does not actually contribute to, certain powerful ideological trends in our culture about human nature and its limitations. The neuroscientists reflect on the relationship between scientific knowledge and "extrascientific" ideological assumptions. The historians are also warned against dogmatism on their part.

JOHN DURANT: I think there is a confusion here. I understand the concept of homology at the genetic level and at the organic level, and I accept it. Every evolutionary biologist needs it. But the evolutionary anatomist doesn't go looking for a reptilian heart in a mammal, or a reptilian kidney in a mammal. He or she goes looking for homologies between the mammalian heart and the reptilian heart. We, however, have been invited to go looking in the brain for a reptilian brain beneath a mammalian brain, and I think that is a different and distinctively recapitulationist view, which I believe has very insecure theoretical foundations.

Can you see why I suggest that, as it were, to a historian's eye, more is going on here than neurological or brain-and-behavioral theory? This is a set of ideas with a very powerful resonance with other ideas in our culture. For instance, there does seem to be in Paul MacLean's writings a thread of

---

[7]The assumption that adaptations that appear later in an evolutionary sequence are added onto complete adult organ systems that otherwise remain unchanged.

thought that implies that our different levels of the brain are somehow either out of touch with each other or in conflict with each other. Why would evolution endow us with a nervous system at war with itself? It's an odd idea, but it does resonate with another powerful nonscientific concept in our culture: that we are, as Saint Paul said, creatures at war with ourselves, with our flesh warring against our spirit.

JEAN-PIERRE CHANGEUX: So all you are saying is that we have to discuss this within the framework of nonscientific beliefs. We have to ask how nonscientific beliefs affect scientific development. What is the effect? Most often, the effect of such beliefs is to interfere with scientific development, as you well know from the history of science. However, there are also cases in which one finds that ideology from the larger culture contributes to the positive development of science. What we are talking about here might be one example of just such a case.

TERRENCE DEACON: I think what John Durant is trying to say is that it is important to recognize that, as neuroscientists, value issues tempt our thinking in a wide range of areas. Consider the idea of "progress" in evolutionary thinking. This was pervasive in our discussions this morning about the size of the human brain and its relationship to what makes us unique. I call this "the bigger/smarter fallacy." Basically, we are hindered by a lot of assumptions that add up to a desire to see ourselves at the center of the universe, or at least at the top of an evolutionary ladder.

So the knife cuts both ways. It's not just (as I said this morning) that neuroscience research constrains or may be able to constrain our thinking about selves and minds and evolution. It's also that our preconceptions about ourselves, our place in nature, and where we stand on some ladder of progression also constrains the research that gets done and the theories by which we organize our data. Such assumptions are many and pervasive, and we need to be able to see through them.

DIANE PAUL: I have no view of the validity of the triune brain concept, but I do have views, strong views, on evolutionary theory. I think it behooves one to be extremely cautious in claiming to speak for the community of evolutionary biologists, to make claims about apparent consensus on the issue of how selection operates. Certainly one of the most striking characteristic of the field at the present time is the depth and the vigor of disputes about the character, the scope, and the optimizing power of natural selection. While John Durant's comments are representative of a school that is a powerful and influential one, there are people who would strongly contest the assumptions that he took for granted. They would themselves, rightly or wrongly, see those assumptions as ideologically loaded and resonant of trends in contemporary culture.

JOHN DURANT: Would you say which assumptions you thought I was bringing to the discussion?

DIANE PAUL: The *All Macht* of natural selection, the assumption that natural selection is all powerful, and that if something is not good for you, selection will eliminate it.

7. Recapitulation of Function in the Human Brain? A Final Look

The Argument: A final attempt is now made to clarify the place of certain controversial evolutionary concepts in the triune-brain model.

FRANK SULLOWAY: I would just like to see if we could clarify this whole issue and distinguish between the bolder Haeckelian notion of recapitulation and the issue of there being conservative trends in evolution. As I understand it, the Haeckelian theory, which involves this notion of terminal additions in ontogenetic development, involves the important assumption that there is a recapitulation in embryos of the adult stages of ancestors. The key word here is "adult." The reason the idea works in the Haeckelian mode of thought is that it assumes a Lamarckian mechanism for acquired characteristics, in which adult organisms retain and pass on characteristics acquired during their whole lifetime. This lifetime of acquirements becomes incorporated into the genome, and eventually these adult "terminal" stages become encapsulated in early embryonic stages.

In Darwinian theory, there is much less selection on embryos than there is on adult stages, which is why vertebrate embryos tend to resemble one another, especially in their early stages. Moreover, characteristics cannot be acquired, and therefore it is hard to see how adult ontogenetic stages can recapitulate adult ancestral stages. In fact, what one finds is *parallelism* in early embryonic stages, with divergence later on, as the organism becomes what it eventually becomes. There is no real recapitulation in any meaningful sense. At most, there is a kind of parallelism in embryonic stages across species. So I would like to ask this question of either Professor Changeux or Professor MacLean: Is there a recapitulation in man of the *adult* stages of reptile brains in any meaningful sense? And if not, then what do you mean by using a term like "reptile brain"? Is it that we parallel *embryonic reptile* brains before leaving this stage behind? How do you reconcile such claims with the fact that Haeckel's old concept of a biogenetic law is considered as dead as a doornail?

PAUL MACLEAN: The embryo would never complete itself if it went through all these phases. That's one of the criticisms, but this is *never a claim that I made.* I also don't use the term "homology"; I use the term "correspondence." All I'm referring to is corresponding structures. My God, I can take anybody into the laboratory and project brain sections, and it's easy to show them the relationships. Anybody can go in and convince him- or herself of these correspondences. I just get so weary hearing about "MacLean's model": these are not my models, they are *nature's* models.

Hell, I mean, the corpus striatum in the human brain is the most beautiful thing you'll ever see, but it reveals primitive anatomic and chemical features of reptiles. And behaviorally it seems to keep on doing the same sorts of things in the higher organisms. Here you have a sort of recapitulation. Elsewhere I have suggested that nature may have utilized the corpus striatum to serve not only as a playback mechanism for ancestral forms of behavior, but also for currently learned performance.[8] Once you learn a piece on the piano, and can do it automatically, it gets shunted down to lower brain regions and if you stop to think about how you're doing it, it may interrupt your ability to play. Language is the same sort of thing. Although the corpus striatum may not comprehend speech, it might parrot-wise be able to process the production of speech.

---

[8]MacLean PD (1972): Cerebral evolution and emotional processes: New findings on the striatal complex. *Ann NY Acad Sci* 193: 137–149.

Morning Roundtable Discussion: Saturday, August 4, 1990

Target Speakers: RODNEY HOLMES, EDWARD MANIER, LONDA SCHIEBINGER

1. Knowledge and Values in the Neuroscience of Sex Differences

The Argument: The question as to whether some forms of scientific inquiry should be discouraged for their potential social misuse starts off this discussion about neuroscientific research into sex differences. There is disagreement over the validity of research results pointing to sex differences, but the main thrust of the discussion ultimately focuses on why this research area is as "hot" as it is, why we as a society choose to invest in it (rather than in other research areas), and what it is we fear about it.

RICHARD HELD: In my laboratory, which has both male and female scientists, we discovered that there are sex differences between males and females in certain visual functions, between the ages of 3 and 6 months. In talking about these results, we have occasionally been told to suppress our results so as to prevent their being used for the wrong purposes. Now I think that is carrying things to an absurdity, and, of course, the main reason I cannot go along with this way of thinking is that the logic behind it is wrong. We cannot make the assumption that because there are sex differences, there are gender differences. That conclusion requires extra assumptions. I do not see any reason to suppress sex differences, when these differences can be validated by both women and men.

LONDA SCHIEBINGER: I would not like to suppress knowledge. I'm more concerned about the priorities of science, how we determine the things we know versus the things we do not know. The world is infinitely complex and our science gives us a narrow path of knowledge through this complexity. If we make research into sexual differences in the visual capacities of babies a priority, then finite resources have automatically been taken away from some other study.

Let me give the most simple example I can think of, one that affects me every day of my life. We know very intricate things about the human body. We can send men to the moon, and do all sorts of very impressive things. Yet we don't really have adequate birth control, a technology that is mostly used by women. I think science could do better. So I do not want to suppress knowledge; I want to look at the priorities of science and say, "Are there not other things we could be funding that may be more fruitful? What do we plan to do with this information?"

JOHN DURANT: I wanted to make a comment on the whole attempt to uncover sex differences in mental performance, and the fears surrounding these studies, as mentioned by Richard Held. In considering these studies as a whole—and particularly their public impact—I would urge us to look very carefully at the question of how we construct our knowledge, technically but also culturally. For example, we are all familiar with the multiple popular articles about "his brain/her brain," how men and women think differently from each other, and so on. What are they all based on? As far as I can tell, they are all riding on the back of one or more statistical studies showing that there are different means in the distribution of certain performances in the male and the female populations. These differences are, I assume, not unlike the differences that have been claimed in the means of IQ for different ethnic groups. In these other cases, we saw a similar phenomenon. We were told that there were "black brains" and "white brains" because the average scores on IQ tests in black and white communities were slightly different.

Now, in both cases, we are confronted with an interesting reconstruction of a knowledge claim: essentially, the transformation of a population study of a phenomenon into a typecasting of specimens. Once you begin thinking in typological terms, then everything has to fit into one of the two boxes. If I'm a man, I can be a good mathematician; if I'm a woman, I can't. In short, cultures take and reconstruct knowledge-claims into meaningful patterns that say a great deal about the culture, but are not in any technical sense inevitable or demanded by the evidence that has been gathered.

ANNE HARRINGTON: I would like to turn our attention to the issue of "nature versus nurture" that I believe is rumbling under the surface of these debates about the research into sex differences. Is it possible that we historians operate with a view of the life sciences, of the brain sciences in particular, that on the one hand confirms our suspicion that all science is in some sense ideologically grounded, but that may in fact be somewhat passé? Are we attacking a rather crude determinist model of biology and neurology that would not in fact be defended by modern neuroscientists today? What does Richard Held personally see as the implications of his work on the functional and morphological differences in newborn infants? What is the relationship between structure and function? What should we be thinking about these kinds of things?

RICHARD HELD: Well, structure/function relationships are very compli- cated, and we're certainly not ready to make raw generalizations about nature and nurture at this point, except with regard to specific species, specific ethograms. We just had a conference in Cambridge on cognitive neuroscience, in which it was suggested, for example, that infants of 3 months or less have "knowledge" of the natural world, of the physical world. Now is that nature or nurture? This remains to be worked out in detail. I find it very hard to answer that question simply.

ANNE HARRINGTON: My point, I think, is in some sense to ask *what* it is we are afraid of when we call something "natural." Where have we developed this idea of an equivalence between something being "natural" and something being destined? Where have we developed this idea of biology as something that is inflexible, and culture as something we can move around in, something we can change? I think many people believe that if something is written from the biological gods above, it is therefore immutable. Hence, I think, the fear Richard Held mentions about the potential misuse of his findings. Do you as a neuroscientist think that that is really the way that people should be thinking?

RICHARD HELD: No, of course not. Biology doesn't dictate our destiny, any more than history or any other science does.

RUTH HUBBARD: I think most or many of us at this point would argue that it isn't nature *or* nurture; rather, this sort of dualistic thinking is a totally mistaken approach. I'm surprised in a way that Anne Harrington wants to frame the question as a dualistic question rather than asking "Why are we thinking in this framework at all?"

ANNE HARRINGTON: Well, my understanding of Richard Held's comment was that, if he had discovered that there were differences in childrearing that then led to sex differences, that that would have been seen as politically correct and published with considerable fanfare. But because of the perceived biological orientation of his research, he is perceived as raising questions about destiny, and therefore we don't want to run the risk that people are going to abuse this knowledge.

RUTH HUBBARD: But I don't think that's what Richard Held said. In fact, one of the interesting things is that when someone argues (I won't say "discovers") that something is social, then it doesn't make the cover of *Discovery* or *Time* or *Newsweek* or *Science*. But when someone finds not just a difference, but hypothesizes that there is a gene for that difference, it becomes an article in *Science*. It also becomes an editorial in that issue of *Science*. It goes on the cover of that issue of *Science,* and it hits every single newspaper and magazine of that month. I think *this* is the phenomenon we need to talk about, not whether his discoveries should or should not have been suppressed. We need to talk about the social context that this research is growing out of, and that it's feeding into.

2. What Would a Feminist Neuroscience "Look Like"?

The Argument: Debate reverts back to the epistemological themes of the workshop, as neuroscientists ponder the feminist argument that neuroscientific knowledge is marked by sexist values. There is confusion as to what in practical terms would have to happen to science to liberate it from its sexist biases. Would the presence of women in the laboratories alone suffice to change the cognitive products of neuroscience? Does the fact that women have been active in sex-difference research argue against this assumption? It is suggested that the feminist critique of science is less about men impugning women and vice versa, and more about revealing the methods and social effects of a certain type of knowledge that we may call "gendered." The institutions within which science is carried out actively obscure the gendered nature of science; a feminist science would thus seek to reform these institutions. We need to demythologize the *Arrowsmith* image of the autonomous scientist–hero in search of high truth, and make visible the contingency and human networking inherent in all scientific work.

RICHARD HELD: Well, granting the sexist nature of science, what would it mean to have a feminine science? People talk as if science would have an essentially different character if women were doing it. I find that very hard to understand, and I would like to have it explicated further.

LONDA SCHIEBINGER: I wouldn't call what I am talking about feminine science or female science, but rather a femin*ist* science. Such a science would, of course, not exclude men, since some of the best feminists over the years have been men. A feminist science would not be exclusionary, it would not depend on what we think of as feminine qualities now. It would reevaluate gender distinctions as we understand them in society and it would acknowledge that one set of values in our society (values such as caring, feeling, nurturing) has historically been set aside and devalued by science and scientists. It would propose that we reevaluate those qualities, and see if it would be fruitful to bring them back into science. The goal is to create a more human and humane science.

MASSIMO PIATTELLI-PALMARINI: I have been living for the past 5 years at MIT among a group of linguists. In the early 1960s, the claim was occasionally made that this kind of linguistics was so abstract, so mathematical, so logically oriented, that it precluded the participation of women in any great numbers. Over the years since, women's participation in this field has been increasing—yet one doesn't see a special kind of linguistics coming from these women. My question then is: How do you justify this idea that science, by leaving out the perspective of women, has left something out or is otherwise skewed? I agree, of course, that there must be more opportunity for women, but I have never seen any evidence to show that they contribute a different kind of thinking.

ANNE HARRINGTON: I would like to follow up on Massimo Piattelli's point, and look at this same issue from a slightly different perspective. This comment is offered sympathetically. My understanding of Londa Schiebinger's argument was that the substance of science is what it is because women have been systematically excluded. However, in the past few decades, as women have come into the neurosciences, I have been struck by the numbers that have taken the forefront in promoting what seem to be arguments for biologically based sexual differences. The requirements for a feminist science would seem, then, not to be exhausted by simply increasing the numbers of the excluded sex in the laboratories.

LONDA SCHIEBINGER: I agree. And my answer to both comments is that our institutions are now structured in a way that requires men *and* women to master certain skills. In other words, the institutions of science and scholarship may, in a historical sense, be called "masculine" because the qualities that we as a society consider masculine — qualities that may not even be inherent to men — are demanded by these institutions.

RUTH HUBBARD: Since nobody has said it, I think it's important that we make clear that a question like "How come women find this, when it is what men should have found or should have wanted to find, etc.?" is the wrong question. In gender studies, we are asking questions about the *genderization* of knowledge, how it got to be gendered historically and where that is taking us. We are not asking what women think or what men think about women or men.

ANNE HARRINGTON: Could I just respond to that, because I think it was directed at me. The reason I made the comment I did was in reaction to Londa Schiebinger's argument that we can change the nature of scientific knowledge if we let women into the sciences. She asserted that women see the world differently, ask different questions. Given that, it was striking to me that the evidence she gave looked at the historical construction of an ideology of female biological inferiority, but drew actively on sources written by women scientists both in the past and today.

LONDA SCHIEBINGER: I think it insignificant that knowledge about sexual differences has been validated by women as long as they are working within present institutional structures. I don't think that we can make any judgments about how women might or might not make a difference in science until they are represented in equal numbers with men in all areas of science and especially in positions of leadership.

SUSAN LEIGH STAR: I would like to add to this discussion of how we think about social change and how we think about women making a difference to the process. It seems to me, in the first place, that a viable critique of science from a feminist perspective must speak to the way work is organized. There is, for example, the important point that women's work in our society is essentially invisible, that a lot of work that women do is behind the scenes, is formally unacknowledged. It's invisible in the sense that the work that lab technicians and janitors do is formally invisible, but in fact ends up having a big impact on science. Now what are the consequences of this invisibility, and what would it mean to make the behind-the-scenes dimensions of science more visible? These are the kinds of questions we should be asking.

The second point I want to make is that the feminist critique of science is talking about exclusion, about the experience of being an outsider. For whatever reasons, we were excluded from the academies of science, almost all of us, until the last few years. Feminism has had a lot to say about what it is to be an outsider, and how outsiders and insiders listen to each other in different ways. For example, there are studies on language and communication that show that we have learned as women to say "yes and," while men say "but." So when you're talking about a feminist science, essentially what you're talking about is interdisciplinary work; you're talking about compromise, about being outsiders to each other and learning to listen.

ALAN FINE: I am still unclear about what all this would mean in practical terms for the substance of science. When you are trained as a scientist, you learn at an early age that science is not an enterprise that is settled by compromises, it is not an enterprise that is settled by give-and-take. Rather it is something that is after some sort of clarification of realities; it is groping after something that is objectively "out there." Whether the scientist happens to be masculine or feminine is in a sense irrelevant. What do you think a science would look like that involved this kind of give-and-take or compromise you are proposing?

SUSAN LEIGH STAR: It's a big question. We had a conversation last night, in which you were saying that your understanding of science was basically that if you didn't do what you were doing, there would be someone else waiting in line to jump in and do it instead. I said that wasn't my understanding of science at all, that I saw you as a person who had been nurtured by your teachers and colleagues, who had all these contacts, and was giving and trading favors with people all the time. You lived in a specific place, had a career. I didn't know if you were married. I didn't know anything about you, really, but I assume that a feminist science would be one that didn't teach this image of the scientist as someone stripped of all his or her social networks, a more or less isolated individual pioneering through under his or her own steam. A feminist image of science would see the scientific enterprise as a lot more burdened down, a lot more contingent and collective.

EDWARD MANIER: If I have been following the conversation correctly, I think we have agreed that the feminist critique is not just a critique of science; it's a critique of all our social and cultural institutions. I think it's a critique, in part, of the ideal of the autonomous self, of the idea that we're each of us supreme lawgivers, that in and of ourselves we can create a frame of light once and for all, and that striving after such high truth is our obligation as sovereign thinkers. This way of thinking doesn't seem to leave much room for friendship, for parenting, for the ordinary mechanisms of evolution in a sense that might apply to human beings. Walker Percy criticizes the set of institutions that encourage us to believe that we can rise above the tide and create ourselves afresh. He wants to call us back to more ancient traditions: the Aristotelian or the Augustinian. The problem is that those traditions have now reproduced themselves in ways we find oppressive. The question then becomes: Where do you go from here? Do you propel yourself into the future you can't clearly imagine, and whose final effects may or may not be equally oppressive, or is there some place to go in the past where there was a viable critique of the autonomous self? I don't think there are any easy answers. We are left with some real dilemmas.

3. Is There a Difference in Science between Error and Bias?

The Argument: It is suggested that historians of science assume too readily that all science is distorted by bias and ideology; sometimes knowledge – claims in science that may *seem* sexist are in fact nonmotivated mistakes that have far more innocent explanations. This remark leads to a discussion about some of the methodological assumptions of the history of science, and whether, like psychoanalysis, it is in danger sometimes of overinterpreting its material.

JAMES SCHWARTZ: I have a rather different sort of question. Even if it is clear that all science has some agenda, isn't it also the case that sometimes scientists just make mistakes? For example, Aristotle said that women have the wrong number of teeth. I'm sure that was not because he felt any particular ambivalence toward women; he just didn't count their teeth because it wasn't the style of scientists of the time to count.

Again, this is not to deny that scientists have social agendas. For example, Jean-Pierre Flourens, an opponent of the localizationist theory in France in the early 19th century, opposed brain localization probably for religious reasons. So there is a social agenda, but there is also the fact that scientists sometimes make mistakes.

LONDA SCHIEBINGER: I'm glad you chose Aristotle, because it's not as important whether or not a scientist simply makes a mistake, as what the culture does with it. *If* Aristotle did make a mistake, the question remains why did his science, including what he had to say about women (a deeply misogynist characterization of their bodies and souls), remain alive in the culture? Certainly, Aristotlean science remained one of the more important determining factors in the collective characterization of women for 2,000 years in Western society, including women's self-characterization.

JAMES SCHWARTZ: I didn't mean to say that he made a mistake in his teeth counting. I meant to say that his acceptance of the idea that women have fewer teeth than men probably stemmed from the fact that Aristotle in fact did not believe in experimental science. Somebody told him or he discovered in some encyclopedia that women had fewer teeth. He just continued the error. It probably did not have anything to do with his feelings about women.

FRANK SULLOWAY: I am struck by what Jimmy Schwartz is saying about Aristotle, and the extent to which scientists sometimes just make mistakes. It occurs to me that an historian of science would probably not have made a statement like that, and I think your remark therefore brings up some interesting questions about the ways in which historians think about science, and how these differ from the ways in which scientists think about science. I think there is an assumption among historians that parallels Freud's conviction about the significance of apparently accidental slips of the tongue in speech and so on. Like Freud, historians of science have a tendency to assume that *everything* has meaning, that *nothing* is really just accidental once you look into the matter and reconstruct the broad context. The historian's goal is to assume there must be an explanation even for something so trivial as to why Aristotle failed to get the number of women's teeth right.

Let me give you an example, and then I will come back to Aristotle. When Darwin visited the Galapagos Islands, he immediately rushed from island to island, not labeling any of his birds by island. Fortunately, he was later able to reconstruct the scientific story from other *Beagle* crew members' accurately labeled collections. We might think that Darwin's collecting behavior was a simple "error." The more you look into this incident, however, the clearer it becomes that Darwin's collecting habits were in fact very much determined by creationist theory. His tendency to collect only a male and a female of each species, for example, is a classic instance of typological creationist thinking. He is confronted with a gold mine of population diversity, and he picks out just one specimen. Fifty years later, the California Academy of Sciences went to the Galapagos with the equivalent of machine guns and procured 10,000 specimens. So we see a whole difference in collecting mentality driven by differences in biological theory.

Let us take the case of Aristotle again. The historian might look at this case and ask, "Well, wouldn't Aristotle have checked something that violated his expectations?" If he or she can find evidence to support this assumption (that Aristotle tended to check dubious facts), then he or she might go on to suggest that Aristotle's belief in major anatomical differences between men and women, of which teeth were one example, was consonant with his overall thinking about the sexes. In fact, we know that Aristotle checked lots of things. His work with marine biology was excellent. So it is not really that he was not an empiricist: it is just that he was not motivated, in some way we do not fully understand, to check something as absurd (to us, at least) as the idea that men and women have a different number of teeth.

This is the way historians of science think; they try to get under the skin of the person of a lost time, and to reconstruct the meaning buried in the little things. What I want now to suggest is that there may be a potential error implicit in our typical historical approach, because it may in fact be possible that, every once in a while, something really *is* accidental. You can end up reading too much meaning into trivial historical details, much the way the entire psychoanalytic profession does with its patients. I hope that we as historians don't go that far. I think we need to develop a kind of a rule of thumb. Initially we should perhaps give the benefit of the doubt to the theory that finds "meaning" in all errors and in all facts. The residue, then, the things we definitely cannot account for, may then be categorized as accidents or "chance" events.

4. Is Science Inherently Self-Correcting?

The Argument: The problem of ideology and bias in science is explored further, now from a somewhat different tack. It is suggested that a society may criticize the ethics implicit to a scientific research program, but these judgments must be separate from all judgments about the science as a testable set of statements about the world. "Good" science is not the same as "ethical" science, and may still remain "good" in the informational sense even if society should judge it ethically "bad." Good science can be recognized because it is inherently self-correcting. Others ask, however, what social values and biases may be present in this process of inherent "self-correction," and whether it might not be important to learn more about that process.

TERRENCE DEACON: I want to shift a little bit away from the feminist issue as a focus here and turn back to a broader issue. What I want to talk about is the notion of good science versus ethical science. I want to distinguish a couple of things. One of these is the notion of value-neutrality, as contrasted with a notion of a self-correcting science. I do not think that most of us, including a lot of the public, still believe that science is value-neutral. I think that myth was lost with the atom bomb. However, many people still have the view that science is inherently self-correcting, and I myself share this view. This is not the same as saying that we're right, not denying that science is biased and human. It is simply suggesting that if you let science run its course, so long as biases change over time, it will become self-correcting, although not necessarily quickly and not necessarily to the good of many people during long periods of time. This is what makes for good science; it means that you've got to have good evidence, and if someone else has stated something else with his or her evidence, then you have to perhaps take the time to test that evidence again, to retest it, to remeasure it in another context to see if your views hold up. The competitive nature of science makes it inherently self-correcting, and this allows us to think of better science, not as truth, but as progressively better or more reliable news.

So that's *good* science. Now what about *ethical* science? You often hear the comment that Nazi science was bad science. I happen to think that it was unethical science, but I'm not sure that it was *bad* science. Let me clarify. I happen to do surgery on animals. There's a lot of animal rights activity these days, and I can imagine a society 20 years or 30 years from now in which it is no longer considered moral to do the work that I do now [cf. the discussion on animal rights, below; also Durant, this volume]. I think what I'm doing is moral now, but I'm just one man who's fallible. Nonetheless, the historical perspective that Londa Schiebinger has drawn, the indictment that's been made against bad science or biased science in the past, would in this case be drawn against me.

My question then is: In what respect will my science 30 years hence be seen in retrospect as bad — biased, unethical — science, and in what respect will it be considered bad science in the sense of representing unreliable news? Even if the moral climate changes, it is still not clear to me that the information I produced, using the tools and methodologies that I did, has nothing to contribute to the self-correcting attempt of science to produce more and more reliable news. In other words, sometimes science is bad because it represents badly selected problems with bad connotations. Sometimes it is bad because of its inappropriateness in a sort of informational sense. I simply want to distinguish these two aspects, and ask Londa Schiebinger if she does not agree that science is self-correcting? Was it not self-correcting even in the 19th century?

LONDA SCHIEBINGER: I think the self-correcting issue is an interesting one. If you think of science as swerving within a certain amount of space between right and wrong on its way to the truth, then one could argue that it is self-correcting. However, if there are systematic biases in the infrastructure of science, then these would not be touched by any inherent self-correcting mechanisms. I think you can't expect science to self-correct on this level until you have done away with systematic exclusions.

TERRENCE DEACON: I think you misunderstand. Let's say I remeasure some biased figures reported in a past paper and report different numbers. Now it's possible that no one will read my work, and that my contribution will disappear. However, someone else sooner or later is going to remeasure those same figures and also find that they don't work. Yes, it's certainly possible that, in the history of the world, a particular bias will never disappear. Yet it doesn't seem to be the way things happen. Enough people remeasure and find that the measurements don't match up, and this troubles enough people that eventually the accepted measurements change.

Now these changes may be only a result of changes in the social context. I don't know. I don't have a causal story to tell me which is causing which to happen. But I do think we underestimate this self-correcting nature of science. I think, moreover, that the self-correcting aspect of science helps explains its continuing power, both for the public mind as well as for those who make it their profession.

FRANK SULLOWAY: You are describing science as a self-correcting process that ultimately results in products that are understood as more or less right or reliable. It struck me that thinking about scientific knowledge as ultimately a correct *product* perhaps misses the issue that Londa Schiebinger was trying to address and that I think Ruth Hubbard was trying to address as well. This is the way I see it:

In the history of science we used to concentrate a lot on the products of science, the great theories, and so on. Recently, there has been a deemphasis of that viewpoint in favor of looking at science more as a process, of understanding science "in the making," so to speak. It seems to me that when you begin to think about science as a process, you realize that we have no adequate theory of how science really works. We just have a lot of philosophical theories that have never been adequately tested by anybody.

One conclusion that seems likely, however, is that we will come to understand the way in which cultural biases and interests are integrated into scientific products only by studying science as *process*. Sure, science is self-correcting, but it could be a lot more self-correcting if we understood the process better and if we could decontaminate that process from some of its internal biases, including but not restricted to gender biases.

5. The Walker Percy Critique of the Human Sciences

The Argument: Discussion now turns to the critique of the Southern novelist Walker Percy. Percy felt that it was language that made humans ultimately "human"—with all the paradoxes and complications that state entails—but that the neurosciences were unable to deal coherently with this phenomenon. It is proposed that Percy's interest in the power of language is relevant to the debates immediately preceding: one aspect of science that may not be inherently self-correcting is its rhetoric or language use. Paying more attention to language in science may not only make science conscious of new biases; attention to the literary dimensions of science may also help to humanize this social institution.

JOHN DURANT: Ed Manier seemed to be suggesting that Walker Percy was intentionally offering an alternative agenda for the study of human beings, or at least a critique of the way human beings are currently being studied within the dominant scientific culture. My question is: Does Walker Percy believe that there are issues that should be addressed about the human condition which science, in its present state, is missing? If so, I have a further question that perhaps takes us back to what Leigh Star was saying about being an outsider trying to talk to insiders. Here you have someone writing novels in the Southern tradition. I have an awful suspicion that most of the shots he is firing at the neuroscientific community are automatically missing simply because people are not paying attention. People in the neurosciences are not going to spend much time reading the kind of literature you are talking about, and probably are not prepared to take a mere novelist seriously, even if he is invited to give the Jefferson Lectures. This is reason for concern. My question then is: How do we give space to the Walker Percys of this world, how can we guarantee that they have their say and their inputs?

EDWARD MANIER: I'm just sorry Walker Percy can't be here to answer instead of me. I wanted to start my reply by saying that I think Terry Deacon's point about looking at science as news is excellent. I very frequently look to science for news myself; that is, news that will help me figure out what it's like to be a human being in the world. I think that's also what Darwin struggled with, and he struggled with it in a way that would have appealed to Walker Percy. Darwin didn't have to worry about whether his article was going to be published in *Brain*. He didn't have to worry about whether Harvard University Press would take a perfectly mundane project and blow it up with lots of publicity in a way that would ensure that it would become a cause célèbre and a center of controversy. His language was under his own control.

For me, his decision to use the expression "struggle for existence" has always seemed to be a kind of poetry—not Malthusian poetry, but Wordsworthian poetry. Beethoven *struggled* against his deafness, Darwin *struggled* all his life trying to decide whether to use expressions connoting war for natural selection or whether to use much softer expressions connoting equilibrium. The very long species book he was working on when Alfred Russell Wallace's paper came in makes very plain that he was digging around for alternative expressions. Now it does seem to me that part of Walker Percy's agenda for science could be construed as suggesting that science become much more cautious and considerate about its language, its rhetoric. What is not necessarily inherently self-correcting about science is the rhetoric, the language, the medium in which the message is cast. For the most part, we do not subject scientists to literary criticism, and the fact that we don't tends, I think, to dehumanize them. I think it tends, for example, to perpetuate the sense that the scientific profession consists of interchangeable parts. Nobody would think of the literary community that way. Nobody would think that if a poet were to die, somebody else would automatically step into his shoes and continue his work.

So it strikes me that there may be something for neuropsychology in Percy's agenda: this phenomenological or existentialist notion that language gives birth to being. It is only through language that we do create that bond with nature, and we re-create whatever we might think about the boundary between nature and culture with our language as well. Again, this is certainly one of the things that made Darwin's work very sensitive, because his language redefined that boundary. Neuroscience faces a similar dilemma.

The other point I would make is that I think Percy is interested in language, not only for its rhetorical functions in science, but also as a *phenomenon* neuropsychology cannot afford to neglect in any science of human beings. Language functions as a medium to connect us with the world, and as a medium to connect us with each other. I don't know if Walker Percy gets the credit, but I had the feeling that we were talking to each other in this session. I don't know whether there's a neuropsychological correlate for this kind of interaction, but if there's not, then that's a piece of news that neuroscience isn't going to be able to tell us about. Yet, so far as we're interested in what it means to be a human being, there's no denying that it is important that people sometimes can really converse with each other.

Finally, I think Percy had an agenda for anthropology having to do with the place of myth in history. He believed certain central religious theses were *historically* true. He thought that a central piece of *news*, of which anthropology had to take account, consisted of the experience of being told, by a perfectly reliable messenger, "You were chosen, you really did screw up, and you really do need help." If we abstract from the meaning of that particular message, it represents the general anthropological problem of connecting myth and history. Percy suggests that anthropology trivializes itself as a human science if it ignores this issue.

Some anthropologists agree.[9] For example, what did the Sandwich Islanders who killed Captain Cook think they were doing? Perhaps, having understood Cook to be a god, they interpreted his untimely return to the island as transforming his divine nature in a way that justified or even required his death, an ominous combination of myth and contingent happenstance. If this is what was going on, we cannot accurately describe or understand Cook's death as resulting from a decision on the part of the islanders to kill another human being. Percy thinks any science of human beings worth its salt must recognize this double layer of meaning in human life.

[9]Sahlins M (1981): *Historical Metaphors and Mythical Realities: Structure in the Early History of the Sandwich Islands Kingdom.* Ann Arbor: MI University of Michigan Press.

Afternoon Roundtable Discussion: Saturday, August 4, 1990

Target Speakers: JOHN DURANT, ALAN FINE, ELLIOT VALENSTEIN

1. What Are the Criteria by Which Experimental Animals Could Have Moral Rights?

The Argument: A heated discussion, dominated by the neuroscientists, breaks out over the problem as to whether it is morally wrong for neuroscientists to use animals in their research if such research will involve animal suffering. Various standards are discussed that we as a society seem to use — inconsistently — when selecting those animals we find most deserving of moral rights: sentience, intelligence, personhood. The neuroscientists find hypocrisy and antihumanistic implications in the claim that so-called species-ism is a form of immorality equivalent to racism and sexism. The neuroscientific imperative to pursue its truths and research questions for the good of humankind is defended. The suggestion is made that much sentimentality and anthropomorphism confounds these debates.

TERRENCE DEACON: In talking about the case for animal rights, John Durant made frequent mention of this notion of sentience. I never heard him once talk about intelligence. Yet I believe that perceptions of intelligence are in fact a key issue, and that the forerunners of the current animal rights movement — Greenpeace and so on — in part grew out of the assumption, without much grounding, that apes, whales, and dolphins are the most intelligent creatures on this planet. Even scientists have told me that dolphins probably have much more intelligence than we do, but are just keeping secrets from us. There is a popular view that intelligence increases the value of the life, and this drives our interest in the welfare of whales and dolphins, and explains why we tend to be more outraged at the killing of whales than at the killing of deer, for example. What part does John Durant see intelligence playing in the case for animal rights?

JOHN DURANT: The reason I concentrate on sentience is that I see this as the issue that is key for the animal liberation movement. More than anything else, people are outraged over the suffering that is alleged to be being caused to other animals. This is not to deny that there is much else going on besides. Attitudes to other animals are full of contradictions.

TERRENCE DEACON: That's what I'm saying. Suffering is treated differently, even by the animal rights people. There is no attempt to treat everything as equal. And in so far as they don't see all suffering as equal, they typically — in my opinion — are guided by the intelligence factor. People scale creatures on the basis of what they think the creatures can understand. All creatures have feelings, they say, but it's especially bad to hurt chimpanzees and whales because of their intelligence.

JOHN DURANT: Okay. I take that on board. But let me make one observation in return. The conservation movement knows — at least the people I talk to — that they are beneficiaries of the symbolic significance of a very small number of animals. So far as I can see, the whales and the giant pandas between them have done more to promote the cause of animal conservation around the world than all the rest of the animal kingdom put together. This fact raises a lot of issues that I think would be worth exploring. Why those particular animals?

I'm told by some of my animal behavior colleagues that there is a tendency for many people to greatly overestimate the relative intellectual abilities of the whales. I'm told that the whales may not be significantly more intelligent than many other animals. Nevertheless, they certainly have come to be seen that way. And I would say we need to understand this phenomenon in terms of a symbolic significance that has become attached to these animals, for whatever reasons.

ROBERT RICHARDS: What contribution do you think larger social movements have made to this movement? For instance, in the early part of this century, the Fabian socialists were in some way connected with the antivivisection movement. And it strikes me that after the Vietnam era in the early 1970s, there was a resurgence of animal-mindedness. This is the era when Jane Fonda got the urge to save the whales, and so on.

JOHN DURANT: I do think there are larger movements. I brought with me a newspaper article about an extraordinary debate that's going on in Britain at the moment about mad cow disease, *bovine spongiform encephalopathy*. It is very interesting for me to see how public perceptions of this issue have developed. One of the concerns people have lighted on is the fact that cows in Britain have been fed offal-based feeds. These are believed to have been the source of the BSE infection, but this is not the fact people have been protesting. Rather, they are distressed by the idea that animals presumed by most people to be herbivores, are being turned into omnivores for the sake of a more efficient industrial agricultural process. I mention all this because I think the most significant other factor playing a role in the animal rights movement concerns a changing sensibility about our relationship with the environment, a change more broadly based and profound than the environmental movement itself, of which saving the whales is a part. The article I mentioned appeared in the *Independent* last weekend, and quoted even farmers saying that they felt pressured to treat cows as walking milk machines in a way that just wasn't appropriate. One has the impression of an almost mystical conviction: "We'll pay a price for this." We're not told how, but there is a feeling that somehow nature will get its own back for this kind of treatment.

MASSIMO PIATTELLI-PALMARINI: Clearly the attempt to liberate the monkeys and other animals being used in experimentation is a major phenomenon, and the populist sentiment potentially quite destructive. It seems inevitable that we're going to have to develop some kind of criteria for decision making. John Durant mentioned the 99-percent DNA correlates between humans and chimpanzees. Maybe DNA should be a criterion. Maybe we should protect the sweetest, most loveable animals. What would be the least *un*reasonable criteria that one could develop?

JOHN DURANT: If you ask me personally, I'm broadly sympathetic with a utilitarian moral philosophy. For this reason, I find the viewpoint represented by Peter Singer, for example, quite persuasive. The point he's making is a point that Jeremy Bentham made a long time ago. Bentham said, "The question is not, Can they *reason?* nor Can they *talk?* but, *Can they suffer?*" If they can suffer, he argued, then they have some moral standing. It's not clear why the suffering of one being should necessarily count enormously less than the suffering of another being. I think the charge of species-ism has to be dealt with, just as the charges of racism and sexism have had to be. Up to this point, I have not discovered any good arguments in favor of species-ism.

ALAN FINE: With respect to this notion of sentience as the determining factor, I would like to suggest that it can't be enough as a criterion, for the simple reason that, if it were, then anything would be permissible on an organism that was painlessly anesthetized. The fact that we don't accept this argument means that there must be other criteria. I think that we ought to define them and that in fact most of us have, in the back of our minds, a much broader notion that can be conveniently lumped under the category of *personhood*. Personhood includes things like intelligence, the ability to make plans, have a history, memories, and so forth.

JOHN DURANT: I feel a sort of great world of debate opening up in front of us, which we will not be able to pursue. I can say that I think utilitarianism is a powerful, but not necessarily a complete philosophy. The problem you've raised is one of the traditional problems of utilitarianism, and I wouldn't pretend otherwise. However, I would also want to be very careful about attempts to define rather abstract terms like "personhood" and would be especially skeptical of any argument that led conveniently to the conclusion that only people can be persons.

DETLEV PLOOG: The neurosciences depend on investigations of the nervous system. In doing this research, you cannot help but do animal experiments. Without animal experimentation, we wouldn't be able to investigate memory, learning, emotions, communication—all the things we care about. All this is very pertinent, and while we don't yet know much about it, we do know that it will be necessary to develop a neuroscientific explanation for these things in animal models if we are ever to stand a chance of understanding the human brain, and also of developing treatments for human neurological diseases. The necessity of animal research, then, cannot be seriously doubted. We need, rather, to come to a moral conclusion on the way we will *continue* to do our animal experiments, while trying to keep the suffering as low as possible.

JOHN DURANT: I think some people are interpreting my talk as if I were actually saying, "We should or should not be doing certain things to animals." That really wasn't the main point of my talk. What I was trying to suggest is that there is a tension between the scientific community's work on animals' brains and behavior and the wider culture. What I was also trying to suggest is that, even within the sciences of animal behavior, there is a tension between neuroscientific work that is distancing us from animals, and inclining us to treat them as objects that advance our knowledge, and ethological work that is bringing us closer to animals and encouraging us to see them as relatives, as subjective individuals that share significant qualities with us. And all of these forces are going on at the same time in the same culture.

I agree with Detlev Ploog that we could not hope to answer many of the questions he mentioned without doing experiments. But that does not mean that there is not a moral dilemma. Let me put it this way: Suppose that we were the only species of animal on Earth. We would still be interested in how our memories and imaginations and emotions work, but I do not think we would approve of experimenting on children or adult people in order to answer those questions. We would all say: "Unfortunately, we're not able to find these things out, because the price we would pay is too high."

If we agree on that, which I assume we do, we then have to ask the question: "Well, if we think differently when it comes to other animals, on what grounds do we think differently?" In other words, we need a moral position on why we would be prepared to do things to other animals that we would not do to ourselves. Do not misunderstand me: there are such positions. We need, though, to think them through. Moreover, I think the arguments are changing as we learn more about the nature of some other animals. I think people are beginning to feel differently than they did perhaps a generation ago, about the justifications you require to do certain things, let's say, to the great apes.

JAMES SCHWARTZ: It seems to me that the issues as to how or whether we should be using animals for research are political or legal issues, not scientific ones. I think we should all agree on that. Because these are essentially political issues, they require political action, which in turn should be based on historical information. So I think one should turn to the historians to explain the reason for what I would call a form of *sentimentality* toward animals. For me, using the word *personhood* is about equivalent to using the word *Bambi*. We're dealing here with a Walt Disney-fication of the animal world. I think this sentimentality isn't honest or innocent, but represents a social diversionary strategy: feed pet cats but disregard human misery. Over the past 20 years the greatest changes in attitude about the use of animals in research occurred in those very countries in which the degree of civility to human beings has decreased dramatically. By *degree,* I mean the *relative* change, not the absolute regard for people. In those countries where the level of civility did not change all that much—for example, Czechoslovakia, Mexico, and Sweden—even though these countries started at very different levels of regard for human life, there has been little popular sentiment against animal experimentation and use of fetal tissue. In contrast, in the United States and Great Britain, where the degree of civility *has* markedly decreased, the uproar is enormous. It is up to historians and social scientists to explain the mechanism, but it seems suspiciously like what psychiatrists call displacement. Only I don't know how unconscious it actually is.

When does this concern for animals emerge? History seems to say during hard times: after the Industrial Revolution in the 19th century, and after Vietnam with Jane Fonda and the whales in the 20th. If one faces the so-called tooth-and-claw aspect of nature, we know that when people are under siege and starving, they will eat cats. And on lifeboats, as we've all heard, people eat people. It may be repugnant, but they do it. My point is that there is nothing inevitable about this sort of sentimentality we are discussing here. It just may be deceptive and comforting.

JOHN DURANT: I think there is a lot of sentimentality toward animals, what you call a "Bambi" factor, and I believe it has to be taken account of. But I would personally think it quite wrong to dismiss all of what's going on as "mere sentimentality." I think there is a proper place for sentiment. We don't say in the pejorative sense that parents are being sentimental when they are concerned for the welfare of their children. We say that they have a concern that is a valid and proper one. I want to argue, similarly, that there is a legitimate question: "What is the level and the extent of the valid concern that we should have for animals that do not happen to be human?" I don't want to try and foreclose what the answer to that question should be, but I would be very uncomfortable about dismissing it as mere sentimentality.

JAMES SCHWARTZ: Yes, but, you know, you can go back to the cave of Lascaux, and you just see that the hunters that lived in that cave respected those bison. Yet they went out and they killed them because they had to. They had to, because they had to eat. And this is essentially the rubric that we have to use. We have to decide what we want and have to do. The morality, the ethics of the situation changes with the situation.

JOHN DURANT: I don't see how anything I've said affects that. I agree with you. You have to decide moral issues in terms of the situation you find.

JAMES SCHWARTZ: All of life is like that. Life is a sliding scale. There's no one absolute context.

JOHN DURANT: I wasn't suggesting that there is.

JAMES SCHWARTZ: And you can't say that monkeys are forever untouchables.

JOHN DURANT: Well, I never said that either. I don't understand why we are disagreeing.

JAMES SCHWARTZ: But what is all this about sentience? What does that mean?

JOHN DURANT: It means that another being may actually be feeling distress. That should be morally significant.

JAMES SCHWARTZ: Yes, but sometimes somebody comes at you, is going to kill you, and you have to shoot him.

JOHN DURANT: Yes, but that isn't relevant. I don't believe that the animals that most neuroscientists work on are threatening them with guns.

JAMES SCHWARTZ: No, my point is that sometimes you want to achieve something that someone else might not like. I mean, this group, on your side of the table, might not like the fact that we on the other side of the table, want to discover the mechanisms of memory in higher animals. Because, in order to do that, we will have to kill higher animals.

2. "Hard" versus "Soft" Approaches to the Understanding of "Animal Mind"

The Argument: The question of anthropomorphism and sentimentality continues to be pursued as the discussion shifts to the validity of those reports purporting to tell us that animals have "personhood," capacity for complex suffering, etc. It is suggested that the methodologies and Cartesian traditions of the neurosciences have hindered these fields from contributing much to our scientific knowledge of animals as subjective agents in the world. It is further suggested (and disputed) that ethologists have provided a much fuller picture of "animal mind." In short, the discussion sets the problem of "animal rights" in a disciplinary debate over the problem of "animal mind," a debate that is characterized by conflicting epistemological and methodological standards. The problems faced by the neurosciences generally in dealing with the mental realm are discussed, and the questions of whether these problems are inherent to the nature of the brain research enterprise, or whether they can be overcome are raised.

TERRENCE DEACON: So what is the responsibility of the neurosciences in all of this? I think we are agreed that we need to make judgments about sentience, about personhood and such, in order to come to moral conclusions about how we are going to treat animals.

My argument would be, though, that ethical and moral judgments need to be informed. They need information. Without information, our moral judgments have little correlation with the real world. John Durant seems to suggest that the information we need should come from pet owners and ethologists, but not neuroscientists for some reason. His explanation for this had something to do with a process of distancing that goes on in the neurosciences. As neuroscientists, we treat animals as objects, as mechanisms, and so on. Certainly, the explanations that we develop are presented in mechanistic terms. However, that is not the issue when it comes to what our responsibility — as neuroscientists — should be in these debates. We must take seriously the question of how to make the translations — translations that we have historically struggled with and sometimes refused to make — about life, about personhood, about self, about emotion and sensibility. These are the values that are required of neuroscientists as we do our work. It's a curious dilemma that it often involves treating animals as mechanisms to some extent, in order to find out that they're very similar to us.

JOHN DURANT: Okay. I think there's an interesting argument here. You're right in your characterization of what I said, and I think I would want to try and defend it. If neuroscientists can and wish to contribute to enlarging our view of the mental worlds of other animals, I shall be delighted for them to do so. But it was my judgment, which I would stick to, that to date far more has been contributed on that question by people observing other animals in close proximity and with great intimacy, in more or less undisturbed situations, or even in an owning situation.

TERRENCE DEACON: But what kind of information is that? If I'm to make moral judgments about life and death, I want clear information. I don't want someone saying "My dog can talk."

JOHN DURANT: No, you misunderstand — and I think this is the heart of the matter. How is it that we know so much about what goes on in other people's minds? It is not because we know much or anything about their brains. It's because we relate to them and it's because we have an empathy with what is going on in their heads. People also have that empathy with animals.

TERRENCE DEACON: There is empathy and there is communication.

JOHN DURANT: Well, but they're linked. Here's George Shaller in his study of *The Year of the Gorilla,* a popular book.[10] He writes, "Since the gorillas are so closely associated day and night, tempers naturally become a little frayed at times. Usually for trivial reasons. Quarrelling is usually confined to the females, with the silver-backed male listening in aloof silence. Bickering goes on. Females harshly screaming." This is the sort of stuff you find over and over again in the writings of ethologists.

TERRENCE DEACON: Exactly.

JOHN DURANT: Well, I'm sorry if I telescoped things. It's true that this is written in a somewhat popular style. Nevertheless, I believe that good ethologists are generally not sentimentalizing, and they're not all doing "Bambi" work. The good ones, and there are many of them, are reporting genuine insights. They know what is out there, and I'll bet that in

---

[10]Schaller GB (1964 [1988]): *The Year of the Gorilla.* Chicago: University of Chicago Press.

hard-nosed terms they can predict better on the basis of their knowledge how these animals will behave than any neuroscientist who's ever lived.

TERRENCE DEACON: I'm not suggesting that ethology is not to be valued.

JOHN DURANT: It's personal knowledge, and I think personal knowledge of other animals is valid; indeed, it's probably indispensable in trying to get to know about other animals.

TERRENCE DEACON: I'm not suggesting that we toss ethology out. I'm suggesting we use ethology, but we don't toss neuroscience out. My responsibility as a neuroscientist is to contribute, not to wait and let the ethologists answer the questions.

JOHN DURANT: Right. But the reason I've become slightly handwaving here is that I detected a tendency on your side of the table to want to dismiss these kinds of claims from ethologists as soft and not reliable. I believe that we've got to come to terms with the genuine contributions of this apparently soft approach to animals.

DETLEV PLOOG: Here I believe you are getting two things mixed up. I think certain questions in neuroethology, for example, are only possible to investigate if you are an ethologist, if you know your animal. You must be in a position to produce descriptions of the sort you have read to us. But then you have to ask a question the neuroscientist must ask: What is the cerebral mechanism behind this? This is what we want to know.

JOHN DURANT: Well, it may be what you want to know, but it's not necessarily to the point of the nature of animal mind.

TERRENCE DEACON: Yes, animal mind is the question here. Do animals have minds? You're telling me that by emoting with my dog, I will understand that this animal has a mind. That's the level of discussion.

JOHN DURANT: I didn't actually say that.

TERRENCE DEACON: Well, you came pretty close. I realize that there is scientific ethology that is very useful. There is also a problem in identifying what is good ethology and what is not good.

JOHN DURANT: I mentioned in passing, and I'll recommend it now explicitly, an article in *Behavioral and Brain Sciences* by Daniel Dennett, on cognitive ethology.[11] What he's asking about, in a very tough-minded way, is what criteria we must use in order to justify the attribution to animals of certain kinds of mental states, such as intentionality. It has always been the position of the behaviorist to try to explain animal behavior in the simplest conceivable terms. And the question Dennett is asking is: "When are we entitled to say, no, the simple explanation is not adequate. This animal must have some awareness of its own position, some insight into what's going on." He claims that the kind of observations you sometimes need in order to answer this question are precisely the types of observations that certain kinds of ethologists make.

---

[11]Dennett DC (1990): Betting your life on an algorithm. *Behav Brain Sci* 13 (4): 660.

Deceit, for example, is a very, very interesting phenomenon in the animal world. There's a famous story many of you probably know from the Dutch literature, where a chimpanzee is trying to put boxes on top of one another in order to reach bananas. Suddenly it sees another chimpanzee coming. It takes the boxes down, scatters them, and metaphorically goes off to sit in the corner and twiddle its thumbs, looking the other way. The other chimpanzee comes in, passes through. As soon as it's gone, back go the boxes, up goes the first chimpanzee. When you see such things, it's very difficult to avoid the conclusion that this chimpanzee has insight into its situation, that it is practicing intentional deception. I see more potential insights into the nature of animal mind from studies like this than I do from direct studies of brains.

RODNEY HOLMES: Let me just see if I can summarize what some of us seem to be suggesting here. It seems we are asking aloud whether the very nature of the neuroscientific discipline is such that it may not be able effectively to study the very thing it finds most interesting, and that is mind. I would like to give people an opportunity to respond directly to this possibility. Is it the case that certain kinds of knowledge are denied our discipline by virtue of its inherent characteristics?

ALAN FINE: It doesn't seem to me that neuroscience is restricted to electrophysiological studies or tracking assessments. Modeling studies, ethological studies, and so on, are all part of the large enterprise we call neuroscience. So the idea that insight into mind is beyond possible attainment using the approaches of neuroscience strikes me as a rather incredible one. The only way to answer it definitively is to go and do the enterprise of neuroscience.

RODNEY HOLMES: Well, that's precisely the point. Are there some things that the methods of empiricism cannot reach?

ALAN FINE: I think that should be an empirical question.

DETLEV PLOOG: May I just add one sentence to that? Rodney Holmes says that neuroscience may be incapable of investigating the problem of mind and subjectivity. I think this is not the case. Do you not think that the investigations on split-brain patients have revealed a lot of new insights about the mind that we didn't know before?

RODNEY HOLMES: Of course, here one has defined "mind" in a way that permits us to study it with the methods we have on hand.

WALTER ROSENBLITH: What other definition of mind would you like to impose on neuroscience?

RODNEY HOLMES: I would simply like to suggest that there are vast other literatures about mind, which are not empirically based.

3. What Are the Ethics of Human Fetal Neural Transplanting?

The Argument: The calculus for balancing the value of new knowledge against the value of ethical restraint is further debated, as the focus shifts to controversies surrounding human fetal neural transplants. Attention is paid particularly to the way in which affective components—especially an unanalyzed feeling of revulsion (the "yuck factor")—may shape social debates over the use of human fetal tissue.

SUSAN LEIGH STAR: I want to address a question to Alan Fine on human fetal neural transplanting. In your presentation, I found it courageous the way you laid open the process that you went through in reflecting on these issues. I was just struck by one phrase that you used, and I have a different reaction to the matter. You said that you're concerned about "protecting" the woman from knowledge of the fact that the fetus she is aborting might be used for other ends. If I were having an abortion, I think that I would be glad if something good could come out of it. That's a very different way of structuring the medical ethics of the situation. I'd be glad to be told even if it were just a possibility, but I don't know for sure.

ALAN FINE: I'm fascinated by what you're saying. I don't remember using that particular word "protecting," and I would not have thought it something that would have sprung to my mind. But indeed we are concerned with the question as to whether a woman should be allowed to make this a component of her decision. I actually was surprised that the clear consensus, as this issue went through the approval process, was that this must *not* be allowed to be a factor—that we had to make sure that the decision to have an abortion would not be influenced by the potential use of the tissue.

What do I think of this myself? Well, first of all, it seems to me to be impossible, in the sense that there's no way of policing it. You can't ever tell what a woman's motive would be, and I think that's a good thing. I'm also curious, however, that we draw these distinctions. Why is it that we're perfectly happy to allow the targeting of a kidney, and actually extol this as a virtuous, altruistic act? What are the salient ethical differences? Yet the clear consensus was that similar targeting of a fetus is repugnant.

RUTH HUBBARD: I can see a reason to be opposed to abortion for the purpose of salvaging the fetal tissue simply because it opens up women to pressures. I mean, suppose her father has Parkinson's disease or her mother or whatever—imagine the subsequent family pressure. We've certainly seen such things with organ donations in hopes that dad or mom can be helped, and I find that distressing. But the issues we are speaking of are not limited to abortion. What about the woman who conceives and carries through a pregnancy for the purpose of using the baby for a bone marrow transplant? A case of this sort has been published in the scientific literature and popular press. To me and a lot of people, such behavior raises a lot of very difficult issues.

ALAN FINE: I think that's fascinating and extremely important. My own feeling about it is that a lot of us have a resistance to these issues that seems to have its roots in the ethical creed that human beings are ends in themselves, and should not be used as means. For me, the crucial point here is: What *is* it about human beings that requires that they should not be used as means? Is it the fact of biological humanness, the fact they have a human genome, or is it the *personhood?* It seems to me that personhood is what's crucial, and this is where a difference arises between a fetus, on the one hand, which is arguably not a person, and a baby which is, or sometime shortly is going to become, a person.

JOHN DURANT: I would like to raise a problem, of which I see the present conversation as a good example. There is a level on which we may have a philosophical debate about the use of fetal material from the point of view of two moral philosophies. The debate is complicated, however, by a problem—and I can't get around using a slightly awful phrase which a British moral philosopher used—of what we can call the *"yuck* factor." In other words, there is a much less rational level on which people respond to the pros and cons of what we are discussing. The actual physical process of what is involved is perceived as very distasteful. And this gut repugnance does become important in shaping the perceptions people have of the ethical issues. How does one deal with that side of things, with that emotional response that says: "That doesn't sound right. Don't argue with me about the right and wrongs. That just sounds awful."

ALAN FINE: I think that's an extremely important point. It's crucial to the way in which the issues are addressed publicly. There clearly is a "yuck factor" with respect to the use of fetal tissue, and it also comes into play in the context of animal experimentation. The fact, as you stressed, that pandas and whales were so important in galvanizing public opinion, draws on this same phenomenon. I think that part of our jobs as scientists, historians, sociologists, people with some sort of dispassionate distance, is to try to eliminate the "yuck factor." It's our job to try to weed out the irrational components, and to isolate the crucial and rational elements of the debate.

4. Experimental Procedures in Testing New Radical Therapies

The Argument: The imperative of ethical restraint in applying the fruits of new research in the medical arena is discussed, with focus on the need to develop better ways to test radical new therapies before allowing patients to be subjected to them. The spread of frontal lobotomy treatment in American psychiatry had been used by the speaker (Elliot Valenstein) as an historical illustration of the risks of therapeutic incaution, with resulting great harm to patients. The ethics of the proposal that new therapies be experimentally tested is questioned, given the fact that calculated nonintervention could also pose risks to patients. It is also suggested that when caution becomes a point of rigid principle, the effects can prove harmful. Finally, the obfuscating functions of the generic term "therapy" is queried; in deciding allowable risk, we need to ask about the interests involved case by case, and it is not always clear who represents an unbiased party acting in the patient's best interest.

JAMES SCHWARTZ: This question is directed to Elliot Valenstein. I am a bit disturbed by what you've been saying because it seems to me that ultimately it's unethical to practice experimental medicine on patients who come to you for treatment. You must treat each patient as you go, and this is why medicine is not a science. In other words, you must offer the treatment you think is in the best interest of your patient; you can't do a controlled experiment on that patient, or your patient might die needlessly.

ELLIOT VALENSTEIN: Well, naturally, we don't want to allow patients to die needlessly. Partly my answer to your comment would be that there is a certain amount of confusion about the word "experimental." The word is often avoided because it obviously has legal ramifications. The word "experiment" in a courtroom is often used to imply that people were used as guinea pigs, which I think is what you are referring to. To me, "experiment" doesn't mean that you don't have the patients' best interests in mind, given our current state of knowledge. "Experiment" means that you run your procedures on a group of patients in such a way that you have a reasonable chance of drawing some kind of valid conclusions about the efficacy of the treatment. This is what makes what you're doing "experimental." In normal observation, you have no way of comparing your results to any other group. You don't know what the outcome would have been, had you not treated that person that way.

Now, as to deciding when techniques are sufficiently controversial, that is a problem in itself. Who's to decide when a technique is officially controversial and when it's just a modification of a technique already accepted? I don't pretend to have the answer to that, and it needs to be addressed. But when a technique is sufficiently controversial, and untested, I think we have to institute some ways of getting it tested, much as we do with new drugs. With drugs you can do double-blind experiments. With procedures that do not involve drugs, it is often more complicated, but we also need to put the techniques through various stages of testing, with procedures that have been approved, so that we have a reasonable chance of drawing valid conclusions about efficacy. There are many legal and ethical considerations here, but I don't think they're insurmountable. I think they all can be addressed.

Let me give you a case in point: coronary artery bypass operations. Without making any judgment about the actual effectiveness of these techniques, I think it is absolutely essential to recognize that they were adopted for many of the wider social reasons I have talked about. The first coronary artery bypass operation was performed in 1964; the second not until 1967. A thousand operations were then performed in 1968, and 35,000 in 1974. It's estimated that the current level is somewhere between a quarter of a million and 300,000 operations annually. Yet the first controlled experiment was performed only about a year and a half ago.

Now, it may turn out that this operation is an absolutely marvelous thing, but we don't know that. Instead, this kind of spread of technology is driven by powerful economic forces, as many people have suggested who have studied the factors underlying the spread of these operations. It has been estimated that we are talking about a $5 billion industry, if you include the companies that are supplying the equipment. I think this is a problem that has to be addressed. We will not be able to avoid mistakes, but we can institute procedures to limit their magnitude.

ANNE HARRINGTON: I wanted to ask whether you might not agree that there is an additional layer of ethical concern behind the whole issue of ensuring that there are scientific safeguards and procedures to test the efficacy of new treatments. I mean here the whole issue of therapy itself, and the different sorts of things that hide behind this word "therapy." You mentioned the case of Rosemary Kennedy, which I found quite striking. It reminded me of the way we talk in the history of psychiatry about the difference between *treating* versus *managing* patients.[12]

---

[12]Rosemary, the first daughter of Rose and Joseph Kennedy, was born mentally retarded. When she was a teenager, her family, under medical advice, subjected her to psychosurgery in an effort to improve her temper and behavior. The surgery so incapacitated her intellectually that she was never able to function again outside of a context of total custodial care.

Attractive and initially sweet-tempered, Rosemary as a child had been both protected and rigorously schooled by her family to behave in ways that would allow her to "pass" as "normal." In fact, the family did not publicly discuss her handicap until Rosemary's brother, John Kennedy, ran for the Presidency in 1960 and rumors began to interfere with the campaign.

In 1941, when she was about 17, Rosemary's behavior began to regress and (in the words of her mother, Rose Kennedy) ". . . her customary good nature gave way to tension and irritability." Rose went on to explain how the decision was made to subject Rosemary to a lobotomy or lobectomy:

She was upset easily and unpredictably. Some of these upsets became tantrums, or rages, during which she broke things or hit out at people. . . . Also, there were convulsive episodes. Manifestly there were other factors at work besides retardation. A neurological disturbance or disease of some sort seemingly had overtaken her, and it was becoming progressively worse. Joe [Kennedy] and I brought the most eminent medical specialists into consultation, and the advice, finally, was that Rosemary should undergo a certain form of neurosurgery.

The operation eliminated the violence and the convulsive seizures, but it also had the effect of leaving Rosemary permanently incapacitated. She lost everything that had been gained during the years by her own gallant efforts and our loving efforts for her. . . . She would need custodial care (Kennedy, 1974 p. 286).

Another biographer of Rose Kennedy, Cameron discreetly suggests (1971, p. 148) that, at the time the decision was made to put Rosemary into custodial care (the psychosurgery is significantly not mentioned), there had been increasing concern in the Kennedy family that, as Rosemary matured into a woman, ". . . a clever opportunist would turn her head." "[T]he Kennedy millions could be a great temptation."

(*Sources:* Cameron G (1971): *Rose, A Biography of Rose Fitzgerald Kennedy.* New York: G.P. Putnam's Sons; Kennedy RF (1974): *Times to Remember.* Garden City, N.Y.: Doubleday.)

The point I am making is that all sorts of complex realities may hide behind the neutrality of our medical vocabulary. Take the case of treating criminals—say, sex criminals—with a form of negative conditioning that may be very, very painful, or the brisk drug-based industry that treats so-called hyperactive children. I wonder if you could speak to *this* ethical dimension of therapeutic innovation, and how we should proceed.

ELLIOT VALENSTEIN: I'm not sure that I know how we should proceed. There are many people who maintain, I think quite correctly, that psychoactive drugs are used in part for management problems. To a great extent, management seems like it is for the convenience of the family, custodians, the hospital staff, and so on. However, management is often also important for the quality of life of the patients themselves; there are certain kinds of restraints, certain restrictions of freedom that are necessary, if someone is suicidal or aggressively unmanageable. These can then be lifted after the situation has changed.

It does seem that we need some kind of review process to determine whether particular procedures are in the best interests of the patient. I mean, a review process that is not a hand-washing, in-house matter, but one that involves competent outsiders. They're difficult decisions to make.

ANNE HARRINGTON: Yes, but who is the "we" in this scenario? Let us imagine we had a group of 19th-century doctors, considering whether to perform a clitoridectomy on an allegedly hysterical woman demanding a divorce from her husband; there were such cases. Maybe this group would come to the conclusion that removal of her organs of sexual pleasure would be in her best interest. We, from a distance of 100 years, might see the matter quite differently. What represents an unbiased view?

ELLIOT VALENSTEIN: I don't know the answer to that. I purposely said the review board should not be an in-house type organization, that there should be some kind of outside influence. What exactly the nature of that would be, I'm not prepared to say. I'm sure, in any individual case, that we're never going to be assured that the particular people chosen have the best interest of the patient at heart, or made the best decision for that person. There are never any 100-percent guarantees.

JAMES SCHWARTZ: One reason why I decided against practicing medicine is that it's not a science. It really isn't, and shouldn't be. I mean, I think it's unethical to do experiments with a patient. My most talented teachers treated each patient as you go—by instinct, intuition, and experience. In other words, if they thought a treatment was the best one for a patient, they would think it absurd to do a controlled experiment on that patient. For me, a most critical point of career decision occurred in the late 1950s in my last year of medical school when I was a clinical clerk in pediatrics at Bellevue. In a lecture to students and residents, a white physician from South Africa reported a set of controlled experiments on children with tuberculous meningitis, then a fatal disease if untreated. The aim of the experiment was to evaluate the effectiveness of intrathecal injections of streptomycin. The patients were divided into treated and control groups. And guess what happened? All the controls died, whereas most of the treated survived. And I still think that was unethical, especially since all of the children were black. I asked: If the doctor's child had had the meningitis, wouldn't he surely have been treated?

GODEHARD OEPEN: This discussion is about applied knowledge and social accountability, and most of us are thinking in terms of experimental therapies that are potentially dangerous and not under operational control. However, my personal experience with neurology and psychiatry has been that *too* much caution, too much fear of imprecision and risk can also be dangerous. I want to give you two examples. The first has its source in the shift in psychiatry nowadays away from psychoanalysis and psychodynamics and toward a much more biological orientation. At least in Germany, where I worked, it has become very difficult to stand up and say that relationships are also important in psychiatry. You are immediately put down as "soft" or "unscientific," and people say: "No, not at all. Just give them the right drug, and that's it." I witnessed the way in which our new director intentionally disrupted the dynamics of our clinic in the interest of efficiency and with no regard for doctor-patient relationships; we had several suicides as a consequence. Yet it was not possible to discuss this in the public forum. So it isn't only the case that one must be prudent in developing certain types of new treatment; one must also be alert to the risks associated with blanket suppression of treatments that happen to have fallen out of favor, perhaps partly for irrational reasons.

On this point, you should know that your book about the risks of treatments like lobotomy[13] has had an important positive impact, but also has resulted in certain problems you might not be aware of. I am sure you are familiar with the political pressure that Robert Heath from Tulane University in New Orleans came under in the 1960s to stop his pioneering work in stereotactic surgery; some of this work was very promising, and I think the reaction against it was simply overblown. Your book plays a role in continuing to fuel this kind of process today. A friend of mine, the Zurich neurologist Gregor Wieser, does work similar to that of Heath but much more sophisticated. Wieser works with patients who suffer from horrendous behavioral problems, problems where you have tried all options using medication, structured settings, whatever. Using depth probes in the brain, he has been quite successful, in selected cases, in locating subcortical epileptic foci apparently responsible for the uncontrollable symptoms. Now, however, he is running out of funding; he's suffering political harassment; he's having departmental problems. I find this irrational.

ELLIOT VALENSTEIN: I think people who know me, know that I never have

---

[13]Valenstein ES (1986): *Great and Desperate Cures: The Rise and Decline of Psychosurgery and Other Radical Treatments for Mental Illness.* New York: Basic Books.

taken the position that we should outlaw all psychosurgery. In my book, I talked about some of the factors, at particular times, that accounted for a dangerous, wildfire proliferation of terrible mutilating procedures. If people have interpreted the horror story I describe to imply that we should never put electrodes in the brain, then they are simply misreading the book. I don't feel responsible for that. I personally have a different evaluative opinion of some of the people that you mentioned, but I won't discuss their contributions.

5. Does the Game of Knowledge Justify Its Means?

The Argument: Having now discussed three specific issues that force the question of ethical restraint and limits in different ways, the neuroscientists reflect on the calculus they use in deciding when and how to proceed with controversial research and applications. There is general agreement that this ethical decision-making process should be made a more visible part of their neuroscientific work.

RODNEY HOLMES: In some sense, all three of the speakers, in different ways today, raised a fundamental question that I never thought would ever be raised by scientists themselves. This is the question of limits. Alan Fine asked the question like this: "Does the game of knowledge justify the subject?" We are asking here deep questions that go to the root of scientific activity itself, and we seem to be coming to a consensus that a blind adherence to the principle of knowledge for knowledge's sake is not ethically permissible. How does one decide whether there is some kind of knowledge in neuroscience that is not worth gaining, or is not justified in principle?

ALAN FINE: I have little doubt that there are indeed some questions, the answers to which can be obtained only at costs that we ought not to pay. Though I do much of what I do out of a sense of curiosity, an interest in knowledge for its own sake, I don't think that that confers upon me any sort of absolute rights. To offer a very trivial illustration: I don't feel that I would be justified in undertaking a neurophysiological assessment of the activation of the anterior thalamus if it would produce extreme pain in my subject. There are some things that I just don't think are warranted, however much I might be interested in the questions.

RODNEY HOLMES: On what ground can you say that? On what justifying principle?

ALAN FINE: I think that is a crucial question, but frankly I'm at a loss to answer it. I'm sympathetic to the notion that some animals have some ethical claims. That is to say, I believe there may be animals—higher primates, for example—that demonstrate properties that I would consider evidence of personhood. And as a result of that, these animals are entitled to certain considerations that should keep us from doing anything we wish to. But what I don't know is how, in a particular practical case, we do the calculus. In particular, how do we factor in a possible future good that may come out of what, for the sake of argument, is pure knowledge right now. In a case like that, when there isn't a clear ledger of plusses and minuses, I don't know what a proper calculus may be.

TERRENCE DEACON: I think the calculus question is incredibly important. We make such decisions every day. It's not the case that we are talking here about some decisions that we *might* have to make some day. It's a very troublesome issue for me. Killing an animal that I have, in a sense, fallen in love with — which is what happens in the laboratory — is a terrible experience. I would not wish it on anyone. Day and night, I struggle with justifications and nightmares. It's not an easy process.

Moreover, there is a calculation that particularly troubles me. The more I learn — as a neuroscientist — about the brains of different animals, including humans, the more I see the similarities. The similarities are the things I want to learn about. Yet it is also the similarities that frighten me, that make the moral issues more difficult for me. It's a conundrum. If I had not done the research, I wouldn't know these things. Information is the critical issue in making responsible decisions, but we also have to figure out a way to get that information responsibly. We start out making flawed judgments, coming up with a flawed calculus. The next time around, we'll hopefully get the calculus better. I think it's a fallacy to think that there is a cut-and-dried single answer, a point at which we can stop and say: "Here's the cut-off point. Here are all the animals we work on; here are all the animals we don't work on." The process is very difficult. It involves issues of judgment and balance and cost and suffering at every stage, and it changes from moment to moment.

ALAN FINE: The fact, though, that you are able to make a decision and I am as well, implies that we do have some rules, whether we know what they are or not. And it seems to me that everybody else in our society has some claim on us in asking us to be explicit. This is a challenge that is on the table right now. If we have rules that we are using, it seems that we must dredge them out somehow, hold them up, and say to the society at large: "Here they are, and we can defend these. We think these are sensible."

ROBERT RICHARDS: Alan Fine suggested one ought to make explicit what the rules are by which one makes ethical judgments. I agree that it's incumbent upon neuroscientists and all of us to do so. In fact, both John Durant and Alan Fine *have* specified what criteria they think are applicable; they are somewhat different. "Sentience," on the one hand; "personhood," on the other. In our discussion at coffee break, I asked you, Alan, to decide a case, to which you gave an answer. It's an artificial case in order to test what it really means to say that "personhood" would be the criterion. I'd like to put the case to you again in this forum, but change the conditions just slightly.

The case has to do with fetal transplants and the situation is this: Let's say that xenographs are perfectly acceptable. And let's assume you have a chimpanzee whose neural tissue would be accepted by someone with Parkinson's disease. Now let's say you have a couple in which the woman is

6 months pregnant. She has a husband who has Parkinson's disease. The couple have a pet chimp. The woman says to her doctors: "I want you to cure my husband. Do it the best way you can. I don't care how you do it. I understand that it can either be fetal tissue or the tissue of our pet." Which of the two would it be: the chimp or the fetus?

ALAN FINE: It's significant that you changed the conditions. When you first asked me, the choice was between a chimp and an 8-week-old fetus. I then had no question in my mind in that particular case, because there it seemed clear that the chimp had a stronger claim to personhood than the 8-week fetus had. When you start pushing it to a much older fetus, or even potential infant, then I have trouble. I don't know how to do the calculus in those gray areas. But this does not mean that there are no cases where we cannot get a fairly clear answer.

WALTER ROSENBLITH: I truly sympathize with all I have just heard, and one thing occurs to me: We have rules for what scientific papers must say and justify. Perhaps editors of journals should "force" scientists and medical people working in these frontier areas to make explicit, in their papers, what determined their ethical choices, the number and types of animals used, etc. I think that might yield rather interesting results.

In addition, I think that when one looks at the things that Elliot Valenstein talked about, one wonders whether we should not further think in terms of an Office of Technology Assessment. I agree, when knowledge becomes technologically, societally important, that there is an obligation, both from the government and from the scientific community, to seek responsible assessments from outside sources. Which is not to take away the responsibility of the individual or group of scientists.

Clearly, there is a great deal of work to be done still. If Norbert Wiener were alive today and could rewrite the book, *The Human Use of Human Beings*,[14] I am sure he would find a great deal of material in the discussions that we have had here today.

---

[14]Wiener N (1950): *The Human Use of Human Beings: Cybernetics and Society.* Boston: Houghton Mifflin.

Evening Roundtable Discussion: Saturday, August 4, 1990

Target Speakers: WARWICK ANDERSON, SUSAN LEIGH STAR, ANNE HARRINGTON

1. Localization and the Mind/Body Problem

The Argument: The centrality of mind/body category dilemmas in the localizationist (modularist) model of human brain functioning is discussed. There is general agreement about the inadequacies of much modern localizationist thinking, and some reasons are hazarded for the persistance of this way of conceiving structure-function relationships. Alternative metaphors are volunteered.

TERRENCE DEACON: You made the suggestion that the problems involved in localization theory in the 19th century remain just as much a problem today; that we have no sense of brain localization today in advance of what was happening then. I found that very surprising, and wanted to ask what you mean by that. Do you really mean it the way I just said it?

SUSAN LEIGH STAR: I only mean there's no consensus, there is ongoing debate, and the terms of the debate are often framed as problems of the mind/brain relationship.

TERRENCE DEACON: You are talking, then, about localization of the mind rather than about localization of sensory-motor function?

SUSAN LEIGH STAR: Those things are often confounded together. Perhaps there's more consensus about localization of function, but there's not complete consensus about that either.

TERRENCE DEACON: If you're talking about mind, I don't have any problems with your criticism. If you're talking about sensory-motor functions, autonomic functions, whatever is involved, I can't see how you can support your argument. So am I right in interpreting this as a "mind" question, or do you really want to make a stronger claim?

SUSAN LEIGH STAR: My sociological observation is that there is an incredible mixup around these terms. It's not as if people say: "Great, we've got the body solved; we know all about function. On the other hand, mind is a mess." People are constantly confusing whether they're talking about function, body, mind, and how these all connect. They mush them together in very different models of how they connect. Moreover, there's a whole wing of neurophysiology that just says: "Well, there isn't any difference." That's my sociological observation.

JASON BROWN: Well, I wonder whether in fact *neuroscience* has created some of the problems that Terry might be alluding to in regard to localization of mental processes. We've gone from the "centers" and "pathways" of the old neuropsychology to operations in smaller and smaller areas, functional columns, command cells, and so on. Simultaneously, you have such a profusion of work in even clinical neurology, to say nothing about neurophysiology. Some of you may know *The Handbook of Clinical Neurology* consisting of 24 volumes, with 600 or 700 pages per volume. A life's work can be one page in that whole series. So things are dividing up to the point where it's almost inconceivable that anybody could come up with a mind/brain theory that could possibly do justice to the complexity of the brain.

Yet, at the same time, the mind theorists talk about mental states and brain states—and never bother to specify what they mean by "a brain state," though perhaps they have some idea of what they mean by "a mind state." I think things are tremendously dislocated today, with the cognitivists on the one hand, and the neuroscientists on the other. It's hard to see how they can come together. In this sense, I agree with the argument for incoherence and lack of consensus. I would go so far as to say that things have gotten even worse.

DETLEV PLOOG: I just wanted to say a word about Norman Geschwind and the localization model he resurrected in the 1960s. The disconnection syndrome of Geschwind is practically an application of the Lichtheim scheme in the 19th century.[15] You have your connecting pathways, and you cut them and you have this or that syndrome. This also doesn't suggest we've made too much progress.

JASON BROWN: We're here with sociologists and historians, so we have a chance perhaps to understand why such a simple account of problems of such complexity managed to achieve so firm a foothold, to appeal to so many people, especially in this country. I think it is still a dominant way of thinking, and continues in a different form in modularity and other current trends. What is it about the way the human mind works that makes it receptive to that kind of explanation?

ANNE HARRINGTON: It's simple: it's easy to visualize and to teach.

TERRENCE DEACON: The atoms of analysis that Geschwind uses are sounds and sights, or are spots in space, line edges in space. All things that we all have no trouble identifying. I go back to William James's comment: really most of thought is this stuff, this fuzzy stuff that is not quite "finished" yet.

---

[15]See Lichtheim L (1886): On aphasia. *Brain* 7:433–484; Geschwind N (1965): Disconnexion syndrome in animals and man. In: *Selected Papers on Language and the Brain: Boston Studies in the Philosophy of Science,* Cohen R, Wartofsky MW, eds. Dordrecht, The Netherlands: D. Reidel, 16:62–92.

WALTER ROSENBLITH: "Buzzing, blooming confusion."

TERRENCE DEACON: That's right. That's what most of it is. And we don't have a nice easy commonsense way of grabbing hold of what all this is.

JASON BROWN: Well, I think we have to understand it partly in terms of the bias toward things as opposed to processes, what Henri Bergson called "the logic of solid bodies," and Whitehead called "the fallacy of misplaced concreteness." There's a tendency to look at these things as if they're concrete, real things — an idea, a representation, a center, whatever — when life and the physical world are actually in constant change.

PAUL MACLEAN: I have one request: would Leigh Star be prepared to restate the metaphor she would accept for the functioning of the mind and brain?

SUSAN LEIGH STAR: Well, I tried, in talking about the mind as a table, to stress the social nature of these phenomena, the importance of the kinds of arrangements that we make with each other. Another metaphor I like very much is about memory and about the fact that our memory is distributed in arrangements that we've made with each other, as well as in all kinds of technologies. An awareness of this exists in computer science as it has been trying to model memory and intelligence, and increasingly realizing that much of the way we remember things is spread out over locations. You're trying to find your way from point A to point B, but you don't have the map of the entire way from point A to point B. Instead, you have a sense of how you get to the next intermediate point where you can ask somebody how to get to the next intermediate point, and then ask somebody again.

PAUL MACLEAN: I like Emily Dickinson's metaphor: "The brain is just the weight of God." What do you do with that one? I feel it's the most beautiful metaphor in the world.

2. Epistemological and Methodological Alternatives to Localization Theory

The Argument: From localizationism, the discussion looks at the rival perspectives of holism. The epistemological issues underlying much of the reaction against localizationist theory are stressed. In interwar Germany, at least, holism was not only about developing alternative models of brain functioning (although it was about that, too); it was also about developing alternative, less empirically based standards of knowledge and method in the mind and brain sciences. These changes in the brain sciences are related to the interwar cultural and political mood.

JASON BROWN: I'd like to move the conversation from localizationism to the other extreme of holism. To me, it's always been a kind of irony that Kurt Goldstein's major work in English, *Language and Language Disturbances* (1948), has been used as an illustration of the bankruptcy of holism. The book is divided into an initial holistic, theoretical half, and a case-study half — and this second part is the half that all professionals read carefully. What is striking is the impossibility of mapping the one half of the book onto the other half. The publication of that book in 1948 almost coincides with the death of holism in the United States.

ANNE HARRINGTON: I am actually relatively happy and prepared to admit that holism is a more or less "failed" scientific movement, at least in terms of how much productive science it could produce. If science is in the business of producing work and in the business of producing useful knowledge, then on this level holism was not really an effective rival or alternative to localizationist theory. There were productive exceptions, notably the experimental work coming out of Berlin. However, generally speaking, the rhetoric surrounding the importance of holism far exceeded the actual fruits it could claim. My main point was, that because there was so much ideology and social value invested in this approach, it nevertheless managed to achieve a certain prominence in Germany and, to a certain extent, on the international scene.

JASON BROWN: Do you believe that one has to take an epistemological stand with regard to the interpretation of brain facts that are themselves neutral? This was certainly the argument of people like Paul Schilder, who I believe to be the greatest neuropsychologist to come out of the holist tradition. If the epistemological stand for the localizationists was more or less one of naive realism, then one might suggest that the holists would take the alternative idealist point of view. Would you be willing to go that far, and would you be willing to say that it was the inability of idealist philosophy to gain a footing in this country that accounts for the waning of holistic ways of thinking?

ANNE HARRINGTON: I don't know what accounts for the waning of holism in the United States. I will talk a little bit about what I think happened in Germany. There is an important element of truth in what you are suggesting, in that holistic science historically has shown an affinity with idealistic philosophical traditions, specifically, the traditions associated with the extension of the Kantian critique. At the same time, the actual epistemological solutions you see emerging from the holistic movement vary across a spectrum ranging from a highly politically loaded irrationalist celebration of "wholeness" and "life" to a rather moderate attempt to reform the scientific method. For this reason, you cannot easily lump all of holism under any single epistemological category like "idealist," at least not without a lot of qualification.

Take, for example, the attempt of Jakob von Uexküll, the ethologist, to "biologize" Kant in the 1920s. His starting point was the *Umwelt,* the subjective reality of the individual animal. Somewhat like the argument Nagel makes in his paper, "What Is It Like to Be a Bat?," he argued that the reality of the bat is profoundly different from our reality and in some sense permanently impenetrable. Even among humans, *Umwelts,* or realities, are profoundly individual.[16] This fact of the relativism of reality served Uexküll

---

[16]Nagel T (1979): What is it like to be a bat? In: *Mortal Questions*. Cambridge:

310 Fragments of a Dialogue

as the starting point for a series of harangues against the epistemological arrogance of the natural sciences, and led him back to an affirmation of the possibility of mystical, religious truth. Still, he always considered himself a natural scientist and respected the experimental method and so on.

Somewhat more extreme was the attitude of the Leipzig school, that was ultimately rather antagonistic toward rationality, looking upon it as secondary to the promptings of the heart, even in science, and in some sense a foreign import from England and France. It channeled its epistemological critiques into *völkisch* cultural traditions and ultimately into political traditions.

Finally, since you brought him up, I will refer again to the approach of Goldstein that saw in human experience and especially human personality phenomena that, in their *wholeness,* could only be perceived intelligently using the methods of intuition and empathy; what John Durant would call, I guess, "personal knowledge." For this reason, Goldstein taught that the neuroscientist of the human brain must ultimately become—as a neuroscientist—more than a natural scientist if he is to come to terms with what is perhaps most important in his subject matter. I must say I am rather sympathetic to this position, but I don't think it qualifies exactly as idealism [cf. Harrington, this volume].

DETLEV PLOOG: Anne, this was a fascinating story you have told. The surprise to me was that you put so much weight on the Leipzig school, whereas I saw much more merit and later development coming out of the Berlin school. If you go through the *Psychologische Forschung* journal— *Journal of Psychological Research*—from 1923 on until Köhler left Germany, it's one experimental paper after the other.

WALTER ROSENBLITH: Hear, hear.

DETLEV PLOOG: Whereas, if you read this Krueger material out of Leipzig, it is mediocre, and really hasn't had much influence in psychology. His writings are full of romanticisms.[17] In contrast, I think the most current discoveries in neurophysiology were very much on Köhler's mind when he claimed, for instance, in Edinburgh at the first international psychological conference after the war, that when we see a cross, a neurophysiological representation of the cross is somehow in our optical lobe. In this sense, the

---

Cambridge University Press; Brock, F (1939): Die Grundlagen der Umweltforschung Jackob von Uexkülls und seiner Schule. *Verhandlungen der Deutschen zoologischen Gesellschaft* E.v. *(Zoologischer Anzeiger. 12. Supplementband)* 41: 16–18.

[17]For an historical assessment somewhat in this debunking vein, see Ulfried G (1980): Die Zerstörung wissenschaftlicher Vernunft. Felix Krueger und die Leipziger Schule der Ganzheitspsychologie. *Psychologie heute* 7(4): 35–43. It is worthy of mention, however, that microgenetic theory, which is discussed and defended in this volume, traces its roots back to the Leipzig school, especially the *Aktualgenesis* work of Krueger's student, Friedrich Sander (see Hanlon R, Brown JW [1989]: Microgenesis: Historical review and current studies. In: *Brain Organization of Language and Cognitive Processes,* Ardila A, Ostrosky-Solis F, eds. New York: Plenum, pp. 3–15).

experiments of Colin Blackmore and others later on turn out to be very close to the ideas of Gestalt psychology à la Berlin. Köhler is the same man, as you know, who wrote *The Mentality of Apes*. His first book, though, was on physical gestalts in the nonorganic realm, a work full of mathematics, with none of this romantic blur.[18]

Gestalt psychology was a scientific attack against association psychology, which was closely connected with Wundt. Wundt, of course, is now modern again. Unfortunately, many roots of the early experimental gestalt psychology never really took hold in the United States. Neither Kurt Lewin nor Wolfgang Köhler nor Koffka nor Wertheimer reached any real importance in this country. In Germany, psychologists saw the beginning of something interesting, and then it was cut off. All these people had to emigrate, and the story of Gestalt psychology ended practically with the war.

ANNE HARRINGTON: I just want to reply very briefly to what you're saying. Again, from the point of view of productive science, I agree with what you're saying about the Berlin school. I didn't come to praise the Leipzig school; I came to bury them, or at least to try to elucidate their historical success in the context of their time. The Leipzig school interests me because of the clarity of the epistemological break you see being advocated among its advocates. The members of the Berlin school were, in a sense, more conservative epistemologically, and were consequently attacked by the Leipzig school as physicalists and phenomenologists who had missed the point of what holistic epistemological radicalism was all about. Because Berlin remained in the mainstream, they fell somewhat out of the focus of the immediate story I wanted to tell tonight, although if I were telling the larger story of holism, they would come back in.

WALTER ROSENBLITH: I think Anne has given a really remarkable account of that period of *squishiness* in philosophical thinking. I also think, however, that Ploog is entirely correct in his remarks, except that he underestimates the influence of Köhler, Wertheimer, and others in the United States. Maybe I'm biased by the fact that Köhler and Richard Held came to work in our laboratory at one stage because they thought I had equipment that might show some of the things that Köhler had predicted.

There's a little historical footnote to all this that I cannot suppress. I lived in Germany in those years before Hitler came to power. I was a young person, not a scientist, but already committed to going into science. But I had quite a few friends, contemporaries of mine, who belonged to what was known as the *Deutsche Jugendbewegung* [German Youth Movement]. The *Deutsche Jugendbewegung* was a form of *holistic* political and quasi-philosophical expression, even for people who were not Nazis. This

---

[18]See Köhler W (1925): *The Mentality of Apes*. Winter E (trans). New York: Harcourt; Köhler W (1920): Die physischen Gestalten in Ruhe und im stationären Zustand. Eine naturphilosophische Untersuchung. Reprinted in condensed English translation in: Ellis WD (1938): *Sourcebook of Gestalt Psychology*. London: Routledge Kegan Paul, pp. 17–54.

movement in some ways expressed these people's *Unbehagen in der Kultur* [discomfort with culture]. Under the circumstances, I think that a study of the way in which holism developed, and the disappointment with it, needs to see its connections even to those who were not scientists as such, but just young people who felt generally alienated and very uncomfortable with what they thought was the bourgeois structure of society.

Concluding Roundtable Discussion: Sunday, August 5, 1990

1. What Can the Brain Sciences Tell Us about Being Human?

The Argument: The ability of the brain sciences to speak to any of the existential issues that validate us as human beings is questioned. Yet there is a general conviction in our society that the brain sciences, in defining the natural limits of our souls, do possess a position of some authority in debates over what it means to be a human being. The usefulness of the human brain's hunger for existential validation is then questioned: through a trick of nature, our brains are indeed programmed to hunger for validation, love, and justice, but our hungers are inevitably betrayed. The world in which we have all been born is in fact characterized by arbitrary cruelty, separation, and endless suffering.

RUTH HUBBARD: I was asking myself this morning why no one this weekend had felt the need to ask the question why a person should *expect* neuroscience to tell us much about the human brain, or, more specifically, about what it means to be a human being who has to live with this awfully big, burdensome brain. I was feeling short-changed. I was feeling that I had not learned anything about being human, or about the significance of having a human brain, that I would not have learned more satisfactorily by reading poetry, or working in a daycare center, or teaching non-English-speakers to speak English, or a whole range of human activities that call on what we assume to be our brain's functions.

ANNE HARRINGTON: Well, I think we will agree that we live in a society that has generally accepted the idea that neuroscientific study of the brain can tell us something about what it means to be human. In principle there is a willingness to look for the sources and ultimate meaning of our humanness in our biology, in our creatureliness, and particularly in the peculiarities of our brain.

Of course this is an article of faith in our culture, and open to debate. This is why we are here. We wanted to hear from the neuroscientists about how far they think human values and basic understanding of what it means to be human is changing or should be changing under the onslaught of new information in their field. But that was not the only goal. One of the dilemmas we also need to address has to do with the fact that we are talking about a human brain trying to reconstruct itself, trying to create a scientific image of itself, but inevitably drawing on its own value-laden human resources. In other words, we are also interested in trying to understand the marriage of knowledge and values that is forged in this process of the brain trying to draw itself. It is not only the brain that is "human"; our scientific drawing of the brain is also inevitably a "human" drawing.

ALAN FINE: Speaking for myself, I didn't decide to be a neuroscientist in order to learn something about what it means to be human. I decided to become a neuroscientist because I thought that the brain was an extraordinary thing, a miraculous thing. It just seemed incredible that this physical thing could do all it seemed to be doing. To me, the brain is the most fascinating object in the material universe. And it is this sense of utter fascination that drew me into the field, and that has dictated my career. I wanted to learn to understand even just a tiny bit about how the brain functioned. But all the same, it would have seemed outrageously presumptuous of me to have hoped that studying the brain would clarify issues of what it means to be human, would clarify issues of value.

PAUL MACLEAN: Professor Hubbard, I wanted to ask you a question, which is also a question I ask myself; it's a question that is going to follow me until I die. This question represents the main reason why I decided to come to a conference like this. I want to ask you, Professor Hubbard: "Why, with all the suffering in this world—human suffering, and also animal suffering (as for example the slow suffocating death of the tons of fish dragged from the sea each day)—why would one want to perpetuate life, either in this world or anyplace else in the universe?" For me, the existence of infinite suffering is the ultimate silencing blow to all debate on the issues we have been talking about, on values and so on.

I've often said I would never assume office in any society, but I broke my vow and did accept office in one society. It doesn't require much of a time commitment because I'm president, and there is only one member—a corresponding member who is a Jesuit priest; he says, "Paul, you can understand why I can be only a corresponding member." He has to pay 25 cents to write to me, and I use government envelopes to write to him, so it's also a very inexpensive society. The name of the society is the Society for the Elimination of Life in the Universe (SELU). As a physician, I learned that my role in life is to alleviate suffering, physical and psychic suffering. If that is my role, then the most assured way to alleviate suffering is to prevent the occurrence of life both here and elsewhere in the universe. Hence SELU! You might call me an animal rightist against *nature*.

I've never heard anybody give me a satisfactory answer to the problem of suffering, and I've asked many people, including scientists. I've never heard anything that I could believe and support in a way that I could think, "Okay, the world makes sense and now I can die." Why is our biology and brain so constructed that we are compelled to build up all this love and affection for the human family—just to watch the people we love suffer, and ultimately be torn away from us? So the wonderful thing is, I've come to the end of my life not being able to believe anything (for the reasons stated in my talk on Friday), yet not being able to shake loose of this limbic delusion that this is all brutally unfair, all because our wonderful paleo-mammalian brain—even though it can't read or write—makes us love and strive for good against cruel odds set up to produce disappointment and suffering.

SUSAN LEIGH STAR: What you're talking about is powerlessness, you're experiencing powerlessness.

PAUL MACLEAN: Yes—the one consolation that I have is that I have something up here in my brain that won't let me rest, that gives me this terrible concern and urge to avert things. And I guess I have remained a scientist and physician because somewhere I still believe that knowledge might be able to avert this situation, might be able to turn the whole thing around.

2. Do the Brain Sciences Lead to a Sense of Human Meaninglessness?

The Argument: Discussion here begins with an outsider's perspective on an apparent paradox in neuroscientific culture: scientists combine great zest for life and personal enthusiasm for the adventure of knowledge with an apparently nihilistic message. The message from neuroscience is that our soul or personality is a product of impersonal physical processes that know and care nothing about our "humanness." The so-called photon effect of science is discussed: the conviction that once we know how a sunset works, its beauty has been somehow invalidated, explained away as an illusion. Others fail to understand the existential dilemma here, and affirm the aesthetic joys of coming to know the world better.

JUDITH ANDERSON: I am a student, an outsider, and not a professional. And I can tell you that one of the questions that trouble those of us on the outside is whether the brain sciences threaten values in a very broad sense; whether they threaten the perception that life is meaningful, that being a human being is more than just being some sort of programmed machine. We've heard some stories at this meeting, some historical and fictional accounts, of people who've been profoundly troubled by the implications of the mind and brain sciences, so troubled that they've taken their own lives.

Yet, as I have observed the interactions of this conference, I have been struck by the extent to which you scholars and scientists who are engaged in neuroscience seem not to be paralyzed by the apparent nihilistic implications of your own findings. All of you are vibrant, passionate, curious individuals who are fascinated by the world, and profoundly engaged in life. Paradoxically, the question as to whether neuroscience calls into question the meaningfulness of traditional human values isn't particularly loaded for you, because your *work* as neuroscientists gives tremendous meaning to your lives. Maybe the content of your work gives us no insight into why we should continue to consider life meaningful, but your example as scientists perhaps does give us insight. By devoting your lives to a passionate pursuit of answers to profound questions, you send the rest of the world a strong message about what it might mean to be human.

TERRENCE DEACON: Where does this belief come from that the thrust of the neurosciences is ultimately nihilistic or destructive of human values? I have had a number of experiences in the past in which I've tried to explain a neuroscientific model of something or other, and the response that I have gotten is the same as if I had just explained why a joke is funny. No one really cares why a joke is funny. If you get the joke, you get it; if you don't, you don't. Neuroscientists are people who are profoundly curious as to why the joke is funny. However, for those who lack this curiosity, to explain a joke is to systematically destroy its effect. You're left with this flat, anticlimactic feeling. This, I think, is what is behind the perception of an existential threat Judy Anderson is talking about. We don't necessarily want to hear why a joke is funny; we just want to have it work its effect on us. Literary criticism might be considered threatening to aesthetics in a similar way. We don't want to hear the ponderous reasons "why" the poem is beautiful; we just know that we like the poem. The prosaic task of explaining why the joke is funny or why the poem is beautiful doesn't really resonate with the beauty of the poem or the beauty of the experience of working with children, or interacting with a chimpanzee. It seems like you've taken all the luster off.

But Judy is also right in sensing that we neuroscientists who are devoted to explaining why jokes are funny are not killjoys or immune to the passionate, experiential side of life. Quite the contrary. We are looking at a different luster, feeling a different excitement. It's a different joke for us, perhaps.

RUTH HUBBARD: I like what you've said very much, but I remain somewhat troubled. There remains the problem of power and privilege. The fact is that our society considers some jokes to be more important than others; some jokes are privileged epistemologically and politically as more worthy of explanation than other jokes.

ANNE HARRINGTON: My sense also, Terry, is that people feel threatened by the neurosciences because they feel you are not just explaining to them why the joke is funny; you're telling them that the joke isn't really funny; the human brain is just programmed to think it was funny. In other words, human values and a conviction of human meaningfulness emerge in the neuroscientific explanation as adaptive delusions that have their ultimate source in blind material processes.[19] Neurosciences tells us that the brain —

---

[19]RICHARD HELD: In reviewing an earlier draft of these discussions, I was surprised not to see a reference to the views of Wolfgang Köhler since the Berlin group was

the seat of our human existence—is a product, an impersonal force that itself knows and cares nothing about our "humanness." A material object has been created with a hunger for personalness and belongingness and existential justice, yet cursed with the fate of living in a blind, deaf, and silent system where these terms have never existed and make no sense.

When I was first starting out as an undergraduate at Harvard, I met a somewhat older student who wanted to be an artist. This student was existentially tortured because he was convinced that science had made it impossible for a person to maintain faith in the aesthetic reality of the world. Once you've explained the joke, you can never really laugh at it again. He and I had a long conversation once about the fact that sunsets aren't really beautiful. People *think* they see beauty out there in the sky, but my friend knew that science can explain all that in terms of photons and refractions. So we both became quite profoundly depressed about the fact that there really isn't any beauty in the world.

This is part of the personal motivation that initially drew me into the history of science at the age of 19 or so: an urge to understand the picture science gives of what it means to be a human being, to understand further what *kind* of truth science represents, and to come to some conclusions about how I should go on to live my life then.

SUSAN LEIGH STAR: I've had analogous experiences, having spent 3 years teaching in a computer science department. I carried out a kind of experiment in which I would vary what I would tell people who asked me what I did for a living. When I would say "I'm in computers," they would respond with: "Well, what should I buy for my kid?" But if I'd say "I'm in artificial intelligence," I'd get the photon reaction: "Oh yes, that's what diminishes and takes away from us."

JUDY ROSENBLITH: I am a behavioral scientist, and thus an outsider in sociology, history, and neurosciences, but I would still like to make a comment. I am struck by the course of this conversation. I don't think it takes one iota of beauty out of something to understand it. I can understand all the physics of the sunset, but that doesn't keep me from appreciating it as a sunset. Quite the contrary. Consider a medical researcher, trying to

---

discussed a few pages previously. Köhler held the view that a kind of value could be discerned in certain kinds of purely physical processes (electric field distributions). This notion is expressed most elaborately in his book *The Place of Value in a World of Facts* (1938), although it occurs in other writings. In his view, values have a rightful place in neuroscientific explanation precisely because the evolved brain allows scope for values in the operation of physical processes. While Köhler's views are debatable, they seem to me to be deserving of mention in this context. (ANNE HARRINGTON: Indeed, in 1923 the German writer Robert Musil, who was briefly involved with the Gestalt movement, would refer to the great cultural promise implicit in Köhler's theory of physical Gestalten: Whoever "has the knowledge to understand it," he wrote, "will experience how, on the basis of empirical science, the solution to ancient metaphysical difficulties is already implied" [cited in Ash MG (1991): Gestalt psychology in Weimar culture, *History of the Human Sciences* 4:409]).

understand how something interferes with cell growth, or with cell biochemistry, and in this way affects the ability of an infant to develop normally. There is beauty in the thought that you may be able, if you learn enough, to do something for that infant. It certainly doesn't take away from anyone's humanity to be concerned with issues of that kind. I don't understand the tension that people are talking about. This is a beautiful world and we should know more about it. Why should there be any conflict?

JASON BROWN: For Escher, whose engraving adorns the conference schedule, the beauty of nature was a *function* of understanding her better. He was always searching for the underlying order, and his drawings celebrated the symmetry and the beauty of the natural order. I think I agree with Judy Rosenblith, that the highest form of aesthetic experience is found in confronting the limits of what's knowable. It's like a musician playing a piece. Some of my musician friends tell me that, even though they may have played a piece 50 times, there occasionally comes a moment when they suddenly disappear in the music. I think that neuroscientists, and scientists generally, can experience the same sort of thing. There is what one could call an "aesthetics of knowing" that motivates me, and I think it motivates a lot of us.

TERRENCE DEACON: I think that would be my answer to Paul MacLean's original question, about suffering. The rose may wilt and become a rotten, brown clump, but all the same there did exist that moment of beauty, of exquisite symmetries and asymmetries. I can't imagine that there's any other reason to go on, to stick around, except for the fact of those few moments where everything just seems to fit.

3. Is Knowledge in the Brain Sciences Ethically Neutral?

The Argument: The argument is made that neuroscientists might do well to insulate themselves from the sorts of ethical debates that dominated much of this workshop. Their judgment on ethical issues is no better than that of anyone else, and they should therefore be left alone to continue the theoretical and experimental work that they do best, leaving more expert individuals the task of deciding on the most ethical means of applying the new knowledge. Others disagree, and stress that ethical dilemmas are inherent already in the process of knowledge construction. Neuroscience is not just about Popperian falsification procedures; it is about technology and competition for resources and large amounts of capital.

JASON BROWN: I wanted to say further that I have had difficulty over the past 2 days, relating many of the issues that have come up to the kind of work that I personally do. I can understand that there is a dimension of brain science research that is very technological and pragmatically oriented, and that the social implications of these technologies and activities need to be addressed. However, some of us are more interested in developing models for the way in which the brain is organized, and I am not sure these models are so dependent on technology. I think that technology is mostly used in cases of theory building to confirm or disconfirm models that are created intuitively by independently working people. These models have implications for our understanding of reality. Values and ethics enter into the discussion when we begin to ask questions about the uses to which knowledge should be put. Scientists don't have any privileged information on this score, because we are not talking about a scientific decision, but a philosophical or a political one.

If, then, it is the case that we're no better than anyone else at ethical decision making, then maybe we should concentrate on what we do best: creating and testing models of reality. I cannot help but think of a remark that Hans Reichenbach made about science. He suggested that too much philosophical analysis may be harmful and paralyze the pioneering spirit. One needs the courage, he said, to walk a new path with a certain amount of irresponsibility.

RODNEY HOLMES: I am puzzled by what you are saying and I could almost not disagree with you more. If one looks at recent scholarship on the problem of facts and values in the scientific process, it becomes very difficult to say that facts are not value-laden.

ALAN FINE: I don't know about that, but I do feel strongly that an ethics of activity is urgently needed. My sense, as a medically oriented neuroscientist, is that the more we learn about how things work—and I'm thinking of the brain in particular—the more power this knowledge gives us. And I worry about the tendency we all have to assume that knowledge and power also confer on us some sort of automatic right to *act*. Take molecular biology: if we were to develop the knowledge to do so, would we have the *right* to make bigger babies, or blue-eyed babies, or whatever else we wanted to order up? I worry about the consequences of our growing scientific knowledge, because I don't see clearly our means for limiting or monitoring the social tendency to want to exploit that knowledge, to intervene.

JOHN DOWLING: What would you propose? Would one stop doing the work, or would one stop worrying about rights?

ALAN FINE: I don't know.

JOHN DOWLING: Either the former or the latter, this is the fundamental choice.

ALAN FINE: I understand, and am absolutely torn here. On the one hand, I am busy contributing; on the other hand, I sometimes want to throw a shoe in the works. I don't have answers and I'm deeply troubled by it. Yet the end effect of all my soul searching is that I keep on working.

EDWARD MANIER: I think Alan is saying something very important. I'm very interested in molecular neurobiology, and one of the things that's very unclear to me is the relation of this powerful science—a science that arguably has something to tell us about the brain and reality—not only to the ethics of technological application, but to the ethics of technological production. It is obvious to any of us who read the cover of *Newsweek* that pharmaceutical companies are very interested in supporting "basic research" in neurosciences, and that products of the new knowledge — antidepressant drugs like Prozac — are a source of great wealth. We're talking here about limited resources available for research and the reasons why one problem is funded rather than another. It's clear that the technological products — or expected technological products — are critically important to such decisions. Competition for resources sets the agenda of the field. We have to move beyond this idea that some of us still try to defend: the idea that our research programs are disinterested quests for truth, quests that sprang fully armed from the head of Zeus, or fully armed from the head of Karl Popper; that we are all just looking around for hypotheses to falsify, and there's nobody here but us chickens.

4. Mediating Knowledge and Power Relations between "Cultures"

> The Argument: Discussion now moves to the general difficulties inherent to the sort of interdisciplinary dialogue attempted at this workshop. Problems identified include the difficulties in communicating one's insights to a naive audience, feelings of asymmetry in authority across disciplines (with a sense — later disputed — that the "hard" scientists tended to command more authority than the social scientists), and lack of sufficiently diverse representation among the invited speakers (especially ethnic representation). The historians and sociologists reflect on the discomfort that may result for all parties when scientists feel themselves made into objects of scholarly study. The general lack of dialogue between the social sciences (history, sociology, anthropology) and the natural sciences themselves is also discussed, and is seen in part as a side effect of critical changes in the formers' attitude toward natural science over the last 20 years. It is suggested that science-studies professionals have an intellectual responsibility to test their ideas against those people — the scientists themselves — most likely to resist them.

ROBERT RICHARDS: We all are aware that there are some difficulties with interdisciplinary conferences; let me just say, though, that I've been to three interdisciplinary meetings in the last year, and believe it or not, this one has worked the most smoothly, and the exchanges have been the most fruitful. When you have any group of professionals — even when the areas of their expertise are somewhat adjacent, as is the case with historians and philosophers — you inevitably have what at a distance looks like people speaking past one another. Each individual brings to the dialogue special- ized interests and very specialized knowledge; and one cannot always count on the ability of the other people to react effectively to an argument where he or she feels inexpert. Maybe some of us would have liked a broader representation of disciplinary viewpoints, but there's also a price to be paid for such expansion — increasing mutual incomprehension — and something to be said for bringing people together whose concerns are a little more tightly integrated. For the most part, I thought the exchanges here were fabulous. We didn't talk past one another, at least not only: frequently we were abutting one another rather hard.

RODNEY HOLMES: I would like to know, though, how people think we can get around a problem that is perhaps inherent to these sorts of meetings: the problem of asymmetry of authority. There is a general reflex tendency at such gatherings, it seems to me, to consider the natural scientists to stand in a position of superior expertise to the social scientists. None of us want this, and all of us want to have faith in a democratic process of information exchange. But it's not clear to me how we can guarantee this.

SUSAN LEIGH STAR: It seems to me that the question you are asking is a question about fundamental social change and not just about how to organize effective conferences. I'm as fascinated by science as many of you neuroscientists are by the brain; what I do for a living in part is to study values. My career is side by side with scientists, studying scientists. This puts us in an odd relationship to each other at an interdisciplinary conference about values in the sciences. In effect, I am the amateur in your domain and you are the amateur in my domain, and there is a strange tension of deference and arrogance that must be mediated in order to get a true dialogue flowing. The politics of interaction in such a setting become very tricky to manage because the things that we're talking about are also things that have to do with our common citizenship and the community we live in. One thing that we discover is that the politics of being a professional—a wielder and promoter of expert knowledge—is often carried out at the expense of certain aspects of citizenship. I think we have a long way to go still before true dialogue has been achieved; but it's nice to think about building a place where it would be safe to explore issues—our deepest values and human concerns—that take us beyond our individual professional personae. Perhaps that is, in part, what successful interdisciplinary work would mean.

ANNE HARRINGTON: As a sociologist of science, Leigh, you say you work side by side with scientists, but I think you are still a rare breed in our profession. One of the things that impressed me very much when I was doing my graduate work in England was the fact that most of my peers had almost no contact with the scientists whose knowledge-claims and social activities they were criticizing, sometimes very radically. They were well trained in sociological theory and subtle epistemological critiques, and they knew that science was a very different animal than it pretended to be. Nevertheless, the scientists in the next building kept on doing what they had always done, and the groups did not talk—did not seem to want to talk. I think this isolationism is part of the fallout of the professionalization of our field. The history of science has transformed itself in the last 20 years; it is no longer prepared to identify itself as a handmaiden to the natural sciences, articulating and reinforcing their self-images and honoring their heroes. There is instead a much greater tendency to think of the history of science as a tool to analyize the nature of science, to look at the processes of science, as you put it, Leigh, as an object of scientific inquiry, and to look at scientists in some sense as experimental subjects. This is a welcome shift, I think, but it has been accompanied by a loss of direct professional contact with the working natural scientific culture. I think that has to change, if our work is to remain ethically credible. If we're not prepared to test our ideas against the people most likely to resist them, then we are abdicating something of our scholarly responsibility.

TERRENCE DEACON: Before I came to this conference, I had never spent time talking about the value of neuroscience. However, I have always felt the tenuousness of scientific truth. Okay, that may not have come out so clearly in all these discussions, but it is part of being a scientist that one defends one's views. Don't think I'm not aware that, caught in time, it won't be too long before my views are forgotten. Theories that I've proposed that I think are *so* true, that beautifully describe the world, may eventually be laughed at. That has always both troubled me and interested me, because when the chips are down you have to ask yourself: "What are you doing? What is the worth of it if 30 or 50 years from now it will be almost certainly clear that you were wrong, and that people will dismiss you?" I worry a lot about biases, both my own and those of my colleagues. I have to admit that most of the biases I worry about are intellectual biases; I can't see, or don't know how to see, social biases or ethical biases. So self-interest in some respect draws me to a conference where I might find people from the outside who can see these other layers of bias better than I can.

I didn't feel that there was such a profound asymmetry of authority at this meeting. As a natural scientist, I have more information of one type, but not of another type. The moment you began talking about science, I become a subject, an object of inquiry. Certainly, if your description of what I do doesn't resonate with what I *think* I do, then I'm going to probably reject your claim to authority and say something like "Look, I don't do that; I do this."

EDWARD MANIER: When, as an historian, I switched from studying the dead Darwin to studying live scientists I did that, in part, because I really wanted to be able to *ask* Darwin about his rhetoric. I wanted to know if I were reading certain issues into his works, or if they were salient for him as well. It turns out, though, that the effort to study living subjects is thwarted by the interest they have in creating and maintaining an image of themselves, in managing the news about their own lives.

It's fascinating, for example, to watch population geneticist Sewall Wright and historian Bill Provine take part in the same panel on the history of developments in population genetics in which Sewall Wright was a major player. Bill detects interesting and important shifts of view, the so-called hardening of the neo-Darwinian synthesis,[20] where Dr. Wright finds

---

[20]"The development of Wright's theory of evolution: Systematics, adaptation and

something like a seamless garment of investigation and theoretical perspective. Dr. Wright thinks Bill is getting it wrong, but Provine's historical command of the literature of population genetics in the 1930s is unparalleled. I would think the resulting tension would be uncomfortable for both parties, but Bill has pointed out to me several times that it's a natural part of work as an historian of contemporary events.

On reflection, I agree with Arthur Danto that such experience is a crucial aspect of what it is like to live a human life — the stories of our lives are continually rewritten, retrospectively by ourselves and by others.[21] It is quite plausible, however, that one might find the prospect that someone else, another person, is becoming a professional, public expert on important aspects of one's career quite unacceptably frightening. After all, we use the stories of our lives to advance our most important projects. Why should anyone willingly share control of such an important resource with a stranger?

SUSAN LEIGH STAR: What you're saying is relevant to another values issue that social scientists studying living scientists constantly face. I come from a sociological tradition that very strongly guarantees anonymity to respondents; I have a lot of arguments with people who come from different traditions about the fact that we sociologists are reluctant to speak about the specifics. They have a point, because if you base your study on anonymity, then you are implicitly denying the specific historicity and situational nature of the problems you are attempting to analyze.

EDWARD MANIER: I agree. The idea of an historian giving anonymity to somebody — well, that's just not what history is all about.

5. Toward a Reforming Vision of the Neuroscience Profession

The Argument: A final vision for the humanization of the neuroscientific profession.

PAUL MACLEAN: I think ultimately that the only chance we might have for making our discipline more humane and more ethical would be to open it up. We'll never get far in this world with just a lot of knowledge; we need

---

drift." In: *Dimensions of Darwinism,* Grene M, ed. Cambridge: Cambridge University Press, 1983, pp. 43–70, Provine, W. (1986): *Sewall Wright and Evolutionary Biology.* Chicago: University of Chicago Press.

[21]EDWARD MANIER: Danto has a great phrase for this experience, which is an important aspect of what it is to live and know and experience life as a human being, "To exist historically is to perceive the events one lives through as part of a story later to be told." Drawing on Sartre, Danto makes it plain that he thinks we cannot have a true conception of ourselves, of the narrative of our own lives, until our sense of self has been transformed by the possibility that others will tell the story of our lives (*Narration and Knowledge,* Columbia University Press, 1985, p. 343). Wright and Provine publically discussed Provine's studies of the origins of the genetics of natural populations at a meeting of the History of Science Society in Madison, Wisc.

wisdom and we need empathy and love. To extend what I said in Friday's talk, it would be hard to imagine a more isolated and lonely creature than a human being with the subjective brain imprisoned in its bony shell. The totality of experience is confined within those prison walls. No one else can ever enter or leave that cell. The only time you even get close to a human being, it seems to me, is when you look him or her in the eye. If you look in the eye, you see the retina, and you know you're about as close to the brain as you can ever get. Regarding the need for wisdom and empathy, the only hope I see for human society is to work to get a majority of women involved in the decision-making process — in politics and in science (including medicine in particular). There must be a correction in the disparity of attitudes toward the two sexes. It seems to have been overlooked that for more than 180 million years, the female has been central to mammalian evolution and the evolution of the family. Responsibility begins in the nest, and that carries over to become what in human society we identify as conscience. I've had a number of women who have raised their families say to me: "I'd love to go to medical school but I can't get in because I'm too old." And I tell them: "Well, that's ridiculous because you know the learning curve gets better every year, and women are naturals at medicine." There's this prejudiced notion that if you show a little gray in your hair, you're not worth investing in. Even someone with all the background of our friend Walter Rosenblith couldn't get into medical school and that's a crime. And continuing with the matter of wisdom: it used to be that society had to depend on young people who must do or die before the age of 50. Now life expectancy has increased to 75 years. One of the hopeful things about this is that as people get older and wax in wisdom, they also become nicer mammals. It has been known for more than half a century that just by restrictions of diet, animals may be kept alive in a healthy condition for periods of 25–30 percent beyond the usual life expectancy. Think how many of the world's vexatious problems could be easily resolved if the lives of healthy people with increasing wisdom could be extended by another 20 to 30 years. That might be the next step in human evolution.

EDWARD MANIER: Paul's just resigned from his society, I think.

17

"So Human a Brain": Ethnographic Perspectives on an Interdisciplinary Conference

MICHAEL FORTUN AND SKULI SIGURDSSON

The appearance of Thomas S. Kuhn's *The Structure of Scientific Revolutions* in 1962 was a landmark in history, philosophy, and sociology of science.[1] It provoked a crisis among those who had constructed an ahistorical and superrational image of scientific practice. In the heady and questioning days of the 1960s Kuhn seemed to be a revolutionary or an irrationalist.[2] Of the many protean ideas embedded in *Structure,* there was an underlying anthropological notion that did not receive widespread attention at first. Some 20 years later he made explicit the hidden anthropological perspective of his philosophy and historiography:

My way of using concepts like "revolution" and "gestalt switch" was drawn from and continues appropriately to represent what historians must often go through to recapture the thought of a past generation of scientists. Concerned to reconstruct past ideas, historians must approach the generation that held them as the *anthropologist* approaches an alien culture. They must, that is, be prepared at the start to find that the *natives* speak a different language and map experience into different categories from those that they themselves bring to home. And they must take as their objective the discovery of those categories and the assimilation of the corresponding language. "Whig history" has been the term reserved for failure in that enterprise, but its nature is better evoked by the term "ethnocentric."[3]

It is a measure of the transformation of the history of science in the last 30 years that anthropology and ethnographic methods have now been accepted in the field. That has happened in Kuhn's more restricted sense by making historians of Western science aware of the "otherness" of earlier scientific traditions.[4] Furthermore, with the increasing emphasis on the study of recent science, ethnography has become an indispensable tool for observing scientists at work in the laboratory. (For some references to this work, see the last section of this chapter.)

Therefore when Anne Harrington invited us to this conference as observers, we were quite excited about having an opportunity to test an ethnographic approach in a different setting, that is, by studying both scientists and the practitioners of those disciplines now grouped under the rubric "science studies" – history, philosophy, sociology, and anthropology of science – at one and the same time. In light of the reflexive turn in ethnography, it seemed sensible to problematize the privileged status of the observer and to observe many "tribes" simultaneously.[5]

From one angle, both of us as ethnographic observers were "outsiders" to this conference's culture: we are neither neuroscientists, nor historians, philosophers, or sociologists of the neurosciences. Yet we were by no means neutral or naive observers of a totally strange culture. We are each making a career of studying the ideas, practices, and development of various branches of modern science (Fortun in molecular biology and Sigurdsson in mathematics and physics), and so we have developed our own repertoires of questions, sensitivities, hypotheses, beliefs, and philosophical assumptions, affinities with other scholars, and working theories – all of which we brought to this conference, and to the subject of the neurosciences. From an ethnographic perspective, this range of personal experience is not a source of bias, but rather a source of questions, hunches, comparative cases, possible interpretations, and provisional explanations.[6]

We tried, in short, to be simultaneously open and sensitive to what was being said and done around us, while drawing on our experience for questions, affirmation, and criticism – to be the type of researcher who "wonders first and judges last."[7] Some readers may feel that there should be no room in ethnography for such judgments, that we should have been more descriptive in our account of the conference. But this is not the way we understand the ethnographer's position. As Liz Stanley summarized in an article surveying some conflicts over definitions of "ethnography":

Ethnographic description is actually not, and cannot be, literal description: rather it is a gloss, a summary which contains and indeed is an interpretation which provides a partial selection of "what was" within the description. The inclusions and exclusions of ethnographic description derive from the relevancies of a perspective or viewpoint – and this is informed by the concerns and issues of "the academy" and not those of "the field."[8]

On the last day of the conference, we presented the preliminary results of our ethographic observations to the other participants. The response in the ensuing discussion was both very thoughtful and supportive of our endeavors.

The structure of the rest of the chapter is as follows. We begin by presenting our observations intertwined with a few comments. This part corresponds roughly to our talk at the end of the conference. We go on to consider how the participants responded to our presentation; the responses resolved a number of questions, yet some silences still prevailed. It is to those silences that we turn briefly before concluding the chapter with a number of more general remarks.

Observations

Our first and primary observation is that there seemed to be three instances when people started talking to each other, instead of past each other, and when the exchanges went beyond simple questions and answers. We have categorized these issues roughly as ideology, gender, and freedom of inquiry.

The issue of ideology first arose on the afternoon of the first day, when John Durant asked if Paul MacLean's model of the concept of the triune brain had a potential ideological content, "the notion of the beast within." We expected to hear — or to be honest, we *wanted* to hear — an exchange of views on forms of ideology, forms of knowledge, and how the two do or do not go together. Instead, the discussion devolved into a very specific and in some sense "safe" debate on recapitulation and "addition terms" at the molecular and cellular level. Ernst Haeckel's well-known maxim that ontogeny recapitulates phylogeny was thus shorn of its ideological connotation and made technical and safe.[9] This demonstrated how reductionism can be an effective strategy for ideological immunization.[10]

The interaction between ideology and different levels of explanation — be they molecular, embryological, or ethological — was never addressed directly. Furthermore, the relationship between an ideology resonating with deeply held cultural values and disciplinary and professional ideologies (raised by Anne Harrington in the first session) was largely left unexamined.

That this was not done we found very strange because here was a chance to explore the question of what goes to make up scientific knowledge. Is it a matter of cognitive models imposed on a sea of ignorance, as remarks made by a number of the neuroscientists would suggest? Or do cultural and political values come either to be incorporated into these models or to make some models more acceptable than others?

The former alternative would seem to imply a possible multitude of approaches dealing with that ignorance and uncertainty. One obvious question is that given the admission by some of the neuroscientists that scientific knowledge constrains this ignorance via models, then why do we have particular models in particular places and particular times? At what point in history, incidentally, did neuroscientists start to talk about "models"? What are the criteria for model choice: simplicity, aesthetics, or utility?

The reason why the constructive role of ideology did not get addressed (as opposed to the generally agreed deleterious effects of linking "ideology" and "knowledge") was that the participants did not articulate clearly what constituted "knowledge." This was as true of historical and sociological knowledge, as it was of the scientific knowledge of the brain. Warwick Anderson's talk on the brain disease kuru, juxtaposing the cultural meanings of the different knowledge generated by anthropologists and neurophysiologists, addressed this issue and was a timely corrective, but there was no subsequent discussion — not too surprising, given the late hour on Saturday night.

Londa Schiebinger's talk on the gendered brain, of the three talks presented in the Saturday morning session, became the focus of most of the discussion. It would seem that the issue of gender is more contentious and unresolved than the issues of either professional ethics (Rodney Holmes) or a humanistic critique drawn from the realm of fiction (Edward Manier), however compellingly these two were presented. The exchange on the issue of gender is intimately related to the earlier discussion on ideology: Do we put the concept of gender inside or outside the knowledge-making process?

We would like to offer two comments on this tangled issue and the verbal exchanges that it inspired that morning. The question of the possibility of a feminist science, and how such a science would differ from the science we now have, is the subject of a great deal of debate — not least among feminists themselves.[11] At the conference there was, on the one hand, a lack of acknowledgment of this debate, and a sort of "unified front" on the question of gender and science was presented; such elision of complexity and internal debate is perhaps a strategic necessity when a minority directly confronts the majority. On the other hand, there was a curious lack of imagination[12] on the part of the scientists, who seemed to find it inconceivable that there could be different ways of organizing knowledge, in spite of the underlying sense (which they verbalized occasionally throughout the conference) of transience, tentativeness, and model-dependency that characterizes the scientific enterprise.

The third issue, which we have labeled as the freedom of inquiry, arose in the Saturday afternoon discussion after John Durant's talk on animal awareness and human sensibility. It was perhaps the messiest of these three issues, involving a number of fundamental notions. Here the notion of values in all of its varied meanings played an important role, and yet the word seemed to be used somewhat indiscriminately.[13]

One question concerned the morality or immorality of certain kinds of research. Medical experimentation under the Nazis was cited frequently as an example of the latter sort. Another question involved the possible social process or language for determining the morality of research, that is, how are such decisions of morality and validity made and justified? Should they be the province of experts — and then *which* experts? — or of a wider community? A third question concerned what kind of knowledge is more valuable and valid than another, for example, is neuroscience "better than" ethology? And what would "better" mean in that case?

Stephen Kosslyn in his talk on Friday afternoon made a statement very revealing of this kind of implicit valuation and demand for autonomy, declaring that at this point in history the only way to get this kind of neurological information was to sacrifice animals. He said that the knowledge thereby gained was important, but without making explicit why it was important.

The discussion on the freedom of inquiry and animal experimentation revealed a deep-seated and unresolved anxiety over the place of humans in nature. Belonging to a post-bomb, post-Vietnam generation that has been brought up to be mistrustful of science and technology, we are both sympathetic to the possibility raised by Kurt Vonnegut that "our brains may be just too fucking big," and that the fascination with the supposedly wonderful and special qualities of the human brain and the underemphasis on the damage that brain has done to the ecosystem and other cultures may just be a case, to paraphrase Walker Percy, of "mind among the ruins."[14]

These were the instances of intense exchanges as we saw them. But there were also noticeable silences—issues and questions left unmentioned or unexplored—which the discussions on ideology, gender, and freedom of inquiry did not manage to break. Taken together, the function of these systematic omissions on behalf of most of the participants (with a few noteworthy exceptions) was to paint a picture of modern neuroscience as an overly autonomous discipline, unaffected by historical social processes. That utopian image reminded us of Kuhn's remark "that the member of a mature scientific community is, like the typical character of Orwell's *1984,* the victim of a history rewritten by the powers that be."[15]

There seemed to be a general silence on what was called in the program "the brain research enterprise." What are the constitutive factors of that enterprise? How many people do research in the neurosciences, where are they located, how are they funded, with whom do they have to compete for resources and for students, and what are the commercial pressures and interests on the field?[16] There was, in short, remarkably little attention given to either the more concrete, day-to-day details of life in the neurosciences or to those aspects studied in more traditional institutional history of science. The best example of this was the surprisingly little attention paid to the technological dimension. We found it striking that it was not until Stephen Kosslyn's talk on late Friday afternoon that we even saw a picture of equipment—a PET scanner. How much does such an instrument cost, how many are there, and who has access to them?

On Saturday afternoon in Elliot Valenstein's talk on therapeutic innovations, it would have been interesting to hear more about how the practices of lobotomy fitted into the technological revolution of the modern hospital, and the belief in "technological fixes" to medical problems. Where does the commercial sector enter, or, say, the insurance agencies in that case? Is there an independent technological momentum, and how does that relate to the autonomy of research?

Put differently, how much of the tacit knowledge in the neurosciences is encoded in the machine environment, or, in Susan Leigh Star's words, how much is the technology forgetting? What assumptions are built into the practices and equipment that neuroscientists use without further thought? What is the division of intellectual labor in the neurosciences and do the various gadgets and tools that are employed increase or decrease the hierarchical control of the field from the top down?

The recent history of the neurosciences was also neglected. We were especially sensitive to this omission, since one of the things we had planned to listen for was how the scientists would construct their discipline's history. So there was to us a marked disjuncture between Walter Rosenblith's self-described "microstorytelling" starting the conference on Friday morning—a tale that included Norbert Wiener, the influence of World War II and postwar technologies and concepts, and the restructuring of the field accomplished by physical scientists who brought to it their skills in solving applied problems acquired in wartime laboratories—and the talks immediately following, from which one got the impression that the context Rosenblith described was simply not there, that models and theories sprang from a Popperian world of ideas. Yet the language used and diagrams displayed, with their emphasis on information processes and networks, were redolent of Norbert Wiener's legacy. Perhaps this indicates a lack of interest by the scientists in historical details, reflecting a belief in the matching of theory to nature unmediated by historical circumstance.

Our own presence at the conference, somewhat a serendipitous function of spatial and temporal location, made us wonder how this particular group of people came to be selected. What were the personal, intellectual, financial, or random factors involved in the choice of them? Was their ability to engage in interdisciplinary dialogues one factor? Why was it that nobody from the artificial intelligence community attended, and was somebody invited? (In that instance, how has a certain technological approach shaped the inquiry?) Therefore we would have liked to hear why people came to the conference. What were their motivations? What did they hope to gain, and what do they see as the possible dangers? What does it take in terms of discipline and commitment of time to further such interdisciplinary exchanges?

We asked ourselves what it was that interested us in coming here. One of us had attended Patricia Churchland's lecture sponsored by the FIDIA Foundation in Emerson Hall at Harvard University in April 1990. Churchland gave a very optimistic and antimetaphysical talk on the rosy future of brain science, but paid no attention at all to the conceivable ethical implications of that research. During the question period the discussion entered into the arcane byways of technicalities, and thus one of us was prompted to ask what would stop the brain scientists in the future from continuing to carry out their research when it would become possible for the state, advertising agencies, and so on, to monitor and manipulate human behavior at the neuronal level. The answer she gave was that, in this hypothetical future scenario, this problem would be solved by democratic means. But what would happen, we wondered, if you had this knowledge in the future, but lived in a nondemocratic and coercive form of political organization?

Finally, the dialogue at this interdisciplinary workshop made us think about the relative autonomy of disciplinary discourses: sometimes the participants seemed to talk past each other instead of with each other, as if not enough effort had been made to gear talks to an interdisciplinary audience. That raised the larger question of how and when disciplines speak to each other: for instance, how do the neuroscientists interact with the "outside world" or look at it through their special "windows"? What assumptions about scientists do historians and sociologists make in their own work? How much are the dialogues in each discipline driven by internal criteria so that contact with other disciplines or a broader community is only occasionally accomplished?

Reflections on the Response to the Observations

The response to our presentation on Sunday morning was extremely insightful, positive, and stimulating. The response underscored our general impression that this conference was most successful and productive during the more unplanned and spontaneous exchanges. People began to articulate why they found the neurosciences so interesting. Yet why it was obvious to study neuroscience rather than some other branch of science was not made clear, nor how the choice was justified, nor how the agenda for research was set, and by whom?

One aim of this conference was to start a dialogue between scientists, on the one hand, and historians, philosophers, and sociologists of the neurosciences, on the other. Thus, the question arises for whom the latter group is writing: the scientists themselves, other people in science studies, or the broader public? This question of audience relationship has been raised by Roger Smith in a thoughtful essay review on recent books on neuroscience.[17] The point he makes is that in the books under review the authors take for granted a fair amount of current knowledge in the neurosciences. Thus, they might be seen as writing from what Kuhn called an ethnocentric perspective—but is that necessarily so bad?

The issue of audience relationships brings to the fore the question of what is gained and lost by encounters such as this conference. During the discussion following our presentation on Sunday morning Rodney Holmes took the "liberty" to quote a brief discussion that we had with him and a few other participants at the local bar the night before. There we had said that we had noticed the authoritative fashion in which the scientists spoke, and how the other participants spoke in deferential fashion to them. He said that he had experienced this recently at other similar meetings. Susan Leigh Star admitted this, and after that there began a long discussion as to how people were invited to the conference, and why there had not been more women and minorities.

This brings up concretely some problems of doing studies of contemporary science and dealing with living scientists. It is not only that they have an interest in presenting their activities as having come naturally out of the past. Being the powers that be, they will be prone to view the history of their field differently from how historians, philosophers, and sociologists of science will construct it, because the scientists have a stake in the future. It is therefore possible that in studying contemporary science people in science studies will have something to lose because they lack the authority and knowledge that scientists have, and they care less about the future course of events.

But while in general there was much less posturing on everybody's part and more open discussion on Sunday morning, some silences still persisted. The role of technology remained for the most part unexamined. Jason Brown's comment about the ethics of technology *use* is only part of the issue. The larger question goes to the possible ways in which technology constrains and constitutes knowledge itself—a question not of *use,* but of *production.* Here, for example, we might find in the limitations of technology some connection to the reluctance shown by a few of the scientists present to define terms such as *mind, consciousness, self,* or *sentience*—or to redefine these concepts in terms of what is physically measurable or displayable. Or we might have been able to unpack and look at the technological foundations underlying James Schwartz's comment on Friday afternoon, that his main problem was to convince colleagues that what he as a neurochemist was observing had anything to do with what he was trying to explain—convincing people, in other words, that he was indeed producing *knowledge.* Given the recent trends in science studies away from the study of theory and toward the study of experiment and practice, we see rich possibilities here for the kinds of recent historical and sociological studies of the neurosciences for which Walter Rosenblith called.[18]

The silence about the "brain research enterprise" also persisted. We are still in the dark as to the scale of this enterprise, the kinds of institutions and numbers of researchers involved, the strength of the links to medicine and industry, and so on. If we are to try, however quixotically, to distinguish between pure and applied research in this enterprise, we first have to know its full extent. What are the goals of this enterprise, aside from the grandiose abstraction of "understanding the mind"? What are the projected therapeutic benefits promised to the medical community? Is there agreement among scientists and medical practitioners as to what to do?[19]

Mention was made twice during the conference of President George Bush's presence at a social function that was part of the recently launched "Decade of the Brain." How are we to think about this display of support, if not partly from the paranoid's perspective that this former CIA director would quite naturally be interested in brains?[20] What makes this knowledge so vital that a president puts his imprimatur on it, so important that significant financial and institutional resources are mobilized and thousands of "lower" animals killed to obtain it? Furthermore, what are the actual and possible uses of this knowledge and technology by advertising or television researchers?[21]

We similarly continue to be perplexed by the meager attention paid to artificial intelligence (AI) at the conference.[22] The presence of someone from the AI community, or someone studying that community,* would have made it possible to address simultaneously the status of competing models, funding and the military, and the role of technology in the neurosciences.[23] In other words, it would thus have been possible to start analyzing the set of factors that contribute to the production of knowledge and to ask why we have the particular kind of knowledge we have now and not some other kind. The critic of artificial intelligence John Searle has remarked: "Because we do not understand the brain very well we are constantly tempted to use the latest technology as a model for trying to understand it." He adds that at various instances in the past the mind has been compared to a mill, a hydraulic and electromagnetic system, a telegraph system, and a telephone switchboard.[24] What are some of the sources, and what are the consequences, of current models of the mind and brain?

Ian Hacking has drawn attention to the extensive military support for research on artificial intelligence in the United States after World War II. He has wondered to what extent the funding and the set of goals defined by the military has shaped the knowledge produced.[25] More generally, he has asked whether it may not be that weapons research has changed the forms of scientific knowledge in a fundamental way, that is, we come to inhabit different worlds for both material and conceptual reasons: "May not new knowledge determine what are the candidates for future new knowledge, barring what, in other 'possible human worlds,' would have been candidates for knowledge?" And he adds: "There is a nagging worry that 'science' itself is changed: not just that we find out different facts, but that the very candidates for facts may alter."[26]

Conclusion

We found it a very rewarding experience to attend this conference and to pass as ersatz anthropologists for a weekend. We would like to conclude with some general comments about ethnography and comparative studies of science. What literature might help further anthropological, historical,

---

*Editor's note: A Harvard professor of computer science active in AI research was invited to the conference, but she was ultimately unable to attend.

philosophical, and sociological studies of the neurosciences? How is it possible in the study of the neurosciences to learn from other fields of science studies?

Ethnographies of science — its theories, techniques, and social influence and consequences — have proliferated over the last decade or so. Bruno Latour and Steve Woolgar's 1979 *Laboratory Life* is certainly an early landmark in the field, but is only one of a number of fine laboratory ethnographies.[27] Other areas of modern science and medicine have been studied, including high energy physics,[28] molecular biology,[29] oncogene research,[30] and the social organization of the hospital.[31] But the science studied need not be the modern institutional forms of the present culture. Steve Shapin and Simon Schaffer's *Leviathan and the Air-Pump: Hobbes, Boyle, and the Experimental Life* (1985) has recently been seized upon for advocating the virtues of anthropological treatments of science.[32]

These studies provide models for the ethnographic studies of the neurosciences.[33] After such studies are carried out, it should then be possible to engage in comparative analysis and suggest which features of the daily neuroscientific practice are shared by other modern scientific cultures and which are sui generis. Another avenue for the ethnography of science has been suggested by this chapter: the analysis of conferences of a mixed sort where historians, philosophers, scientists, and sociologists come together.[34]

The transience of present forms of communication and scientific collaborations (cf. this meeting!) is another reason for devoting increased efforts to ethnographic investigations. Letters and other personal documents now face a stiff competition from a wide variety of new communication technologies. Thus the intimate and personal archival documentation that historians cherish so much will be unavailable in the future, while on the other hand they may be paralyzed by the sheer volume of the scientific enterprise. It might also be useful to combine ethnographic studies of the neurosciences with an oral history project as has been done in the physical sciences (i.e., the Archive for the History of Quantum Physics created during the 1960s and the Recombinant-DNA Oral History Collection at MIT in the 1970s).

Another rationale for ethnographic studies would simply be, that for professional historians of science who want a better understanding of the sciences, anthropological fieldwork may give them an intimate feel for how the subjects of their studies behave.

Acknowledgments

We would like to thank Pnina G. Abir-Am, Mario Biagioli, Joan Fujimura, Diane Paul, and Silvan S. Schweber for helpful discussions as we prepared our fieldwork at this conference, and Mi Gyung Kim for careful reading of this chapter.

Endnotes

1. See Thomas S. Kuhn, *The Structure of Scientific Revolutions,* 2nd ed. (Chicago: University of Chicago Press, 1970). [Original work published 1962.]
2. See Skuli Sigurdsson, "The Nature of Scientific Knowledge: An Interview with Thomas Kuhn," *Harvard Science Review* (Winter 1990): 18-25, esp. 21-22.
3. See Thomas S. Kuhn, "Afterword: Revisiting Planck," in his *Black-Body Theory and the Quantum Discontinuity, 1894-1912,* 2nd ed. (Chicago: University of Chicago Press, 1987), 349-370; quotation on p. 364—our emphasis. [First published in *Historical Studies in the Physical Sciences* 14, no. 2 (1984): 231-252.]
4. See Mario Biagioli, "The Anthropology of Incommensurability," *Studies in the History and Philosophy of Science* 21 (1990): 183-209.
5. See, for example, Paul Rabinow, "Representations Are Social Facts: Modernity and Post-Modernity in Anthropology," in *Writing Culture: The Poetics and Politics of Ethnography,* ed. James Clifford and George E. Marcus (Berkeley and Los Angeles: University of California Press, 1986), 234-261.
6. See Leonard Schatzman and Anselm L. Strauss, *Field Research: Strategies for a Natural Sociology* (Englewood Cliffs, N.J.: Prentice-Hall, 1973), 53; and Anselm L. Strauss, *Qualitative Analysis for Social Scientists* (Cambridge: Cambridge University Press, 1987), 10-14. We also found the following useful for our preparation: James P. Spradley, *Participant Observation* (New York: Holt, Rinehart and Winston, 1980); and Danny L. Jorgensen, *Participant Observation: A Methodology for Human Studies* (Newbury Park, Calif.: Sage, 1989).
7. See Schatzman and Strauss, *Field Research,* 65.
8. Liz Stanley, "Doing Ethnography, Writing Ethnography: A Comment on Hammersley," *Sociology* 24 (1990): 617-627.
9. See Stephen Jay Gould, *Ontogeny and Phylogeny* (Cambridge, Mass.: Harvard University Press, 1977), esp. 115-166.
10. See also Edward Manier, "Reductionist Rhetoric: Expository Strategies and the Development of the Molecular Neurobiology of Behavior," in *The Cognitive Turn: Sociological and Psychological Perspectives on Science,* ed. Steve Fuller, Marc De May, Terry Shin, and Steve Woolgar (Dordrecht, The Netherlands: Kluwer, 1989), 167-198.
11. The concurrences and conflicts between various feminist perspectives on science are discussed by Sandra Harding in her *The Science Question in Feminism* (Ithaca, N.Y.: Cornell University Press, 1986); "Feminism, Science, and the Anti-Enlightenment Critiques," in *Feminism/Postmodernism,* ed. Linda J. Nicholson (New York: Routledge, 1990): 83-106; and "The Method Question," *Hypatia,* 2(3) (1987): 19-35. Also see Ruth Bleier, ed., *Feminist Approaches to Science* (New York: Pergamon, 1986), and the contentious and detailed exchange between Evelleen Richards and John Schuster, and Evelyn Fox Keller, in *Social Studies of Science,* 19 (1989): 697-729.
12. A colleague who gave us many helpful comments on this text suggested that we were pulling our punches by using the phrase "curious lack of imagination," when in fact we knew enough about feminist critiques of science and the dominance of a male culture in science to make a much more critical judgment.

She is right. We decided, however, to keep the wording as we read it on Sunday morning.

13. See the cynical, but nevertheless pertinent, discussion in Langdon Winner, "Brandy, Cigars, and Human Values," in his *The Whale and the Reactor: A Search for Limits in an Age of High Technology* (Chicago: University of Chicago Press, 1986), 155–163.

14. We would like to thank Edward Manier for conveying these metaphorical possibilities from Vonnegut and Percy to the conference and to us.

15. See Kuhn, *Structure*, 167.

16. It was therefore interesting to learn during the discussion on the freedom of inquiry about the use of animals for the testing of cosmetic products because it took the conference out of the realm of knowledge for knowledge's sake—where it so often stayed—and confronted it with more mundane matters.

17. See Roger Smith, "Origins of Neuroscience," *History of Science* 26 (1988): 427–437. Here Smith discusses Edwin Clarke and L. S. Jacyna's *Nineteenth-Century Origins of Neuroscientific Concepts* (Berkeley: University of California Press, 1987) and Anne Harrington's *Medicine, Mind, and the Double Brain* (Princeton: Princeton University Press, 1987).

18. See, for example, the programmatic introduction by Timothy Lenoir to a recent issue of *Science in Context* devoted to this question: "Practice, Reason, Context: The Dialogue between Theory and Experiment," 2 (1988): 3–22.

19. See "Scientists Disagree Over the Goals of a Federal Program That Stresses Research Aimed at Curing Brain Disease," *Chronicle of Higher Education,* August 8, 1990, pp. A5–A6.

20. See John Marks, *The Search for the "Manchurian Candidate": The CIA and Mind Control* (New York: New York Times Books, 1979).

21. See Rose K. Goldsen, "The Great American Consciousness Machine: Engineering the Thought-Environment," *Journal of Social Reconstruction* 1, no. 2 (1980): 87–102.

22. As an introduction to this cluster of ideas, see Steve Woolgar, "Reconstructing Man and Machine: A Note on Sociological Critiques of Cognitivism," in *The Social Construction of Technological Systems: New Directions in the Sociology and History of Technology,* ed. Wiebe E. Bijker, Thomas P. Hughes, and Trevor Pinch (Cambridge, Mass.: The MIT Press, 1987), 311–328.

23. What we have in mind was demonstrated in an article that we read on the way to the conference: "Out of Control! At MIT, Robots March towards Independence," *Boston Globe,* 30 July 1990, pp. 29, 34. At the MIT Artificial Intelligence Laboratory instead of looking at the human mind, researchers have begun to construct small "creatures" that can perform a variety of tasks without having to possess higher intelligence. This is an interesting example of different forms of knowledge.

24. See John Searle, *Mind, Brains, and Science* (Cambridge, Mass.: Harvard University Press, 1984), esp. pp. 42–56; quotation on p. 44.

25. See Ian Hacking's review of Marvin Minsky's *The Society of Mind* (1986) in the *New Republic,* April 20, 1987, pp. 42–45.

26. See Ian Hacking, "Weapons Research and the Form of Scientific Knowledge," *Canadian Journal of Philosophy,* suppl. 12 (1986): 237–260, quotation on p. 238. We first became aware of Hacking's writing on these topics after reading

Paul Forman, "Behind Quantum Electronics: National Security as Basis for Physical Research in the United States, 1940–1960," *Historical Studies in the Physical and Biological Sciences* 18, no. 1 (1987): 149–229, esp. on pp. 224–225.

27. See Bruno Latour and Steve Woolgar, *Laboratory Life: The Social Construction of Scientific Facts* (Newbury Park, Calif.: Sage, 1979; reprinted by Princeton University Press, 1986); Karin Knorr-Cetina, *The Manufacture of Knowledge* (Oxford, England: Pergamon Press, 1981); Michael Lynch, *Art and Artefact in Laboratory Science* (London: Routledge and Kegan Paul, 1986); Steve Woolgar, "Laboratory Studies: A Comment on the State of the Art," *Social Studies of Science* 12 (1982): 481–498.

28. See Sharon Traweek, *Beamtimes and Lifetimes: The World of High Energy Physics* (Cambridge, Mass.: Harvard University Press, 1988).

29. See Max Charlesworth, Lyndsay Farrall, Terry Stokes, and David Turnbull, *Life among the Scientists: An Anthropological Study of an Australian Scientific Community* (Melbourne, Australia: Oxford University Press, 1989).

30. See Joan Fujimura, "The Molecular Biological Bandwagon in Cancer Research: Where Social Worlds Meet," *Social Problems* 35 (1988): 261–283.

31. See Charles Bosk, *Forgive and Remember: Managing Medical Failure* (Chicago: University of Chicago Press, 1979); Anselm L. Strauss, Shizuko Fagerhaugh, Barbara Suczek, and Carolyn Weiner, *The Organization of Medical Work* (Chicago: University of Chicago Press, 1985).

32. See Bruno Latour's essay review of that book, Michel Serres's *Statues* (1987), and Traweek's *Beamtimes:* "Postmodern? No, Simply Amodern! Steps towards an Anthropology of Science," *Studies in the History and Philosophy of Science* 21 (1990): 145–171.

33. Indeed, Lynch's *Art and Artefact* is a study of "shop talk" in a neuroscience laboratory.

34. This has already been done for strictly scientific conferences. See Pnina G. Abir-Am, "Towards an Historical Ethnography of Scientific Ritual: The 50th Anniversary of the First Protein X-Ray Photo and the Origins of Molecular Biology." Paper presented at the joint meeting of the British Society for the History of Science and History of Science Society, Manchester, England, 11–15 July 1988.

Index